Evolution Fact & Fantasy:

The Psychogenic Theory

LAWRENCE L. HORSTMAN

Finished January 15, 2011

ISBN: 1466469412
ISBN 13: 9781466469419

Preface to the Internet Edition

This book was originally drafted in the 1970s, intended as the first sequel to *The Lotka Hypothesis*. Neither found a publisher at that time, and wound up gathering dust.

Some 30 years later, a series of books about consciousness by high-profile authors – two were by Nobel laureates – got me fired up again because they were so hopelessly inadequate. Lotka's hypothesis is the obvious and the only solution. Accordingly, I dusted off the old manuscript, updated it, and sent it off, confident that it would be recognized. Again, no publisher could be found. Because I feel it is important, I ended up publishing The Lotka Hypothesis at my own expense.

That was another big disappointment. It turns out that nobody takes any notice of self-published books, probably for the good reason that the great majority are junk. Then again, the same is true of the great majority of commercially published books. Be that as it may, it now appears that the internet holds promise for leveling the playing field. Just as politics is too important to leave to the politicians, so are academic matters too important to leave to the academics, much less book editors.

Coming back to this book, *Evolution,* it was intended as "book II" of a planned series to demonstrate how the principles set forth in book I can be usefully applied to real-world scientific issues, more specifically, to biology, vis a vis its central pillar: the theory of evolution. That is to say, this book presents an alternative theory, or more correctly, a complementary theory, which might be called the *pyschogenic* theory of evolution. This requires exposing the considerable deficiencies of the existing theory.

How did this book come to pass? Well, after *The Lotka Hypothesis* failed to make a dent, I reasoned, with my usual boundless optimism, that perhaps the topic of consciousness was a bit too esoteric for the popular palette; but the topic of evolution was hot and very public, and besides, my take on the subject was quite original and, I daresay, engaging to read. It couldn't fail! So I set about updating my old manuscript.

Well, it did fail. No publisher expressed the slightest interest. It took me a long time to realize that the editors are business administrators alongside hot dog vendors, not intellectuals.

Internet to the rescue! Enjoy.

Contents

1.

Point of Departure

1. TWO PROBLEMS, EASY AND HARD. Among the great unsolved problems in modern science, second only to that of consciousness, is that of biological evolution. Of course, the establishment people will hotly object, claiming this problem to be essentially solved, but they are preaching to the choir, and a small one at that. Most people of common sense, including many if not most scientists, remain skeptical of the glib explanations, and more than a few are downright contemptuous. Indeed, as we shall see, even among the professional evolutionary biologists there is considerable discord about fundamental principles. The reasons for our doubts will emerge.

But first, it is essential to recognize that "the theory of evolution" is actually two theories, addressing two distinct questions, (i) *whether* it happened, and (ii) if it did, *how?* It is most unfortunate that this distinction has not been more forcefully distinguished in popular accounts, since otherwise we might have been spared much of the rancor of the ongoing debates about teaching "the theory" as fact.

Regarding whether it happened, the evidence is overwhelming that it did happen within the time frame commonly given. The most obvious evidence is the clear succession of fossil forms found in strata of sedimentary rocks, often in sheets like the pages of a book open to all[1, 2]. The genealogical relations implicit in the succession of fossil-bearing strata correspond closely to those inferred from comparative anatomic and genetic studies of still-living forms.

On top of that is the indisputable commonality of all living things so impressively revealed by their shared biochemistry, physiology, and molecular genetics. Thus, *whether* it happened is the easy question, now answered with confidence in the affirmative. To deny that it happened is no more tenable today than to deny that the earth is round. Indeed, the notion that evolution really happened was widely entertained for at least fifty years before Charles Darwin's book of 1859, but was not so widely accepted until Darwin proposed an explanation for how it happened: the theory of natural selection.

That is to say, the second part of the theory – the how or the why of it – is a very different matter, much harder and, we submit, still unsolved. This distinction was fully appreciated by early commentators such as T.H. Huxley[3]. On this question, the doubters and skeptics are entirely within legitimate intellectual bounds.

> "That evolution happened is a fact. Just why and how it happened is, in part, still a matter of debate, and a number of evolutionary biologists are currently attempting to rethink the problem of causality in their discipline." [4]

Accordingly, throughout the remainder of this book, when we speak of the theory of evolution we mean only the how and the why of it.

2. THUMBNAIL SKETCH. The theory of evolution is often said to be the cornerstone of biology, and so it is, or more accurately, so they might; for we shall see that a great many rival theories are now in flux, despite a generally accepted foundation of orthodox principles. Whether the current accounts are right or wrong remains to be seen; and if they are mostly wrong, or weak, or significantly incomplete, then goodbye cornerstone. We will need to look elsewhere for a more satisfactory foundation, such as to Lotka's hypothesis, which brings to biology a set of principles heretofore ignored and suppressed but which can be denied no longer, namely, *psychological* principles.

By way of background, it is common knowledge that the theory of today stemmed from Darwin's book of 1859, *On The Origin of Species*[5], a time when the philosophy of Newtonian mechanism was highly influential. Darwin's theory seemed to explain how the great variety of living things around us could have arisen by the same dispassionate mechanisms assumed to govern the rest of the cosmos. Thus, one might argue that Lamarck's earlier theory of evolution (1802) failed to win much support in scientific circles precisely because it posited mental factors driving living existence, a notion incompatible with the Newtonian worldview. Darwin was fully aware of this intellectual climate and by his own admission, consciously strove to couch his theory in terms agreeable to his scientific peers; that is, in terms of dispassionate mechanisms, despite indications of his later misgivings. He often worried about "making a fool of myself."

In particular, the main mechanism of evolution that he proposed was *natural selection,* a process in which environmental forces such as climate, predation, and competition for resources favored the survival of those variant individuals which happened by chance to be better adapted to survive and reproduce than their littermates – "the fittest survive."

This was a theory already in the air, though. In recent decades, dozens of books about Darwin and his times have appeared, many said to be excellent, in numbers matched only by those about Einstein. A take-home message of both groups of books is that neither of these lions was quite so original as the public imagines. The great novelty of Darwin's theory boils down to having supplied a mechanism, natural selection, for the already-popular notion of the "transmutation of species," along with a more detailed and scholarly defense of the whole idea than anyone else. Indeed, Alfred Wallace got the same idea at the same time[6].

Among the many inspirations to Darwin's thought was the work of plant and animal breeders, seen to yield dramatically improved strains of crops and livestock by crossing selected offspring with desired characteristics. In a nutshell, Darwin proposed that the same principles used by the breeders must have been operating naturally over geological time, constantly selecting those most fit for survival and reproduction. This remains the core and essence of the theory today.

That is to say, among any set of littermates, some would happen to be better able to survive a particular challenge such as a harsh winter, disease, or attack by a predator, owing to some fortuitous advantage such as greater swiftness, stronger jaws, larger fat reserves, etc., rendering it "more fit" or "better adapted" to survive and reproduce.

The great appeal of this idea to the scientific community was that it appeared to explain evolution in terms of *impassive mechanisms,* it was consistent with the prevailing assumptions about how the larger cosmos works. For no *mental* qualities were required. Random accidents of birth alone determined the most fit, and only the fittest survive from each generation, each round of natural selection.

Although this account is an oversimplification of Darwin's own views and did not take hold fully until the twentieth century, it was exactly what the mechanists had been searching for: the missing link between the Newtonian worldview of physical science and the otherwise magical realm of biology. The emerging sciences of biochemistry and physiology had already supplied such a link, but these applied only to the inner workings of organisms. Darwin's theory seemed to explain their variety and outward features. No "vital principle" unique to biology and certainly no *mental* qualities, were required. Conveniently ignored were the mental qualities so obviously implicit in such phrases as the "struggle for survival."

3. SUPERFLUOUS THEORY? Although it was believed that Darwin's theory had brought biology into the tent of Newtonian science, we must point out that it was only partially successful. For if living organisms are indeed just an extension of ordinary inanimate chemistry then we must wonder why a theory of evolution should be needed at all. For if one is committed to the view that living things are not quintessentially different from the nonliving, then it must be possible to view the process of evolution as a kind of long-term chemical reaction. In fact, the origin of life is believed to have been precisely that: a fortuitous chemical reaction. As clearly indicated by Lotka, and as we learn in Chem 101 today, chemical reactions proceed by a series of "competitions" between various molecular "species," the end result being the "natural selection" of those which are "most fit to survive" in a given test tube.

Granting this perfectly logical parallel, if we then find that the existing theory of chemical reactions is inadequate to embrace the "evolution reaction," then we must conclude that our theory of chemical reactions, and our notion of how the world works, is deficient and in need of revision.

At the moment, however, no such solution is available. We have instead the current theory, in which life remains a puzzling exception to the larger cosmic order.

4. DUMBING-DOWN DARWIN. Darwin was aware that the scientific legitimacy and appeal of his theory would rest upon his principle of natural selection, and accordingly gave it foremost prominence. On the other hand, he never intended that it was the *entire* explanation and often intimated additional factors (see later). Yet, even in Darwin's day, his work was widely and grossly simplified by his epigoni, intent on showing that life could be completely explained in terms of natural selection alone. Darwin intensely resented this over-simplification, insisting repeatedly that natural selection was *not the only* modality of evolution, as in the following quotation:

> "But as my conclusions have lately been much misrepresented, and it has been stated that I attribute the modification of species *exclusively to natural selection,* I may be permitted to remark that in the first edition of this work, and subsequently, I placed in a most conspicuous position – namely at the close of the Introduction – the following words: 'I am convinced that natural selection has been the main but *not the exclusive means* of modification.' This has been to no avail. Great is the power of steady misrepresentation; but the history of science shows that fortunately this power does not long endure." Ch. XV[5], and similarly in Commins and Linscott[7]. Italics added.

We hope in this book to vindicate his words. We submit that the dumbing-down of his ideas stemmed from a simple motive: the desire of the scientific community to eliminate doubts and uncertainties, chiefly *mentalism*, from the theory. One detects a curious compulsion of the rank-and-file, as distinct from the more thoughtful innovators, to possess a complete explanation at any cost. This is a mindset that cannot abide any element of uncertainty.

> "From a modern point of view Darwin's evolutionary genetics must be considered defective in one important respect. He believed in the inheritance of acquired characteristics and in inherited habit arising from originally intelligent behavior. This sort of mechanism he thought existed in addition to natural selection. ... This complicated his researches. Our work is now much easier." [8]

Easier, yes, but does this lead us closer to the truth or farther from it? To illustrate by concrete example, Darwin saw clearly that sexual selection (meaning preferences in selecting a mate based on fancy feathers, big ant-lers, song, colors, dances, etc.) must have played a central role in the shap-ing of the species we see. But as recently as the 1970s, when this book was first drafted, the topic of sexual selection was taboo, forbidden from the serious literature.

Why? Sexual selection implies *mental* factors, even aesthetics. Furthermore, the fancy features and ornaments associated with it – the peacock's plum-age, say – represents a waste of valuable resources, anathema to the assump-tion of strict economic utilitarianism presumed to operate in the natural world. Since that time, however, the topic of sexual selection has exploded back into the theory – it couldn't be ignored forever – where it now rests most uncomfortably (taken up in Ch. 7). One is reminded of the uneasy sta-tus of motivation theory in behaviorism around the same time (see book 1).

Despite Darwin's misgivings, his theory was eventually stripped of all el-ements of doubt and uncertainty, leading to the position we speak of as neo-Darwinism, a.k.a. the New Synthesis, asserting that natural selection is *the only* law or principle of evolution, as in the Ghiselin quotation above. This departure from Darwin's position was said to be justified by the rise of genetic science and remains today essentially the position of the orthodox camp, if such a camp any longer exists.

Paradoxically, however, this simplification of Darwin's work has not re-sulted in a simplified theory today. On the contrary, for at least the last 30 years practically every issue of the leading general journals (*Science* in the U.S.A. and *Nature* in the U.K. and Europe) have carried at least one article purporting to clarify one or another controversial topic in the theory. The overall result is a train wreck of obscurantism, conflicting principles, uncertainty, complexity, high-flown jargon, and general confusion. One is reminded of Kepler's teacher on the Ptolemaic system of the planets – "It is too complicated!" – inspiring Kepler to vindicate the much simpler Copernican hypothesis. Likewise, are we inspired by this mess to vindicate Darwin, by application of Lotka's hypothesis?

Darwin himself was partly to blame for the dumbing-down of his thesis that so rankled him, for as noted above, he was motivated in large mea-sure by a craving for the respect and admiration of his peers, leading him to scrupulously avoid mentalistic or teleological factors in evolution. Yet, like Newton, Darwin was a deep-thinking man in search of the truth and was fully aware that his theory barely scratched the surface of the mystery. The rank-and-file who subsequently rose to dominate the field had no such scruples, being more interested in tidy logic and clever arguments. Upon close inspection, however, their tidy logic is anything but.

This is not to say that no efforts were made to embrace mind and consciousness within the theory. Many such books have appeared and continue to appear, some of them discussed later, and many take cognizance of Darwin's later writings that more explicitly dealt with such issues, such as his *Descent of Man*[9], *The Expression of Emotions*[10], and his notebooks and posthumous works. The following is from a review of Richards' book in this vein:

> "Certain intellectual heirs of Darwin and Spencer [Romanes, James, Morgan, and Baldwin are mentioned] used the theory of natural selection to prove that consciousness and mind have been dynamic forces in evolution, that nature is neither mechanistic nor amoral, and that morality, religion, and human freedom have a basis in nature." [11].

Mention might also be made of the deep reflections on such questions by Alfred Wallace, co-discoverer with Darwin of the principle of natural selection, and of the fact that neither Darwin nor his champion, T.H. Huxley, saw any deep incongruity between the theory of evolution and religious or moral convictions. But such efforts lie well outside of the scientific mainstream today.

5. NEO-DARWINISM. Today's theory of evolution did not arise overnight nor spring directly from Darwin's pen. According to Ernst Mayr, as late as the 1930s there was much confusion about Darwin's theory, as well as continuing sympathy for Lamarck's theory of 1802, which twentieth century textbooks often ridicule. (As we shall see, Lamarck deserves more respect, as accorded by Eiseley[3] and more recent historians.) The key to the modern theory lay in enthusiasm for the then-new science of genetics, laughably inadequate though it then was.

Darwin himself did extensive experiments on garden plants in an effort to discover laws of inheritance, but failed. It is a sad irony of history that at the very same time, the more mathematically inclined monk Gregor Mendel was doing similar experiments but enjoyed much greater success, discovering his eponymous laws – Mendel's laws of inheritance. Recognizing the importance of his findings, Mendel repeatedly attempted to meet with or otherwise communicate with Darwin, to no avail. Mendel died unsung except by his devoted parish. His notebooks were unceremoniously burned as the worthless scribbles of an eccentric hobbyist. Only a small few of his writings survived [12].

Thus, Darwin's opus was written in ignorance of any clear laws of genetic inheritance. He believed that parental characters (traits) were blended in the offspring, whereas Mendel demonstrated that many or most are discretely inherited in non-blending units. When Mendel's laws were rediscovered in 1900, a reformulation of Darwin's theory began. By the 1940s, there had emerged what came to be called the New Synthesis, a.k.a. neo-Darwinism, due to work by T. Dobzhansky, R.A. Fisher, J.B.S. Haldane, E. Mayr, G.G. Simpson, S. Wright, and G.L. Stebbbins, to name most of the key players, each with specialized expertise[13-16].

In addition to its great emphasis on genetic theory, the New Synthesis downplayed or ignored many of Darwin's ideas, especially any which might suggest a mental component to evolution, for example, sexual selection. Some authors discern various stages of neo-Darwinism between then and now and even forecast a Final Synthesis in the near future! [17] Fat chance. However, insofar as the fundamental precepts have not changed, despite spectacular progress in genetic science, we see nothing that resembles systematic advance of evolutionary theory, no progressive march towards scientific maturity, instead only a series of shifts of fashion leading to the present disarray, as essayed in following chapters.

It must be added that the New Synthesis of the 1950s was heavily dependent on what appears from the modern perspective to be a very primitive knowledge of genetics. Indeed, as discussed later, even today our understanding of molecular genetics is only dawning, suggesting that from the perspective of a century hence, our present knowledge will seem as primitive as that of the 1950s does to us now. Konner, writing in 2002:

> "One thing is certain. The sequencing of the genome will soon look like the easiest thing that biologists ever did. And what the sequencers euphemistically call 'annotation' and the rest of us call

development – what the genes actually do – constitutes the real code of living systems. To crack that code will take centuries, but getting there will be more than half the fun." [18]

Note the word centuries. The reality is that the genetic science deployed in the 1970s against critics and skeptics was no more than a smokescreen, a pretense at knowledge that didn't exist, a tactic for defending the bastion of mechanism, for slamming the door on alternative thinking, for presenting the theory as if it was proven, solid, settled. Future historians will have great fun teasing apart the sociocultural forces that have for so long propped up this poor imitation of science.

6. CORE ISSUES. The central problem in the theory of evolution is, of course, *speciation:* how and why do new species arise? The bottom line is that nobody knows. Instead of a clear answer to this problem, what we have is a huge corpus of forbiddingly technical literature on numerous peripheral or accessory topics such as genetics, phylogenies, paleontology, hybrid theory, taxonomy, population dynamics, sociobiology, developmental biology, ecology, evolutionary psychology, game theory, etc. that are all under the rubric of "evolution." This functions to intimidate any citizen, including many scientists, who might wish to delve into the subject to evaluate for himself the cogency of it. It conveys the illusion of a highly sophisticated and well-established system, like rocket science, but without anything that really flies.

In the original draft of this manuscript, an effort was made at this point to summarize the main divisions, topics, and approaches in evolutionary theory to orient the reader to the chapters following. But the subject has since become so complicated and hazy that we had to abandon that approach, and must instead let the chapters which follow speak for themselves. On the one hand, this eruption of diversity in evolutionary studies reflects the rich interplay of the many specialists now wearing the hat of "evolutionary biologist," a promising trend away from naive oversimplification. In contrast to the 1970s, when the theorists presented a fairly united phalanx and the general principles were easy to summarize, the field is now more open-minded, or, one might say, fractious and fragmented. Many topics then taboo, such as altruism, sexual selection, and the role of behavioral innovation, are now openly debated. On the other hand, this new atmosphere also illustrates the unsettled if not totally confused state of the theory, since many of these topics or issues represent completely alternative approaches to, or even fundamentally different theories about, the same problem. It's the seven blind men trying to describe the same elephant.

Yet, the basic precepts remain essentially unchanged. Although each leader formulates them slightly differently, the following is according to Bell, abbreviated here from his list of the ten foundational principles with which he opens his book:

> "Heritable variations in replication rate causes evolution ... this arises from random alterations of genes, and does not direct evolution ... This is the only attribute that is selected ... Modifications are contingent upon the fortuitous occurrence of variants ... Outcomes may depend on presence of particular competitors. ..."[19, 20]

That's the flavor of it. In this view, organisms are passively and helplessly molded by external forces alone, acting to select the most fit from the naturally occurring variants in each generation. In this view, reproductive fecundity is the only quality that counts as "fitness," and superiority in this quality arises solely by natural selection of random variants. And that is the core and essence of the orthodox position today, as it was in the 1970s.

7. SUMMING UP. The present proliferation of theories and approaches, to be sketched in later chapters, stands in sharp contrast to earlier efforts to simplify Darwin's thesis. Thus, Darwin's inspired theory has become a jargon-laden mish-mash of all sorts of theories in which only insular cliques of self-appointed experts are allowed to tread.

This proliferation of theories and jargon was initially rather daunting to the present writer's effort to update his manuscript of 1976. However, after some initial qualms, a reasonably secure grasp of the modern landscape was attained. It turns out that the fundamental aspect of the theory in which we are most interested, its total exclusion of psychological factors, has not changed.

Finally, if this book seems at times a bit polemical, adversarial, or confrontational, this is not because of disrespect towards the many admirable workers in the field. There is no more fascinating literature than that about ecology, ethology, speciation, etc., and we admire the many unsung heroes whose tireless work in jungles and deserts and at laboratory benches brings to our attention so many splendid new facts about the natural world. But, when certain self-appointed authorities claim to have all the answers, and pontificate in smug pomposity that doubters are either ignorant, blind, or stupid crazies, that rankles, and demands a trenchant reply.

2.

What is Evolution?

1. THE PROBLEM OF PROGRESS. The first and perhaps most fundamental problem to engage our interest is the meaning of the word *evolution.* To the layman, the theory of evolution addresses the rise of humanity from a succession of prior forms (apes, reptiles, fish, sponges, amebas, bacteria) bespeaking some kind of overall direction, a clear progression. Not so, say the experts, for the reason that nobody has been able to give a satisfactory definition of progress. For example, here is G.G. Simpson, one of the architects of the New Synthesis:

> "In sober inquiry, we have no reason to assume, without other standards, that evolution overall, or in any particular case, has been either for the better or the worse. ... Progress can be identified only if we first postulate a criterion for progress" [1] (ch. 7 of).

This position is often reiterated endlessly indeed, lest we forget it, by the late Steven J. Gould, and here by Henry Gee, senior editor of *Nature*:

> "We've all seen the commercials. A line of figures walking from left to right, first a shambling ape walking on all fours ... [and finally] a tall proud man carrying a spear. ... These advertisements reflect

the popular view of evolution as a progressive force that drives an inexorable improvement. But in reality, evolution is based largely on natural selection ... has neither memory nor foresight; it works only in the here and now ... not a force, like gravity. It is directionless with respect to history ..."[2]

Just for good measure, the following is by the eminent evolutionist, Francisco Ayala, digressing in the course of writing a rather negative review of a book dealing with the historical rise of the concept of progress *vis a vis* evolutionary theory, a book which other reviewers found admirable. After repeating the old questions, he dishes up the old answer:

> "The point I am leading to is that in order to decide which organisms are more or less progressive, we need first to settle the matter of what we mean by the term. ... deciding on a criterion of progress is a matter of personal preference, or, to put it more formally, the concept of progress implies value judgments, and thus transcends the standard mode of scientific discourse ... For this decision science is hopelessly incompetent." [3]

We shall find otherwise. Recall that in the previous volume we ran up against another problem, free will, that all the experts said was hopelessly undefinable, a wishful delusion, a meaningless term alien to science; and yet, we had no trouble rigorously defining it and proving that it really operates in everyday life. That is to say, we are dealing here with culturally entrenched dogmas, a kind of metaphysics behind and within what passes for science, and it is really quite easy for an outsider to see the whole thing very differently, to arrive at clear alternatives, including satisfactory definitions of "evolution" and "progress."

Meanwhile, the orthodox position contradicts another major theme of modern biology, which is how humanity allegedly sprang up so vastly advanced over the apes (reviewed in book one), despite a total absence of clear definitions or rationality for that belief. Obviously, there is a double standard going on here. It seems to me that the theorists know in their hearts that some kind of directed long-term change has happened, and their language shows it. Any fool can see that some sort of "progress" has occurred in the succession of forms from bacteria through jellyfish, worms, fish, birds, mammals, apes, and people.

If we are to believe that no progress occurred, then what can be the meaning of the word *evolution?* Webster informs us that this word is from the Latin *evolvere,* to unroll, and unavoidably implies a systematic progression. Indeed, biologists often use the word appropriately in other contexts: *The Evolution of Theodosius Dobshansky*[4], *Richard Dwain's Evolution*[5], *The Evolution of Darwinism*[6]. Thus, just as the ideas of individuals evolve, so do scientific theories like Darwinism. Therein lays a hint to the solution: the evolution of life has proceeded much like the evolution of our technologies and related theories about, say, chemistry.

Be that as it may, the bottom line is that evolution, like so many other key terms in biology, is an undefined word, whereupon the so-called "theory of evolution" is a theory about, well, who knows? Accordingly, it is fair to conclude that the theorists literally don't know what they are talking about.

2. ADAPTATION? Since at least the 1970s, the word *evolution* has come to be used interchangeably with *adaptation*, or even with any kind of heritable change at all. This can be seen in hundreds if not thousands of publications. I was aghast when I first came across that usage in a 1978 article[7]; but as early as 1953 it was used to describe results of forcing microbes forced to adapt to modified conditions[8], similar to more recent experiments of that kind[9]. What's wrong with that? What's wrong is that an adaptation need not have any particular long-term direction. When an animal becomes adapted to a colder climate, say, then it can presumable re-adapt back to a warmer climate. Can that properly be called evolution? Even the influential and archly conservative G.G. Simpson, an architect of the modern school, balked at this equivalence:

> "The concept of [evolutionary] progress as involving improvement in adaptation is rather obvious and apparently promising ... But on closer inspection it seems to be quite limited as a criterion of progress. On the basis of adaptation alone, there is no reason to consider one adaptive type higher than another. ... Every organism is adapted." [1]

Very true. Lions prey on zebras but in turn are preyed upon by fleas, ticks, worms, and microbes. Worms, flies, and herring are as well-adapted as we are. Actually, they are better adapted since they reproduce much faster and so adjust far more quickly to environmental exigencies. Moreover, owing to their great numbers, they are less prone to extinctions. So why did life advance beyond worms, or even get that far? On this, the theory is silent.

3. PREMATURE EJACULATIONS. Among the many socio-historical puzzles about this muddle is the paucity of effort that has gone into trying to honestly confront the problem of evolution, as compared with that expended trying to evade it. Thus, it appeared to G.G. Simpson in the mid-twentieth century, when genetic science was primitive by modern standards, that the problem was settled:

> "It does seem that the problem [of evolution] is now essentially solved, and that the mechanism of adaptation is known [in terms of genetics]. It turns out to be basically materialistic, with no sign of purpose [or meaning] as a working variable in life history."

Richard Dawkins, the animal behaviorist turned best-selling popularizer of evolution theory, is bolder yet, opening his Preface to *The Blind Watchmaker* thusly:

> "This book is written in the conviction that our own existence once presented the greatest of all mysteries, but that it is a mystery no longer because it is solved. Darwin and Wallace solved it, though we shall continue to add footnotes to their solution for a while yet." [10]

Go tell it on the mountain. Claims only slightly less sophomoric were commonplace among the leaders, e.g. by Stebbins and Ayala[6] in trumpeting the "triumph" of the theory[11]. Such immodest rhetoric is not designed to win friends and influence people. At any rate, we shall see in subsequent chapters that statements of this kind were more than a little premature, not only because of the problem of the meaning of "evolution" but also in view of the still-widening confusion and splintering of the subject into conflicting schools, partly because of the unexpected complexity of genetic mechanisms, the intractability of ecosystems, and the endlessly amazing new discoveries by field observers of the natural world.

4. CREEPING MENTALISM. On the other hand, let it be said that the orthodox position on nearly all of these matters has softened considerably in the decades since this book was first drafted. But the word "softened" is not far from "disintegrated," which is to say that the classical dogmas have become blurry. One is reminded of the decay of behaviorism, once so epistemologically crisp, into the present non-science of psychology. In particular, the old taboo against any kind of mentalism associated with the theory is showing signs of erosion. To illustrate, we read the following concerning kin selection theory (explained later), a hot topic:

> "Recent research from the laboratory of Strand and Hardy on sibling conflict among parasitic wasps sheds light on that most puzzling of social behaviors, spite. ... Upon hatching, soldiers distribute themselves throughout the host [caterpillar] and launch aggressive attacks on other larvae, murdering their unfortunate victims. This has the potential to be spite ... Giron and colleagues demonstrate that soldiers are indeed capable of recognizing their kin ... Nonetheless, the existence of an aggressive soldier caste among parasitic wasps provides evidence that spite does exist in the real world, as Hamilton predicted it would." [12]

One must certainly applaud such work. But at the same time, one must object to the double standard of biologists who on the one hand teach that tiny insects are insensate mechanisms and that evolution is purely impassive, while on the other reporting as above on what can only be described in terms of very human passions. Again:

> "Genomic imprinting may result in a fractured self in which maternal and paternal subsets of genes in an individual acquire their own identities and work at cross purposes ... Yet, clan nepotism appears not to happen. Instead, there seems to be a 'veil of ignorance' cloaking clan identity, forcing individuals [honeybees] to behave for the greater good. ... But why should an individual acquiesce to the masking of her own clan identity...? Lifting the veil of ignorance from matrigenes and patrigenes should have many consequences. ... Haig suggested that imprinting would affect a female's fundamental social choice between reproducing and helping a mother to reproduce. ... The many individual bodies that make up a colony are still there, but their self identity may be erased, leaving strange hybrid puppets animated from above by the strings of a marionette master but also from beneath by the rods of a pair of shadow puppeteers." [13]

These examples, rife with human attributes worthy of Shakespeare (*spite, murder, nepotism, acquiesce, purposes, veil of ignorance, choice*) could be multiplied by hundreds, for there is much literature on these matters (see later). The point to be made, however, is that such descriptions are now in unabashedly human terms, that is, in terms of mental motivations, whereas such terminology was absolutely taboo in the 1970s. Such intrigues are now widely believed, with good evidence, to persist all the way down, not just to insects but to individual cells, as in sperm competition[14, 15], sperm cooperation[16], and even to DNA and the genes within it — the "selfish gene" principle made so famous by Richard Dawkins[17]. Dawkins' dissembling notwithstanding, this implies mentality (motivation, ambition, purpose, zeal) at the molecular level. Consider, for example, the motions of the DNA molecule:

> "Gasser and her colleagues have shown the molecule to gyrate like a demonic dancer. ... she's fascinated with the significance of its endless acrobatics. ... the double helix, it has emerged, regularly morphs into alternative shapes and weaves itself into knots. [Chromosomes] form fleeting liaisons with proteins, jiggle around impatiently and shoot out exploratory arms. ... These mysterious movements may be *just as important as the genetic sequence itself* in deciding which genes are switched on and off." [18] Italics added.

We would view this creeping mentalism as a positive trend except that it remains covert, subversive, not openly admitted or confronted as such. The reason is simple: there is no place for mentality in the foundational assumptions or epistemology of the theory. Indeed, it would be very difficult to make such a place, for mentality is alien to the entire fabric of western science and is allowed only for the special case of mankind and some higher mammals (see book one).

Meanwhile, it seems inevitable that the notion of motivation must be openly introduced to evolutionary theory. It is already implicit. How else is one to account for the facts of observation? Why else would organisms persevere so doggedly in the face of constant struggle? The very word *struggle* in Darwin's classic is meaningless in the absence of motivation or sentience.

To further illustrate, kin selection is now an accepted principle of the theory (though it, too, is rapidly softening, see later). Rarely broached, however, is the question attendant upon it: *why* is allegiance to kin such a dominant force? There must exist motives to drive it. The nature and origin of such motives hold the real key to understanding evolution. To play on Dobzhansky's famous remark, that nothing in biology makes sense except in the light of evolution, we may say that nothing in evolution makes sense except in the light of motivation. Meanwhile, what we have is covert mentalism, reminding us of a government which promises to uphold its constitution, yet whose officials and police ignore it.

5. THE WRIGHT STUFF. Coming back to the matter of progress, Robert Wright grappled with this in a *New Yorker* article, portraying S.J. Gould as an extremist opponent of anything resembling the usual meaning of evolution[19]. The drift of his argument is that Gould's brand of evolutionism unwittingly lends support to the creationists, as well as to secular anti-evolutionists such as Phillip E. Johnson. Johnson was noted for his incisive article in *Atlantic Monthly* and his attack book, *Darwin on Trial*, in which Gould is oft-quoted, especially Gould's *Wonderful Life* (1989) and *Rocks of Ages* (1999), the latter ironically claiming to "reconcile science and religion."

Wright, whose review panned *Wonderful Life,* writes that "anti-progressivism is the grand unifying theme in Gould's oeuvre," and quotes Gould: "'The vaunted progress of life is really random motion away from simple beginnings, not directed impetus towards inherently advantageous complexity.'" Wright intimates that Gould is outside the mainstream on this and other matters, quoting Maynard Smith as saying that Gould "is giving non-biologists a largely false picture of the state of evolutionary theory." He cites Michael Denton's *Evolution: A Theory in Crisis*, as another "favorite of creationists," the arguments of which also draw extensively from Gould, whence the title of the article, *The Accidental Creationist*[19].

Wright attempts to map out his own antidote to Gouldian nihilism, arguing that evolution may indeed have a general overall direction, necessarily leading to human-like qualities: "In the game of evolution, I submit, it was just a matter of time before one species or another ... raised its hand and said 'bingo!'", meaning attaining the putatively lofty state of mankind. He thus appears to harbor some of the same doubts about the standard theory, if there is any standard theory, as card-carrying skeptics. He claims that other evolutionists, E.O. Wilson and W.D. Hamilton in particular, also "believe that the evolution of great intelligence [sic] was likely from the start," i.e. that evolution is indeed progressive and goal-directed.

But there is a problem with his argument: while Wright et al. may well be right, such a claim is antithetical to the existing theory. There is no place for it in the epistemology, it is heretical, inconsistent; Wright's arguments are vague, based on allusions to the bombardier beetle, such "reciprocal altruism," and "positive feedback," whatever that is. He offers no theoretical basis, no fundamental principles, no explanation for the rise of "great intelligence," yet suggests that it was inevitable, predestined. Well, if it was inevitable, then the cause of this inevitability must certainly be counted among the most important principles of living existence, deserving a place in the foundations of the theory. Meanwhile, Wright and colleagues are in bed with the likes of Johnson, Denton, et al. By the way, we see here again a parallel with the struggle in behaviorism against insistent demands for "values," as by Maslow and others (see book I).

Though it is true that Gould is unusually emphatic about the apparent randomness and caprice of evolutionary pathways, he is better described as ultra-orthodox or far-right than radical, for his views on this accord with those enunciated at least since G.G. Simpson. Accordingly, we must respect Gould, right or wrong, for his intellectual integrity. If Gould is wrong and Wright is right, then that can mean only one thing: that the existing theory is wrong.

In any event, Wright's piece further illustrates a simmering dissatisfaction with the existing theory, even within the community of its defenders. Other indications of dissent sporadically appear, such as the recent controversy over an article expressing skepticism that somehow made its way into a respected journal[20]. Horrors! The major general journals, *Science* and *Nature* included, have yet to get the message, treating all dissenters as ignorant cranks, dismissing them out of hand by tactics of ridicule, sarcasm, and epithets. This is bad news for science.

6. LOTKA AND EVOLUTION. Alfred J. Lotka also wrote on evolution but, curiously enough, his views had their main impact in anthropology, not biology. Specifically, anthropologists have long been interested in human *cultural* evolution, and Lotka's ideas are at the root of the leading theory. (Lotka's ideas on this were actually inspired by W. Ostwald, 1902.) Since that is grist for another mill, we will limit this article to the question of defining evolution.

Lotka recognized in 1922[21] that a definition of "evolution" was critical to the development of a coherent theory. He therefore proposed such a definition, namely, that it is an *irreversible* process in the thermodynamic sense, as in the following from 1945:

> "It has been remarked by various writers (including the present [1925]) that evolution is an *irreversible* process: Now the term *irreversible process* has acquired a very special meaning in the vocabulary of the physicist, and not every process that might be [so] described in ordinary speech ... falls into the category ... as the physicist understands the term." [22]

Unfortunately, we cannot easily explain for the general reader the thermodynamic meaning of irreversible[23, 24], but it has to do with the impossibility of constructing a perpetual motion machine. The main point is that Lotka sought to supply this crucial missing piece of the theory and did so in terms of energy, recognizing that the struggle for survival of living things is basically a struggle to capture energy, which plants do from the "flood of sunlight on this planet" and which animals do by eating plants or other animals. He proposed that evolution proceeds to maximize the total energy passing through living systems.

We cannot fully agree with this proposition since it appears that solar energy has been fully utilized for at least the last half-billion years and so cannot explain the "progress" made in that time (The cultural evolutionists, however, find the concept useful since *human* progress has been measured in terms of technology, which depends on ever improving stewardship of energy resources.) We do, however, agree with the proposition that evolution is an *irreversible* process, leaving that term to be clarified. The clarification we have in mind is that *increased knowledge* is irreversible in the sense that scientific knowledge, for example, is clearly progressive in the long term; and this may be cast into the physical dimension of thermodynamics via information theory.

Backing up for perspective, it may be noted that Darwin's theory in large measure stemmed from the rich literature of his day about the rise of human civilization from Stone-Age savages. By the twentieth century, that topic had become the bailiwick of anthropology, but the same key question, defining progress, stubbornly blocked agreement on a theoretical account of this progression. Thus, in a fashion perfectly parallel to the theory of bioevolution, it was widely doubted if anything like "progress" had happened at all since, in the absence of knowing exactly what is progressing (or evolving), it was impossible to say that modern England, for instance, was any more evolved or advanced than any band of naked hunters ten-thousand years earlier.

To the extent that any consensus on this can be said to exist, it was the theory of Leslie White[25] and his pupils, Sahlins and Service[26], that rose to dominance. The essence of it, inspired by Lotka and Ostwald, is that cultural progress is to be defined in terms of technological progress. Whether or not one buys into that (next volume), the key point here is that cultural evolutionists recognize two kinds of change: minor historical vagaries, which they called *specific* evolution, and the long-term direction from the Stone Age to the Atomic Age, which they called *general* evolution. Only the latter is true evolution (irreversible). The former resembles what bioevolutionists call "adaptation" and the cultural evolutionists call "mere history."

We find it useful to make a similar distinction, albeit with different terminology, in bioevolution, for it confronts us with the same problem: innumerable threads of development or adaptation peculiar to each lineage, but also an overall long-term progression. But exactly what kind of progression? The answer to that demands that we recognize that behind our advancing technology lays advancing *knowledge,* which is truly irreversible. The same is true, we submit, of the biosphere.

7. THE SIMPSON SOLUTION. G.G. Simpson, that archly conservative but astute exponent of the New Synthesis (recently biographied[27]), although ultimately negative or non-committal on the question of progress in evolution, was nevertheless keenly interested in it, carefully pondering various proposals. After squelching virtually all of the usual efforts (increased complexity, adaptation, etc.), he ventures his own opinion, beginning with a quote from the then-eminent neurobiologist C. Judson Herrick:

> "'Change in the direction of increase in the variety and range of adjustments of the organism to its environment,' as Herrick puts

it. ... It is *increased awareness and perception of the environment* ... and to develop appropriate adjustments.'" [1] p121.

Of course, Simpson hedges on this, making clear that this criterion is "unofficial" (informal, off the record), for it is plainly heretical to orthodoxy in being subjective, not objective. Yet, he is emphatic about the point and carries on: "It is indeed progress in the perception of and reaction to the environment that underlies and makes possible such quite limited degrees of independence and control as have been achieved"[1] . Actually, he is more or less committed to attempt such a definition by the very title of his book, *The Meaning of Evolution* – a title later appropriated, by the way, by a critic[28].

Taking the liberty to clarify his idea, one might fairly say that "increased awareness and perception of the environment" amounts to *increasingly accurate and comprehensive knowledge of the world;* and we propose that the pursuit of this quality, along with "independence and control," is very nearly the essence of bioevolution – its definition, motive, and cause. It remains only to indicate exactly how this pursuit operates and how it can be integrated into the wider fabric of science. Some headway on the latter was made in book I..

8. THE ASCENDING. A closely related feature of the long sweep of evolution is increasing communication, cooperation, and socialization. For example, the human body is a vast empire of trillions of cells organized into hundreds of specialized tissues and organs, analogous to a colony of social insects or a nation. Even free-living bacteria are increasingly recognized as social[29-32], as are many protists like the amebic slime molds (which periodically form social aggregates, often entailing the transient specialization of subgroups of cells) and other intermediate levels of multicellularity, such as *Volvox* or some jellyfish[33]. The rise of the higher prokaryotic cells (defined partly by their content of organelles such as nuclei and mitochondria) had their origins in cooperative (symbiotic) alliances[34], as first proposed for the case of mitochondria[35].

The evolution of multicellular life is marked by parallel trends or transitions. Thus, the social insects such as bees, wasps, and ants arose from non-social solitary forms. Similarly, a key characteristic of mammals as compared to reptiles is their increased range and specificity of communications by gestures, cries, facial expressions and other signals. That is to say, increasing *communication* appears to be a central underlying feature of evolution overall, and is a prerequisite for socialization. Witness human language, the printing press, or the cell-phone.

Of course, this general movement has always been opposed by trends such as parasitic exploitation, the endless splintering of one species into many, internecine rivalries, even antagonism between the sexes (see later). All of these opposing forces are also seen in human cultural history, e.g., the endless splintering of religions, languages, and nations into subgroups at the same time as other forces are acting to unify them into ever larger blocs. These parallels, we submit, are not mere analogies but reflect the selfsame processes that engendered the living world before us. As Lotka phrased it, "we have an inside view of the processes under discussion"[22].

Many have attempted to see increasing *complexity* as the hallmark of long-term evolution, which is with some merit, but this was dismissed by Simpson and others for various reasons, notably, observed trends towards simplification. Despite a thousand books and articles on complexity theory, it is not yet clear how complexity is to be measured (as pointed out by Lotka). Among the surprises of modern genetics is that the human genome is not significantly larger or more complex than the genomes of much earlier animals, or plants. In any case, complexity by itself is not very illuminating as a criterion for developing a theoretical understanding. For even if it is true that organisms get "more complex," we are left with the question of what that might mean.

Thus, it appears to us more heuristically fruitful to recognize a drive towards communication and socialization, for these can be seen as serving the end of acquiring more truthful knowledge of the world, and as means for advancing one's ends. Knowledge is power.

At the same time, another major feature of evolution overall lies within the same tent: improved "technology" for apprehending the world and working one's will such as the animal eye, the snake's venom, the human hand. This is what Simpson and Herrick were talking about. Certainly the rise of such equipment was promoted by predator-prey dynamics, intra- and inter-species competitions, and all the other winnowing pressures documented in orthodox theory. But those factors alone do not drive such acquisitions, any more than war alone drives human inventions.

9. GEDANKEN EXPERIMENT. To convey the subjective motivational side of evolution, let us perform a thought-experiment. Imagine a set of concentric spherical cages, each separated from the next smaller one by a labyrinthine tangle of passageways — a maze — where each successive cage is larger and brighter with a wider horizon, and containing a larger number of interesting things. Let us now inoculate the innermost cage with the first living thing and observe as it creeps about, exploring and reproducing. Eventually, one of the progeny stumbles onto the passageway to behold the next larger cage, awed and enchanted by the new vista, the greater light, the larger perspective.

It is a higher form, more advanced, and so does not hesitate to cannibalize its former brethren, its primitive ancestors. This is an *irreversible* transition because no mind can willingly return to a state of smaller horizons, dimmer light, lesser knowledge. As we know from first-hand experience, the expansion of consciousness, the growth of knowledge from personal experience, is *absolutely irreversible*. It is this feature of evolution that defines progress. Nobody can reject what he knows to be truth, until faced with a higher truth.

Coming back to our experimental pets, our latest "higher organism" has basked for a time in the light of its greater comprehension of the world, quickly filling the space with copies of itself and again displacing prior forms. But then another individual arises, a still more advanced form, who finds the passageway to the next larger and even more entrancing cage. So on it goes, accelerating in pace as means are acquired to more efficiently explore and learn, to explicitly seek out new doors to higher perceptions as we now do quite deliberately, having discovered the merits of "progress."

10. OPPOSITION. It is a truism of physical science that all driving forces are countered by opposing ones, for in the absence of a counter force each would proceed with infinite rapidity. Thus, Newton in his wisdom gave us *inertia*, the innate property of mass by which it resists acceleration. Accordingly, there must exist a kind of force opposing bioevolution, so plainly evidenced by its glacial slowness over geologic time, moving in fits and starts, and leaving so many ancient forms unchanged today. Indeed, the long-term persistence of so many forms constitutes another serious challenge to the standard theory, for their very existence contradicts all the putative mechanisms by which organisms are supposedly enabled to continuously change, to constantly adapt to ever-changing circumstances. That is to say, today's orthodoxy leads us to expect continuous change in all living things, but what we actually see are "living fossils" as often as not.

The solution to this again requires looking at the subjective side of change, the idea being that once an incipient species has begun to pursue a particular lifestyle (think of the ancestor of an anteater, say, or of Darwin's finches), it has launched itself on a developmental trajectory and will become increasingly adapted to that particular niche, until at last it can do no better, having risen from a "valley" of generality or marginal success to a "peak of adaptive perfection," to paraphrase Sewell Wright's terminology[36], recently re-invented[37, 38] . That is, one envisions a fitness landscape with peaks representing perfect adaptation to a given niche; and having reached such a peak, the form cannot conceive merit in any further change.

But why not? If natural selection got it up there, why can't it bring it back down and then up another peak, as orthodox theory would lead us to expect? The answer is that the organism actively climbed that peak, driven by a vision of perfection. Thus, just as any long-established human culture regards itself as perfect and cannot see the rationality of any alternatives from within its own logical sphere, so is the specialized organism, stranded on its lonely peak.

3.

Some Core Issues

1. ADAPTATIONS. Before getting into the nitty-gritty of conventional theory, it is well to lay out for the general reader some of the core principles and disputes. As good a place as any to begin is with the notion of *adaptations,* for Darwin's theory may be seen as essentially an explanation for how plants and animals became so superbly *adapted* to their environments (to local climate, resources, predators, disease, and so forth).

For example, Arctic mammals such as polar bears, foxes and hares have numerous specialized adaptations equipping them for their habitat, and the theory of natural selection seems to offer a plausible explanation: individuals of any litter which fortuitously possessed an advantageous feature (a whiter pelt, say, for concealment against snow) would be better enabled to survive, reproduce, and to range still further north. A recently documented example of this kind are certain mice and lizards which inhabit sand dunes and so have become whiter, presumably by natural selection since the more visible ones will be picked off more frequently by predators such as birds[1]. Similar cases of "cryptic coloration" have long been studied, as cited, for example, in Sewell Wright's 1968 survey article[2] and as recounted below for the case of a famous moth caught in the act, so to speak.

One of Darwin's favorite cases was the group of species of finches on the Galapagos Islands, known as Darwin's finches, all of which clearly descended from just one or a few pairs of a mainland species, yet each of which was found to be very specifically adapted, most obviously in beak form but consonantly in many other traits, to pursue a particular lifestyle – that is, to occupy a distinct niche: to eat certain foods, to dwell in higher branches *vs.* lower ones, etc. The observed result is a diversity of species which together utilize available resources almost as completely as do the more differentiated mainland species. These finches have been repeatedly studied since Darwin's account, notably by Grant[3, 4], Weiner[5], and related cases more recent[6] and ongoing.

However, the diversification of bird species is not so easily explained as is cryptic coloration, for here we may suspect an element of *active choice* of lifestyle is involved rather than mere elimination of the most vulnerable individuals. The existing theory ignores this clear distinction, instead viewing such 'choices' as the result of *genetic compulsion*, hence not essentially different from the genes causing accidental differences in coat colors. Accordingly, we shall propose a classification of adaptive types, only some of which are plausibly explained by natural selection. Meanwhile, the question of exactly how and why Darwin's finches diversified as they did remains perhaps the most basic unresolved question in the theory today. That question being how new species arise, the *problem of speciation*.

A number of well-regarded books purporting to clarify these issues have appeared in recent decades such as two by G. Bell in 1997[7], one of them a simplified version of the other. But according to reviewer J.F. Crow, neither places much emphasis on adaptation *per se*, which is not surprising since, for reasons given below, the word *adaptation* has become politically incorrect in the evolutionary coterie. Another widely noted book on general theory is by the eminent G.C. Williams[8], said to be an update of his influential work of 1966, *Adaptations*[9], even though the word *adaptation* fails to appear in the title of the new one.

2. ADAPTING FAST AND FASTER. While the principle of natural selection as the explanation for both speciation and adaptation is appealing in its simplicity, it is nevertheless just a hypothesis, and so demands supporting evidence. Unfortunately, well-documented examples have been hard to come by, as might be expected from classical theory, holding as it did that evolutionary changes must be exceedingly slow and gradual, on the order of millions of years, and thus impossible for us to witness. Indeed, a major challenge to the theory in its early days was whether the earth was old enough to allow the process as then envisioned, e.g., creating an eye by random changes. But the old doctrine of 'gradualism' has been steadily eroding first by Gould and Eldredge who persuasively argued for 'punctuated equilibrium'[10, 11], and more recently by observations taken to mean that significant changes – "evolution" – can occur in mere decades. If true, then we are faced with quite a different question: for if species are really so malleable, then what keeps so many of them so fixed and unchanging for tens of millions of years?

The now-classical observation of "evolution-in-action" was that of a certain moth in England which grew darker in color over some decades.

> "The genetic basis for specific features that are polymorphic – that vary among individuals in a population – is usually impossible to determine with precision. ... Unfortunately, of all of the phenotypic variation in nature, the case of industrial melanism in [the moth] *Bistron* is not only a classic example but is one of the few examples ... The polymorphism that causes sickle-cell anemia in humans is the counterpart of industrial melanism in moths, in that it is one of the few examples of molecular polymorphism that has a known evolutionary consequence." [12]

Let us summarize this and some other examples.

(i) <u>Industrial melansim</u>. In a now-classical study, it was observed that certain moths in wooded areas around Birmingham, England, changed color from creamy white to a darker color over several decades. This change was related to the facts that (a) the bark of trees on which they alit also changed color from creamy white to a darker color, owing to soot from industrial pollution; and (b) the main predator of these moths was birds[13].

<u>Conclusion</u>: The genes of the less visible darker moths – better camouflaged by more of the pigment, melanin – enjoyed a differential

survival advantage, and so rose to numerical dominance. Some aspects of that work have been challenged[14], yet it endures[15].

(ii) <u>Sickle-cell</u>. The classical human example concerns the sickle-cell gene, common among blacks in or from African regions where malaria is endemic. When this gene is homozygous it causes sickle-cell disease, entailing much suffering and premature death owing to abnormal red blood cells. However, it was noticed that when this gene was present in recessive form, not only was it largely harmless but protected against malaria[16]. More recently, other malaria-protective polymorphisms have been found in other parts of the world[17].

<u>Conclusion</u>: This mutation of the hemoglobin gene, although lethal to a fraction of offspring, affords a net benefit to the population as a whole, whence natural selection maintains it in a state of balanced genetic polymorphism.

(iii) <u>Microbial experiments</u>. It is easy to produce in the laboratory by deliberate selection of many successive generations of microbes such as bacteria or yeasts, strains which are adapted to a variety of conditions (high temperature, poor nutrients, antibiotics, etc.) that would be lethal to the original inoculum[18, 19].

<u>Conclusion</u>: Natural selection will presumably act in the same fashion over vast spans of time.

These constituted the main classical evidence supporting natural selection as the cause of adaptation, and more controversially, of speciation, on the assumption that extensive adaptation will eventually lead to a new species. More recently, numerous other examples have been tendered. Weiner observed significant changes in the beaks of some of Darwin's finches in response to a dry spell lasting only about 12 years. This was misleadingly reported under captions like "Evolution in Real Time"[5] and "Evolution Made Visible"[20]. Misleading because, as pointed out in letters to the editor and acknowledged by the authors, the observed changes almost certainly relied on latent genes the birds already possessed. Similarly, a claim to have demonstrated "punctuated evolution" in a long-term experiment with bacteria[21] prompted a rather heated exchange of letters between heavyweights[22].

In a related vein, a delightful report on a ten year study by Hori vividly illustrates the balance of polymorphisms for left- *vs.* right-oriented jaws in a kind of lake fish that lives by eating scales off other fish, this asymmetry facilitating attack from one side or the other. Remarkably, however, the frequency of the two forms oscillates with a period of about two years, their time to maturity, explained by the prey becoming wary of attack from one side or the other[23]. Again, in the news around 1998 was the experiment of releasing a certain lizard on several small Bahamian islands where they had been absent. Observing them over about 12 years, clear adaptive changes were seen[24, 25], including "convergent evolution", meaning parallel "adaptive radiations" to similar niches on different islands[26].

Also in the news was the effect of transplanting guppies from river regions where their predators were common to upstream locations where predators were rare; changes were seen in just four years and more dramatically at seven years[27]. This reminds us of reports that certain snails will grow too large to be eaten when in the presence of predators[28], although as I recall, no claim was made that this was the result of natural selection (it might better be classified as adaptive plasticity, see later). A short survey of claims for "fast evolution" recently appeared in *New Scientist*[29].

Whether or not the above examples of rapid change constitute true evolution by any definition, many authorities now seem to accept that adaptation and speciation can proceed much faster than was earlier thought. This trend began with the recognition by Eldredge and Gould, based on paleontological data that species tend to arise in rapid bursts followed by long periods of little change (punctuated equilibrium). The hypothesis of the "Cambrian explosion", a.k.a. the Big-Bang theory of animal origins, holds that nearly all of the basic body plans of subsequent animals appeared quite suddenly, geologically speaking, around 550 million years ago[30], and this idea continues to gain support[31]. Well-documented examples of "fast speciation" include tree species in the Amazon, 2 million years[32]; cricket species in Hawaii, 4.2 million years[33]; and land-adapted crab species in Jamaica, 2 million years[34]. But the whole question of just how rapidly *evolutionarily relevant* adaptations can appear remains unsettled, not only because of the controversies noted above or because of the many ways in which they may occur (see later), but also because of such fudge words as 'microevolution,' as in reference to the suddenly altered yet already instinctive routes of migratory birds[35]. We shall find more than one occasion to revisit this fuzzy but central question later on.

3. BUT IS IT SCIENCE? Despite catchy headlines like *Evolution in a Test-Tube*[19] and *Evolution Speeded Up*, such claims are clearly controversial in their relevance to the real world. For example, in the exchange of letters mentioned above[22], touched off by test-tube experiments on bacteria, the contestants ended up arguing about punctuated evolution. They argued about why new forms appear suddenly and then remain unchanging for long periods, but neither side wins because neither has any decisive answers. It's just hand waving against hand wringing.

In no experiment on forced adaptation was a new species ever observed to arise. The authors of the guppy experiment could find no sign of emergence of a novel species in their transplanted population, not even incipiently[27]. Likewise, the authors of the heroic 11-year experiment on bacteria (24,000 generations resulting in genomes "riddled with changes") ruefully admitted no indication of a novel species arising, they remained plain old *E. coli*. One must conclude that adaptation alone does not a new species make, though expert opinions on this vary all over the place.

Organisms are known to possess special mechanisms facilitating adaptations[36, 37], discussed later. The fast changes seen in the experiments on guppies and lizards and the draught-induced effects on Darwin's finches, while nicely illustrating adaptive potential, mostly resulted from selection of pre-existing genes rather than new mutations, as pointed out in a letter critical of a book review about rapid changes in Darwin's finches[5]. Thus, these outcomes were but adaptive variants (comparable to our domestic breeds of sheep, swine, foul, equines, canines, felines, crops, and ornamental plants), all made in the same essential way: repeated selection of favored offspring, hastened by intensive inbreeding. But dogs are still dogs, all one species.

On the other hand, truly new species can and do arise "in the blink of an eye," literally overnight, as discussed later. More conventionally, they emerge in a couple of million years, as exampled above (trees, crickets, land-crabs) – but this is still fast in terms of geologic time; indeed, so rapid that the hypothesis of punctuated equilibrium was initially highly controversial. Even faster speciation is now well documented, such as the many species now adapted to deserts such as the Sahara, which enjoyed ample rainfall as recently as ten thousand years ago, probably explained at least in part by natural selection of adaptive variants[38]. The numerous species of cichlid fishes in many African lakes also arose very quickly, geologically speaking, but are not so readily explained by natural selection since they are all together in the same lakes, and therefore stand closer to the case of Darwin's finches (sympatric speciation). At any rate, mere adaptations are not what is ordinarily meant by *evolution*, certainly not what we mean by this word in this book.

Thus, the big question nowadays is not that of explaining how novel forms could have arisen in finite time, but is the opposite, trying to explain why so many have remained fixed and unchanged for so long despite frequent upheavals in climate, resources, competitors and enemies.

4. JUST-SO STORIES. Getting back to the role of adaptations in the theory, by the 1960's, literally thousands of books and papers had appeared defending adaptation as the core of theory. However, those books and papers had become a bit too easy to write; for virtually every feature of every living thing may be seen as adaptive. Thus, simply pointing out the "utility" of any such feature was taken as further support of the theory. One is reminded of the too-easy business of naming drives in motivation theory, around the same time (see book I). This weakness, this Achilles' heel, had often been pointed out by critics outside the field, but such objections were shrugged off by the cognoscente, who scorned such critics on grounds of their ignorance of the technicalities.

It took a pair of recognized experts to put an end to the charade. In a now classic paper by Gould and Lewontin, *The Spandrels of San Marco*[39], the whole genre was lampooned as "Panglossian"[40] or likened to Kipling's *Just-So Stories* – children's stories giving fanciful explanations for how the various creatures became as they are – wittily written and dripping with acid sarcasm. Indeed, a whole book was devoted to analyzing the literary style of that influential paper[41]. (Gould's huge literary output is at once appealing for its erudition, insights, and wit, and obnoxious in its high-hatted, long-winded, and sometimes shallow analyses, an opinion widely shared[42, 43].)

That telling critique, so long overdue, brought to a screeching halt the flood of adaptationist works. Unfortunately, it also brought to a halt nearly all writing on adaptations, the heart and core of the theory. Here we are reminded of Boas' critique of anthropology, in which "the baby was poured out with the bath." It is said that the word *adaptation* practically disappeared from the vocabulary. Only in 1996 has a new book on the subject dared to appear, frankly entitled *Adaptations*[44]. Mercifully, the topic is once again respectable, as seen for example in a recent piece, *How the Horned Lizard Got Its Horns*[45]. A play on Kipling's titles (such how the leopard got its spots), clearly intended as a thumbing-of-the-nose at Gould's complaint. Yet, the child had been chastened.

5. FITNESS, ETC. Our only caveat is that *The Spandrels* didn't go far enough or deep enough. As remarked in the previous volume, the hallmark of real science lies in its logical systematics, which in turn demands clear definitions of interlocked terms. Evolutionary theory is notably deficient in this regard, despite heroic efforts to the contrary – efforts often backfiring, resulting in a labyrinth of hazy and overlapping technical jargon. The problem is you can't have well-defined terms unless you have a well-defined theory to place them in. Evolutionary theory is not a well-defined theory. It has been pointed out that theory alone would lead us to expect the entire earth to be covered in homogeneous green slime.

Thus, while the word *adaptation* temporarily vanished from the vocabulary, the word *fitness* promptly replaced it. Granted, fitness is now generally employed with a modicum of rigor, e.g., "relative fitness" is often defined by the number of surviving offspring compared to other mating pairs, and "inclusive fitness," attributed to Hamilton[46] (but arguably[47]), embraces kindred groups, discussed later. But abuses and lapses continue and not everyone is satisfied. Thus, R.E. Michod, in the course of writing a book review, carries on at length about the uncertain meaning of fitness: "What is fitness? How do we measure fitness? ..." We see this word or concept at the heart of the theory, yet it remains very unclear, its proper formulation being "a task for the future"[48].

As another example, the term *niche* is often employed in a manner implying that a definite number of niches exist in a given environment into which a population may expand, e.g., finches in the Galapagos and cichlid fishes in Lake Tanganyika. One frequently reads how such-and-such "filled almost every available niche." As I see it, niches may be discovered or invented, exactly as human entrepreneurs find new niches. There is no way to predict how many might exist, or even to count how many do exist (except in the tautologous sense that it equals the number of species). Could anyone have predicted the niche for fish that eat the scales off other fish? Keller and Lloyd have drawn attention to the ambiguities of 37 key terms in evolutionary theory[49].

6. MORE DIFFICULT CASES. Most will agree that the classical example of the peppered moth's color change plausibly illustrates the action of natural selection in producing modified forms. In other words, to quote Darwin, we cannot doubt that natural selection is "an important principle," but we must also agree with Darwin that it cannot be *the only* important principle because such examples constitute only a small fraction of the features and behaviors of living things. Yet, orthodoxy today asserts that natural selection is the sole and entire explanation.

In reality, the vast majority of living things exhibit complex *behavioral* strategies involving coordinated adjustments of morphological, physiological, and biochemical features which, we submit, cannot plausibly be explained solely by natural selection of random variants. Consider spiders and their webs, for example, or the archerfish, which shoots down passing insects by spitting a well-aimed water droplet, capable of extinguishing a glowing cigarette at 12 feet[50]. This fish possesses specialized eyes to compensate for the refractive index at the air-water interface, etc., but equally importantly, is obviously *motivated* to hunt in this fashion. Is it plausible to assume that this behavior arose solely by natural selection of some random variant? We think not, holding that *mental* factors are indispensable to a credible theory of evolution; and that skepticism is justified until and unless somebody can otherwise explain in persuasive detail exactly how this fish came to be, *sans* motivation.

Similarly complex forms are seen in thousands of species of fish, mussels, snails, frogs, birds, mammals, snakes, spiders, beetles, bees, wasps, ants, bats, etc., and even in numerous plants, perhaps especially those which eat meat. Let us consider a few examples. The larvae of the green lacewings (family *Chrysopidae*) are carnivorous and some cloak themselves in debris, being known as "trash carriers." Eisner tells of one kind, *C. slossonae,* which disguises itself as a wooly aphid (*Procifilus*) by plucking off pieces of the aphid's waxen wool and carefully applying them to its own bristles, thus enabling it to feed on the aphids with impunity, undisturbed by the ants who guard the aphid colony (since they covet the sweet secretions which they milk from the aphids). If the wooly disguise is removed by forceps, *C. slossonae* is immediately set upon by the ants, thrown down from the tree, or killed[51]. Another trash carrier is described on p307 of that reference.

Again, the same author tells of a small blue leaf beetle and its larva (*Hemisphaerota cyanea,* of the *Chrysomalid* family), living in the scrub brush of Florida, such as on palmetto fronds, which conceals itself by weaving a loose "straw hat" or thatch made of sausage-like chains of its own fecal pellets, which it artfully glues to its body in such a way as to provide total coverage. In tests with predators, those which were denuded by forceps were promptly eaten but not those whose thatch was intact. If artificially damaged, "the larvae seemed aware of what needed to be done" and promptly set about filling in the gaps in its cover, requiring considerable gymnastics[51]. Other such cases from the subfamily *Cassidinae* are known. For one more example from the same book, a certain *Synchlora* caterpillar, a kind of inchworm,

> "... disguises itself with bits of flower petals which it cuts to size and deftly attaches to its spines with tiny bits of silk, rendering it practically invisible to birds and other predators. They add fresh pieces as the petals wilt and dress up anew with each molt, and if denuded by forceps will immediately begin fashioning a new cover, even before feeding. When it comes time to pupate, they incorporate the flower petals into their cocoons, camouflaging them, too, ensuring the emergence of a "beautiful green moth""[51].

Numerous other examples come to mind such as the decorator crab, the cadis fly larva, or the basket mite[52]; and see also books by Donald R. Griffen[53, 54] or Mary Stamp Dawkins[55]. Or, we could think about the skillfully made nests of birds[56], e.g., the weaver bird, the ovenbird[57], or the houses of hermit crabs or sandcastle worms[58], beaver dams, etc. Not to mention the innumerable variety of superbly engineered nests and other structures of bees, wasps, ants, and termites. Striking new examples regularly come to light, such as the clever death traps laid by certain ants[59], and just a while back I heard of an ant which builds its burrow in the clay banks of streams and which keeps a large pebble near the entry to seal the entrance if attacked by other ants[60].

Such cases, we submit, *cannot* plausibly be explained in terms of natural selection of random variants. The only explanation is the obvious one: that they originated as *behavioral innovations which subsequently became heritable,* as Darwin wisely insisted. We further submit such cases as a new and different kind of challenge to the assumptions of orthodoxy. The classical challenge consisted of pointing out fine anatomical structures such as the eye (and more recently, sub-microscopic structures such as motor proteins), but these were always explained away in terms of orthodox dogmas (see for example Richard Dawkins[61, 62]). Oddly, however, the types of cases we cite above have rarely been raised to challenge the theory, nor have the theorists ever seriously confronted them.

Why not? The answers must lie in social psychology, or rather, in the cultural philosophy, for these examples demand that lower organisms must possess elements of will, intelligence, and consciousness, whereas our Christian heritage teaches that only humans have such attributes, denoted by soul, imbued by God.

A more contemporary example is the recent observation that many birds have changed their migratory routes within living memory, yet the new routes are already genetic, i.e., "instinctive"[35]. Obviously, there must exist mechanisms by which innovative behaviors can become genetic, and indeed, modern genetics has brought to light mechanisms by which this happens (see later). It is noteworthy that Eisner, that great impresario of insect behaviors and chemical strategies, politely sidesteps any discussion of evolutionary theory in his book[51], leading us to suspect that he, too, is skeptical of the prevailing theory.

7. FAITH VS. FAITH. To me, it is flatly incredible to imagine all those strategies arising by the usual simplistic mechanisms. Indeed, as we shall presently see, there is a quiet revolution in biology going on right now, in the direction of radical new theories, directly stimulated by increasingly frank skepticism of the orthodox explanations. Unfortunately, however, they are not yet radical enough.

The common-sense explanation for the archerfish, say, is simply that its ancestor accidentally downed a passing insect by spitting – an invention, an innovation, a *discovery* – and then deliberately adopted and honed that skill, which was subsequently copied by some of its fellows and/or potential mates, whence it eventually became heritable and perfected in subsequent generations. Darwin himself considered such a process to be self-evident. Now, what is the logic by which such a conclusion is denied?

Briefly, our common sense account is forbidden because (i) it admits an element of *mentality* underlying the rise of such forms, (ii) it implies a degree of social learning in lower forms, and (iii) it requires that behavioral innovations can become heritable and instinctive. All three are deemed impossible under current dogmas, though as we shall see, a few evolutionary biologists are inching away from those dogmas back toward a theory closer to Darwin's original, and even to Lamarck's. They are deemed impossible for a very good reason: to admit these things would require a revolution in *basic science*, by which I mean physics and chemistry, because the minute you admit to mental factors in lower organisms, you are raising fundamental questions about how the cosmos really works, issues that reach beyond biology.

It really boils down to a matter of faith. You either believe the orthodox account or you don't. I don't; and that brings us to the next chapter.

4.

Interlude: On Faiths

1. THE ATMOSPHERE. When this book was first drafted, Christian creationism was irrelevant to the matters under discussion, and still is. The Scopes trial was ancient history. However, in view of the recent tumult about teaching evolution in the schools, and because this book is critical of the existing theory, some comments are in order. We hope in this chapter to offer some novel perspectives, as opposed to the hackneyed inanities about "science *vs.* religion," most of which merely rehash old arguments from Darwin's era[1,2].

First, to be clear about where we stand, we do not doubt that evolution happened and that it happened by natural, not supernatural processes. That said, the phenomenon of life is so astounding and mysterious that it may well have arisen by "supernatural" means, maybe by a higher intelligence from outer space[3], or from another dimension, or by decree of God. But such explanations are intellectual dead ends. They shut the door on our curiosity, they do not lead to productive inquiry, and they are heuristically sterile. Hence, we are better advised to operate on the assumption that continued investigation may someday lead to deeper understanding, perhaps even to knowing our Creator, as Dorothy was led to the Wizard. Why else would we be invested with such insatiable curiosity, if not to solve this puzzle? Perhaps then "the trumpets will sound" and we will sing, "O death! Where is thy sting?"

2. THE MONKEY WARS. By the 1920s, Darwin's theory had collided with Christian fundamentalism in the USA, resulting in laws against teaching it in many states. The law was challenged in Tennessee by John Scopes under the aegis of the much maligned but invaluable ACLU, producing the so-called Scopes Monkey Trial, prosecuted by William J. Bryan, pitted against defender Clarence S. Darrow, July 10-21, 1925. The sensational case was followed daily by millions worldwide. Although Scopes was ultimately convicted of violating the law, upheld by the Supreme Court, it was plain that Darrow, arguing for Darwin, had won the hearts and minds of the great majority. A demoralized Bryan died just five days later, which, by the way, is an interesting example of the mind-body connection.

There things lay for some fifty years. Most states repealed their laws and evolution was widely taught, albeit with some delicacy for fear of arousing further confrontations. It seems there was a tacit understanding that evolution could be taught as science, but not the biblical account because that would violate our lofty constitutional principle of separation of church and state. What nobody seems to have noticed is that there are parts of science that are more religion than science, and evolutionary theory is one of those parts.

The fight broke out anew in the 1980s when political conservatism gained new power, heavily supported by a popular resurgence of Christian fundamentalism (despite repeated exposures of evangelical frauds and scoundrels) with its own agenda for America, taking on such now everyday issues as abortion, stem cells, and of course, evolution; this time with more clout. The popularity of this resurgence was certainly due in large measure to widespread disgust at growing moral dissolution, probably correctly attributed in large measure to the erosion of Christianity as a moral authority, for which science might be blamed, in particular, social Darwinism, an ugly spectre. Declining educational standards also doubtless played a role, as did growing disenchantment with the promise of science and technology. In any case, the previously disenfranchised "silent majority" was silent no more.

3. THE RECENT CHRONOLOGY. There is little need for commentary here since the issues and arguments are common knowledge. Instead, our focus is on the response of the scientific community to this new challenge. The first grounds started rocking in the editorial pages of periodicals such as *Natural History* in the early 1990s[4]. By 1996, such articles were appearing more frequently,[5] and some were touching off floods of letters to editors with opinions all over the place – left, right, pro and con, some impassioned, some blase[6]. New books on the subject and paperback editions of old ones appeared with rising tempo[7-9]. TV shows related to evolution became hot-button issues[10]. Creationism was on the march elsewhere too, even Australia[11].

By 2002, things were really heating up and the scientific community began expressing genuine alarm, such as at the effort in Ohio to get "intelligent design" (ID) into the classroom[12]. The *Scientific American* listed "15 ways to answer creationist nonsense"[13], along with a page by anti-pseudoscience crusader, Michael Shermer[14]. In 2004, Olson looked over the whole affair[15] in the course of reviewing three more books on the matter[16-18]. The scientific community was galvanized, reporting with alarm that creationism had become a key issue even in presidential politics[19, 20] along with stem cells and abortion, all of them trivial compared to real issues like health care, affordable housing, jobs, and a vanishing middle class.

By 2005, editorials were darkly forecasting "The End of Science" and "The End of the Enlightenment"[21], and even "The End of the Age of Reason." A ruckus about a PhD thesis concerning ID at Ohio State University garnered heavy press coverage, much of it apparently misinformed[22]. The conflict even opened old rifts within the Catholic Church[23]. But most of the attention focused on courtroom battles about teaching evolution, notably in Kansas[20], on the heels of the Ohio flap, and then in Pennsylvania[24-26], the latter affair reviewed in depth in the *New Yorker*[27, 28].

In an obviously political gesture, the high-prestige journal, *Science,* named "Advances in Evolution" as the *#1 Breakthrough-of-the-Year 2005*[29], although I could see no advance to warrant such an accolade, especially since six months earlier the same journal listed several problems related to evolution among *The Top Unsolved Questions*[30].

These off-hand notes are, of course, only a tiny sampler and will be out of date by the time this book gets printed (if it ever does). The enduring point being that the scientific community is absolutely frantic over this unseemly assault on its cherished theory by mere commoners.

4. INTELLIGENT DESIGN (ID). Having grasped that creationism was a bit too blatant an affront to our constitutional separation of church and state, the fundamentalists switched gears, proclaiming that living things are just too complex, too finely made to be explained by orthodox theory, ergo, ID deserved to be taught as an alternative. The same argument was advanced by William Paley in 1802, with which Darwin was intimately familiar and which was allegedly refuted in modern times by Richard Dawkins[31]. Although Dawkins' book is so arrogant, so naive, and so misinformed in some ways, one must suspect it helped kindle the flames of opposition. (A more sophisticated appraisal of the *Watchmaker* metaphor had been given by a pair of astronomers[3].)

It seems that the ID movement was in large measure the brainchild of retired law professor Phillip E. Johnson of the lavishly funded Center for the Renewal of Science and Culture at the Discovery Institute, a right-wing think tank based in Seattle. William Dembski is among its more articulate and prolific spokesmen. This group crafted a new strategy to advance their cause, leaked to the press in 1999 as the "Wedge Strategy"[15, 16], perhaps in reference to Darwin's metaphor that species are like wedges hammered into the ecosystem, displacing competing species[32]. One such book from that organization, *Darwinism, Design and Public Education*, is a collection of essays by various ID boosters, along with a few by scientific defenders of the orthodox theory to give the impression of balance but edited by fellows at the Institute[18]. It was scathingly reviewed in *Science,* along with two opposing ones said to "demolish" ID[16, 17]. The authors of which include physicist Paul E. Gross, best known to many of us for his gallant service in the Science Wars (see book I). Another notable attack on the ID camp is Robert Pennock's *Intelligent Design*[9], following his popular *Tower of Babel*[8]. Numerous related books have appeared, pro and con, including a putatively non-partisan historical analyses[2] and a torrent of articles, letters, and editorials in the scientific journals warning of the perils to science education, represented by the ID attacks on science. "Science" was now being equated with evolution theory, which is probably the worst possible example of science. The defenders of ID were unmoved.

5. FRANK APPRAISAL. Having briefly reviewed the dire warnings and perils of the "flight from reason," we end up sympathetic to the doubters and skeptics. Yes, public education is a disaster and has been for 50 years. Whose fault is that? The kettles are calling the pot black.

Resisting the temptation to digress, let us evaluate a few of the common attacks on the ID camp. For example, in the course of panning a book defending ID, evolutionary biologist Brian Charlesworth remarks that Dembski's theory "can explain anything and therefore explains nothing"[33]. However, exactly the same could be said with some justice about orthodox evolution theory, whose axiom is that natural selection can explain everything but in reality explains nothing beyond the trivially obvious. Someone, perhaps Denton, has called the principle of natural selection a "universal acid" of explanation. Granted, disposing of the problem of evolution by saying that "God did it" is not very fruitful either, but is hardly worse than the orthodox position that it was just luck and blind chance.

When Charlesworth states that Dembski wants to "turn the clock back ... to the Middle Ages," we must point out that the existing theory, too, with its tangled mish-mash of jargon and obscurantism, would be quite at home among medieval theories. Attempting to argue with an evolutionist is like trying to argue with a Freudian: both have an answer for everything based not on evidence but on dogma, the same as the creationists. Actually, as we shall see, the evolutionists are worse because they don't really have a clear position, they can't be directly confronted, they are "not even wrong," as physicist Feynman famously complained about fuzzy theories. It's like trying to wrestle with a fog. Furthermore, one could very well argue that that Evolution Theory is itself a religion, replete with its High Priests, jealously guarding their authority.

Take another example. Biologist Kenneth Miller, who testified in court against ID, states that "if there was genuine scientific evidence for intelligent design it would simply become a part of normal science"[34]. But, to put the shoe on the other foot, if there was genuine scientific evidence for the orthodox theory, there wouldn't be any room for argument. For it so happens that many creationists are keenly interested in scientific matters and are perfectly logical and rational people, albeit having adopted certain "non-scientific" premises which they believe offer a better or more complete explanation for things. That takes us to the Galileo affair.

6. THE EVIDENCE QUESTION. To illustrate, we noted in book I that the church opposed Galileo, not out of sheer dogmatism but because he had no real proof for his radical claim that the earth moves. Like modern evolutionists, he could not comprehend how his opponents could be so blind and stupid as to deny his evidence. In reality, however, his evidence was rather weak and inconclusive and could be explained in other ways. To settle the matter, the church solicited the opinion of Tycho Brahe, the greatest astronomer of the day, and Brahe listed numerous reasons for doubting Galileo's defense of Copernicus. Indeed, we now know that the Copernican theory itself was haywire. So there was "reasonable doubt" – until, that is, Kepler finally and conclusively mapped the paths of the planets around the sun. Opposition then melted away. The "reasonable doubt" was gone.

The point is, the general public is not stupid, nor are religious beliefs necessarily naive, or at least, not more so than those of the evolutionists. The bottom line here is that without clear understanding of exactly how the species arose, doubts are legitimate. Until this situation changes, and this book aims to contribute to that, we must agree that the current theory of evolution should *not* be taught in the public schools as if it were fact, no more than astrology should be. As for Galileo, well, it turned out that he was right, roughly speaking, but at the time this was by no means certain.

7. THE RUSE TAKE. Among the issues repeatedly surfacing in the creationist debates is the distinction between science and religion. The standard line is that ID is not real science. It seems that Michael Ruse has put the shoe on the other foot, pointing out that evolutionism is itself a kind of religion, but he didn't go far enough.

> "... leads to a radical conclusion. Although we are used to speaking of a conflict between science and religion, to do so misses the point: it is rather a conflict between religion and religion, he claims. ... it is an intra-family feud, and this explains its bitterness. Superficially this may sound paradoxical, if not perverse. Surely scientific theories of evolution cannot be paraded as examples of religious belief? Of course not. But Ruse has in mind a distinction between evolution as a fact, evolution as a theory that offers mechanisms for evolutionary change, and 'evolutionism' – a metaphysical, naturalistic worldview imbued with values ... It is evolutionism that has repeatedly functioned as a secular religion, offering seductive images of progress

and translating naturalistic methods of enquiry into doctrinaire assertions about what can and cannot be believed about the meaning of human existence." From review[35].

He observes that from the early evolutionists to modern ones such as Richard Dawkins, many defenders of the theory have "self-consciously rejected Christianity only to replace it with a substitute system [evolutionism] that presumes to answer the same basic questions." He quotes Richard Dawkins:

> "All the great religions have a place for awe, for ecstatic transport at the wonder and beauty of creation. And it's exactly this feeling of spine-shivering, breath-catching awe – almost worship – that modern science can provide." [35]

It seems that Ruse admires E.O. Wilson, of *Sociobiology* fame, whose "call to repentance" on the subject of biological diversity (conservation) reminds him of an old-time preacher. Sharp observer, Ruse[36, 37].

8. SCIENCE? RELIGION? The word *science* has come to be used very loosely. We provided a definition in book I, which bears repeating. The first real science was Newton's mechanics, which should serve as the archetype. Newton's physics was explicitly formulated on the model of Euclid's geometry, being a logical system of interlocking definitions and theorems based on a few axioms. This was, in fact, Galileo's vision, i.e. he wanted to extend Euclid's geometry to include the physical world. By and by, the nascent field of electricity and magnetism came to be formulated in the same terms, and then chemistry entered, by way of thermodynamics (concepts of the heat and energy involved with chemical transformations, and the momentum and energy of gas molecules [kinetic theory of gases]). This unified system defines "real science"

It was an impressive system, and a "theory of everything" was already in the air. Science had already outgunned the church on several key issues, first being astronomy (the earth's motion), another being the age of the earth, at least a couple of billion years, not a few thousand. So there came to be this idea that science could actually replace religion as the ultimate authority on all things. But there was a big problem, actually several big problems, with this ambition. The first was the problem of explaining life, and human life in particular. Another, which we addressed in book I, was (and still is) the problem of explaining our consciousness, a.k.a. the human soul, and how it is possible that by a sheer act of will, we are able to stand up and walk.

So religion still had its trump card: explaining all of that in terms of God and Satan, in terms of good and evil forces, in terms of moral struggles, etc. Science had to dream up answers for all of those things if it hoped to displace religion. Thus, the quasi-sciences arose: psychology, sociology, and of course, evolution theory. But they do not deserve to be dignified by the name, science, for they have none of the earmarks of the true sciences defined above.

They are no way directly connected to the system of physics and chemistry, and have none of the key earmarks. Where is the system of interlocking definitions and theorems? It's a house of cards. Early efforts at defining science often appealed to the "scientific method", but that went out the window when it was realized that this boils down to ordinary reasoning, or the Sherlock Holmes genre. Slightly more distinctive is the doctrine of *objectivity*, being that science must confine itself to the visually verifiable. However, in the long run, that could be self-defeating because it will never be possible to address in objective terms alone those very elements of religion that science aims to displace: passions, moral struggles, willing, longing, the appetites, the sentiments – *consciousness*, in a word.

This writer speaks as a patriot of the nation of science, holding that it is indeed a "candle in the darkness," as astronomer Carl Sagan memorably termed it. It is a beautiful and powerful system to which we owe so much of modern life, such as good health. But it is a duty of every true patriot to speak out when he sees the principles of his beloved nation compromised, corrupted, and tainted.

Now, what about religion? What is an acceptable definition of religion? If one examines closely the world's religions, including those of preliterate societies, one eventually grasps that each one is fundamentally *a complete logical system of world explanation*. This turns out to be the same as the aims of the classical western philosophers, from Plato and Aristotle onward; each was seeking to outline a complete logical system of world explanation. Indeed, there are preliterate societies that lack many of the trappings of religion, but none lacks a complete system of world explanation, usually in terms of their system of myths. (There is also reason to believe that everyone develops his own personal logical system of beliefs, which dictate his interactions with the outside world; a religion is a *shared* system of such beliefs.)

And where does science stand with respect to religion, a.k.a. philosophy? It is an *aspiring* complete system of world explanation, but it has a long way to go. It is thus one prominent example of a (partial) philosophical system. It sets itself apart from religion, of course, because all peoples of all times regard their own religion as The Truth and categorically dismiss all the others as "illogical," for the simple reason that they are based on different axioms.

9. THE NEUROPSYCH CAMPAIGN. Some have argued that new developments in neuroscience raise new challenges to religious beliefs, presumably on the supposition that the "mind" is now being clearly revealed by such studies, as in the following remarks from a letter:

> "Sir – The argument over evolution versus intelligent design, discussed in your News Story, 'Day of judgment for intelligent design' [438:267] is a relatively small-stakes theological issue compared with the potential eruption in neuroscience over the material nature of the mind. ... The truly radical and still-maturing view in the neuroscience community, that the mind is entirely the product of the brain, presents the ultimate challenge to nearly all religions. ..."[38]

He goes on to talk about dualism, Descartes, etc. as if that stuff was excitingly new and revolutionary. Maybe it is to him. The reality is that nothing in modern neuroscience has yet revealed anything that wasn't obvious for centuries (see book I), apart from a lot of brain mapping, which is analogous to gene mapping. The writer of the letter belies his ignorance of the subject by speaking of "the material nature of the mind," which is a contradiction in terms.

10. THE BIG CHILL. It is a sad commentary on the state of science that anyone who questions the existing theory is frozen out, labeled an ignoramus, a crank, or of the lunatic fringe. For example, the career of an outstanding astronomer was placed in jeopardy because he dared to doubt the theory of evolution[39]. Have they all forgotten that their eminent colleague, the late Sir Fred Hoyle, was also a doubter?[3] It is said that if we do not learn from history, we are doomed to repeat it, and that is particularly true of wrong-headed theories in science. Scientists are fond of pointing out the remarkable persistence of wrong theories and resistance to new and better ones, especially when the new ones issue from outsiders or misfits, e.g., Ignaz Semmelweiss, Ludwig Boltzmann, Alfred Weggener; and to the present day, as often as not, they are themselves guilty of the same. For example, the discoverer of the bacterial cause of peptic ulcers was ridiculed for a decade, mainly because he was working at a small Australian hospital, as opposed to the big-name institutes like NIH, Harvard, or Scripps[40].

In summary, we submit that accepting the current dogmas of evolutionary theory requires a leap of faith scarcely more or less far-fetched than the Biblical account of Creation; and that natural selection circumscribes only an outer bound of viability outside of which the organism will surely perish, but within which there exists practically limitless latitude for creative variation and invention.

11. ADDENDUM. It is worth noting a spate of relevant books that appeared between writing this chapter and now proof-reading it (2006), chiefly a cluster which argue that religion is not only a silly delusion but is dangerous and corrosive as well, causing endless violence and bloodshed and poisoning the minds of our young people against reason.

These include *God is Not Great: How Religion Poison's Everything*, by Christopher Hitchens; *The God Delusion*, by Richard Dawkins[41]; *The End of Faith*, by Sam Harris; and *Breaking the Spell; Religion as a Natural Phenomenon*, by Daniel Dennett[42], reviewed as a group in the *New Yorker*[43]. This addendum might well have been titled "Taking the Gloves Off: Backlash Against the Religious Right", for that is what it seems to be.

It's a safe bet that these books are a direct response to the rising political power of the Christian Right and its challenge to the teaching of evolution. Thus, it appears that the scientific intelligentsia has gone on the offensive. So we have this big power struggle going on, but why? How did things come to this? I put the blame squarely on the scientific community. Two big failures top the list first, by overstating their claims to knowledge, as in defending evolution theory as real science, which is like throwing down a gauntlet; and second, the disaster of science education. It is tempting to digress on wrong-headed textbooks and teaching approaches, but it is time to get back to the main topics.

5.

Speciation Introduced

1. THE PROBLEM. The core and essence of evolutionary theory is, or certainly should be, explaining why and how new species arise, a process called *speciation,* and why they occur in the numbers they do.

> "In 1959, in a thoughtful essay entitled *Homage to Santa Rosalina, or why are there so many different kinds of animals?,* G.E. Hutchinson set much of the agenda for the next several decades of community ecology. ... Now, 35 years after Hutchinson's 14 page essay, we have a wonderful 414 page volume summarizing the extent to which modern ecology has succeeded in explaining biological diversity." J.H. Brown, in a book review[1].

He then makes two points about the new book. First, that it shows how far we have come – he praises its depth and scope, daunting in wealth of detail – and

> "... the second thing the book does is to make clear that we still have no general, satisfying answer to Hutchinson's question. ... It is almost as if the more we have learned, the more difficult it has become to answer"[1].

This may indicate that the approaches taken in the last fifty years are fundamentally misguided, that the hounds are barking up the wrong tree. Again, from a book review of *Speciation and Its Consequences:*

> "The assemblage of about 40 evolutionists who have produced this book was supposed to help bridge these differences ... [Instead] the book is an incautious attempt to deal with the hardest problems ... Why does life come in so many forms? Why do these forms come in related clusters, some sharing special characteristics? What is a species and how do new ones arise? ..."[2]

Thus, not only is the word *evolution* undefined but its central problem, speciation, remains unsolved. It therefore seems a bit premature for the theorists to proclaim as they do, to *crow* (to borrow a pejorative often used by them) the "triumph" of the theory to the world[3]. Not only is such immodesty in poor taste (imagine proclaiming, say, "the triumph of Christianity") but it tends to rankle and incite those who might otherwise be silent, like me.

Most of the contending theories are united under the principle that natural selection is the entire or major explanation. Or at least they *were* so united, for this dogma is increasingly challenged[4-6]. Still, as a group, they continue to present a united front to the outside world, lest the barbarian hordes pounding on the gates – the dreaded Creationists – come pouring in.

2. SYMPATRY? The classical theory was enunciated in 1942 by Ernst Mayr[7,8] and holds in brief that essentially all new species arise by the isolation of a small splinter group, after which they diverge from the parent population due to new selection pressures, different resources, genetic drift, or the founders effect (a.k.a. peripatry, meaning that a small group of original founders of the new population will happen to have some distinctive features). This is known as *allopatric* speciation: new species arise whenever a fragment of the original breeding community becomes isolated, as by wandering north, or chancing upon a far island, or crossing a mountain range or large river.

Allopatric speciation has been defended with remarkable zeal despite obvious and widespread evidence for the contrary scenario, called *sympatric* speciation, which is the rise of new species within and amidst the parent population. It is easy to guess that the reason for resisting the reality of sympatric speciation is the difficulty of explaining it in orthodox terms. But, like the reality of sexual selection, such an obvious fact could not be ignored forever, despite the hazards.

> "The alternative [to allopatric speciation], 'sympatric speciation,' in which new species are created within a single population, has long been seen as a heresy – to the extent that young biologists would risk their careers if they proposed such a mechanism." [9].

Tautz goes on to explain that "one of the strongest arguments against sympatric speciation" has been that "there are no convincing mechanisms" to explain it; but then adds that in recent decades "the problem has been solved" by mathematical models, the net result of which, however, is "giving to the individuals an active role in choosing their mates"[9]. At last, the cat is out of the bag. An element of *choice* is now part of the game and with it, *mentalism*. Nobody seems to have noticed that this contradicts the former strict assumptions of dispassionate mechanism. (By the way, mathematical models can do anything you want them to[10].)

That is to say, when this book was first drafted (1970s), any defense of sympatric speciation would have appeared outrageously heretical. (For more comments on and attitudes towards this mode of speciation, see notes at the references[11-16].) One eminent commentator is particularly self-congratulatory:

> "The last two decades in particular have brought major advances in molecular genetics, comparative analysis, mathematical theory, and molecular phylogenetics; speciation has consequently matured from a field fraught with untestable ideas to one reaching clear, well-supported conclusions." [12]

It would be easy to compile a dozen such quotations from every decade since 1950. It seems that every generation feels compelled to announce anew that "at last, the problem is solved!" Whistling in the dark. One senses that such men of science are every bit as passionate about their beliefs, and just as insecure in them, as are the religious fundamentalists they so abhor. Why? That is an interesting question in itself.

Among the more obvious problems with classical allopatric speciation is the fact that isolated populations, rather than diverging, tend to *preserve* ancestral features as often as not. Think of the many "living fossils" of Australia, Madagascar, deep-sea retreats, and numerous islands. In fact, it is well known that large and homogeneous regions such as the Amazon basin are the most speciose[17-19]. Granted, there are many complexities, yet it is self-evident that speciation readily occurs in aquatic environments where there are no physical barriers whatsoever, most obviously in open oceans[20], but also in lakes. The most recent and well-studied examples are the cichlid fishes of African lakes (and in Nicaragua) and the sculpins of Lake Baikal[16, 21]. Or we could mention that a handful of soil can contain up to 100 species of mites (earlier cited). In sum then, sympatric speciation is no longer ignored and is widely discussed, yet remains in a kind of theoretical limbo because no clear explanation for it yet exists apart from "mating preferences," of course, which inescapably imply a *mental* dimension to evolution.

3. LANGUAGE ANALOGY. Although human evolution and the origin of language is a later topic in this series, it is relevant to note that exactly the same question arises in historical linguistics: why and how do new *languages* arise? Darwin himself observed the parallels between the diversification of languages and species[22]. The now traditional explanation for language divergences, dating back to the 1940s, was clearly derived from the New Synthesis of bioevolutionary theory: it is said that new languages arise by the *isolation* of a group from the parent speech community. However, this is the exact opposite of what is in fact observed, which is that languages diversify much more rapidly and frequently on large landmasses compared to isolated groups of speakers, which tend to preserve their ancestral tongues. For example, the many long-isolated and widely separated cultures of the Polynesian islands all spoke essentially the same tongue when first encountered by Cook. Similarly, Icelandic is practically unchanged from Old Norse, French Canadian is antiquated French, certain regional variants of American English resemble Elizabethan speech, and so on to numerous other isolated language pockets – Hungarian, say, or Basque. Clearly, drift does not occur. On the contrary, it seems that intimate contact, such as adjacency and contiguity, drives language diversification. Think of the hundreds of languages of India, New Guinea, Africa, native America, or for that matter, the dialects within a single city such as New York or London, all cheek-by-jowl. We propose that linguistic differentiation is more than analogous to biological speciation; it is identical in its motivational dynamics. Likewise are the diversifications of religions, ethnicities, etc. Yes, human evolution is ongoing right now but not by genetic mutations; the genetic changes follow the behavioral changes, not *vice versa*.

4. GENE THEORIES. Symposiums on speciation in 1986 and 1996 revealed a great diversity of approaches but no breakthroughs, no consensus, no unifying principles[16, 23]. The sampler of theories or approaches given in the following paragraph supplies a short introduction, refined in later chapters.

The classical theory holds that new species arise from random genetic mutations and this "central dogma" still predominates, even though in light of modern knowledge it is steadily eroding (see later). Thus, mutations in selector genes might give rise to classical "hopeful monsters," would-be progenitors of new species[24], reminiscent of conjectures in the 1960s that "pleiotropic genes" could give rise to major and multiple effects, i.e., new species, which hypothesis was commonly invoked to respond to skeptics posing awkward questions. This is often called the "big gene hypothesis." For example, Løvtrup is described as "an avowed epigenetic macromutationist"[25]. It is true, of course, that many genes do indeed act in concert, or control a suite of other genes, as shown in many experiments, as by Elena and Lenski[26]. Mutations of the Hox genes, which affect development, have been proposed as a trigger of speciation[27], as has a set of sister genes[28]. Another candidate is called *Odysseus*[29]. Among many interesting examples is the case of two similar species of snail which cannot mate because one has a right-handed twist, the other left, apparently due to a single mutation[30]. A genetic defect in a DNA repair system might also result in the "reproductive isolation" of a group of bacteria from the parent stock, and this is tendered as an explanation for sympatric speciation since it would work "even if there are no geographical boundaries."[31]

However, many such theories are incongruously blended with elements of "choice" and "preference," in the sense that certain genes are said to govern preferences where sexual selection is the mechanism, as in this headline from an article entitled *Searching for Speciation Genes:*

> "The formation of new animal species often results from divergence in male sexual behaviors and female preferences [sexual selection].

> The genetic basis of this sexual isolation in fruitflies is gradually being revealed." [32]

In other words, the tacit assumption is made that behavior (mating preferences) is wholly a matter of genetic compulsion. There is, of course, some support for that. For example, mating habits in fruit flies can be transferred by gene technology from one species to another[33]; in this case, by a gene called *per*, and even from a male to a female. Then again, I'm sure that a brain transplant would do much the same and genes presumably code for the brain.

It's a big can of worms. Among other things, it raises the question of how such "preference genes" arise in the first place; or, in other words, how genes and behavior are related, a hot-button question at least since 1975 when E.O. Wilson published *Sociobiology*[34], on which we commented in book I of this series.

5. MUTATION PROBLEMS. The big problem with mutation theories of speciation is that nobody knows exactly how a genetic mutation could bring about a new species, or even what controls mutations. It's 99% hypothetical. Although genetics is a later topic, it is not too soon to mention that mutation rates are highly variable from one species to another and even from gene to gene. Thus, some bacteria (e.g., *Radiodurans*) are extraordinarily resistant to high doses of X-radiation owing to their unusual ability to perfectly repair segments of their damaged DNA. Many viruses, on the other hand, actively exploit rapid mutations as a means for evading host defenses, as we well know from annual flu outbreaks. But those mutations do not result in new species. Our own immune systems, too, generate mutations by the millions for the very good purpose of producing antibodies that bind specifically to an invading microbe. Granted, there are some regions of genomes which do appear to mutate "randomly," at least sufficiently so to allow estimations of times of divergence of species; but this method of phylogenetic dating, although better now than earlier naive efforts, must be done cautiously, and they remain controversial.

The bottom line? Random mutations as a cause of new species are just speculation. New knowledge of genetics has made this putative mechanism less likely, not more likely, as compared to the 1960s when knowledge was so hazy that practically everything could be glibly explained by appeal to "random mutations." *Smoke and mirrors* is the operative phrase.

6. CHCKEN OR EGG? Here we ask, which came first, the genes or the behavior? In other words, are genes the cause of behavior or is the behavior the cause of the genes? There is only one viable solution to this impasse, and that is to recognize that behavioral innovations may become habitual and heritable, as Darwin himself insisted. Orthodoxy holds that the genes cause the behavior and that variant behaviors, presumably including the caterpillars which decorate themselves, etc., arise from genetic mutations. But that seems so far-fetched that only the opposite can be true, that initial choices can launch new evolutionary trajectories in which the choice (innovation, preference, new strategy, etc.) can become heritable, i.e. instinctive. Such instincts need not be immutable, of course, but may be reversed or overridden at later times in light of new circumstances. In other words, the genes responsible for behavioral paradigms may be likened to long-term memories or to learned behaviors that have become habitual, except that here we are speaking about trans-generational "memories." Solid support for this will emerge in later chapters, e.g., in considering modifications of bird migratory routes, novel snake venoms, etc. Granted, certain details must be cleared up before this view will gain acceptance. But they are not insuperable and modern genetics, such as epigenetics, is pointing that way.

7. CHROMOSOMES, HYBRIDS. A related set of theories for speciation is based on chromosomal rearrangements[12, 31, 35, 36], on grounds that certain changes in them might block cross-fertility to result in a new species. However, it is essential to remind the reader that genetic science is still in its infancy despite populist hype to the contrary, whence all such theories are highly speculative. With regard to chromosomes, for example, new insights on the complexities of their interactions are constantly coming to light[37, 38], as for example:

> "A three-dimensional examination of gene regulation suggests that portions from different chromosomes 'communicate' with each other, and bring related genes together in the nucleus to coordinate their expression." [39].

Another popular group of theories holds that new species arise by hybrids of related species. The classical definition of a species, by the way, is a group of similar organisms which can reproduce only among themselves. Thus, horses and donkeys are two different species because the hybrid offspring, mules, are sterile. (Of course, this definition doesn't work for extinct animals known only from fossils but the experts have ways of working this out[40], notwithstanding ongoing debate[41]. Laboratory evidence of genetic exchange is now an accepted surrogate criterion, with the advantage of being applicable to lower organisms such as bacteria, known to engage in genetic exchange comparable in extent to sexual organisms[42-45].)

The general theory here is that some hybrids are viable and might result in new species. This is not a new theory, having been discussed at some length in Wright's 1968 survey article[46], except that new tools in molecular biology now allow better evaluation of the degree of genetic exchange. Many hybrid plants exhibit "hybrid vigor." A recently claimed new natural species is a hybrid sunflower[47]. The same has been claimed for some animals such as water fleas[48] and a crayfish[49].

8. INBREEDS, OUTBREEDS. Related to the above is the controversy about the importance of genetic diversity in a population. This is of special interest in the conservation of endangered species. It has long been reasoned that diversity is important to the viability of a species, whence insufficient diversity must lead to decline and extinction owing to inability to dispense with harmful mutations or to cope with new adaptive requirements (such as new diseases or climate change) because of loss of heterozygosity needed to cope with such exigencies[50, 51] . This issue is related to the theory of why sex exists (see later) and to the alleged human fear of consanguineous (incestuous) marriages, which tend to perpetuate genetic diseases, e.g., hemophilia[52, 53]. However, this fear is much exaggerated[54, 55], although complicated[56]. The reality is that in many if not most primitive human societies, cousin marriages were not only allowed but were obligatory, and the same custom continues in many parts of the world today, without evident harm.

Turning to the natural world, it is hardly possible to imagine a new species arising without intensive inbreeding of its founders. Surely all of Darwin's finches, for example, arose from just one or a small few breeding pairs. Furthermore, many healthy animal species have been studied showing very limited genetic diversity, e.g., cheetahs[57], monkeys[58], beavers[59], cattle[60], and bacteria[61], leading to a reappraisal of conservation strategies[62]. In the wild, of course, deleterious mutations that might come to the fore through inbreeding are quickly weeded out by natural selection.

Recent work on the theory of speciation by hybridization has often centered on so-called "ring species," being those at the margins around the range of the root species where matings with variant outsiders is more likely[63-67]. Despite all this work, however, most authorities believe that this mode of speciation is rare, and unlikely to result in major radiations of novel species. Hybrids are usually no more than blends of closely related species. In sum, then, neither is the hybrid theory of speciation a convincing general explanation.

9. EVO-DEVO. Evo-devo abbreviates "evolution and development" (taken up in a later chapter, but briefly). Acolytes hold that new species arise in consequence of modifications of the genetic program governing embryonic development. The rationality is that development determines whether an organism will be a dog, sparrow, fish, elephant, pumpkin, or man. Thus, a fin becomes a foot; a foot becomes a hand or a wing, etc.

Attempts have been made to forge a new New Synthesis purporting to unite all the disparate views based heavily on evo-devo. (Another claimant for a new New Synthesis was mentioned earlier, namely, the "Final Synthesis"[68].) For example, we read in a book review that Carrol

> "... declares Evo Devo to be the third revolution in modern biological thought, after Darwinian evolution and molecular biology"[69].

Similarly, Pigliucci comments on a book by Jablonka and Lamb, which heavily features evo-devo, speaking of it as another effort towards a new New Synthesis:

> "The clamor to revise neo-Darwinism is becoming so loud that hopefully new practicing evolutionary biologists will begin to pay attention. It has been said that science often makes progress not because people change their minds, but because the old ones die off and the new generation is more open to novel ideas." [70]

Amen. The book purports to be mildly radical, notably in defending a cautiously limited kind of Lamarckism. The review is favorable and informative but stops well short of proclaiming the book to be anywhere near an ultimate solution.

10. SYMBIOSIS AND CO-EVOLUTION. Another interesting approach to speciation and to evolution in general focuses on the formation of symbiotic relationships among and/or between species. This goes back to Darwin's observations of the exquisitely refined relationships between flowering plants and their pollinators, which include, by the way, not only bees but numerous bats, hummingbirds, ants, etc., a topic of enduring interest[71-73]. It was a major theory in Cold War Soviet biology and still has western adherents[74]; the general idea being that the form of nearly any species is determined in large measure by those it depends on, or which depend on it, such as parasites or predators, so that any significant change in one should force change upon the other, a concept called co-evolution[75-77]. The evolutionary arms race between predators and their prey is sufficiently well known that it needs no comment, except to mention that it is not limited to what you see on TV about gazelles *vs.* big cats, but is pervasive, extending all the way down to the insects, bacteria, and plankton of the seas[78], and to the constant battle between our immune systems and infectious organisms.

Viewed in such broad terms, however, the concept becomes unmanageable since in the extreme, virtually every organism is interdependent with others, resulting in a hopelessly complex web[79]. A further complication is the absence of any clear line between relationships of mutual benefit (symbiosis) *vs.* exploitation (parasitism, predator / prey); see, for example, any good ant book[80, 81] or article[82-86]. That, in turn, raises the knotty but profound question of cooperation *vs.* competition, beyond the scope of this overview.

Accordingly, theories of this genre tend to be more restricted, such as to endosymbiotic events giving rise to major groups, most famously the mitochondria and other organelles within the cells of higher organisms (eukaryotes), now extended to many algal plankton[87-89] and other attributes[90]. In a like vein, mention could be made of lichens, fungal relations with plants, and so forth[91, and 92].

Lynn Margulis, that "maverick biologist"[93, 94] who first established the phylogenetic history of mitochondria[90], has written a book, *Acquiring Genomes: A Theory of the Origin of Species,* arguing that such relationships have been the central motif of speciations[95]. Although the review by DeDuve was luke-warm (predictably), her stature demands that she be taken seriously, e.g., her *Handbook of Protoctista* has been called "her second great contribution to biology"[92]. Her minor writings on microbial communities are always fascinating[96]. However, Margulis is by no means alone in propounding such a theory. Frank Ryan, author of *Darwin's Blind Spot,* is identified by reviewer Steven Frank as another "radical symbiologist"[97].

Nobody can doubt the importance of species interaction as a determinant of speciation outcomes. Nor can we resist the fascinating allure of various symbiotic relationships, e.g., the luminous bacteria in squid[98] and other sea creatures, or the gut bacteria (and yeasts[99]) in nearly all animals, including some 400 species in our own gut[100]. Yet, despite the pervasive existence of such interactions, a general theory of speciation based on symbiotic and other mutually obligatory relations has yet to be persuasively enunciated.

11. GROUP SELECTION? There are plenty of other theories, all disputed. To mention one, there is the question of the level at which natural selection acts. At the one extreme are those who hold that even individual genes are "selfish," ever striving to increase their numbers, rather like an infectious disease, made famous by Richard Dawkins in *The Selfish Gene*[101]. At the other extreme is the concept of group selection, meaning that an entire breeding population, not its individuals, is altered and optimized by natural selection. This theory, attributed to Wynn-Edwards in the 1960s, was subsequently "discredited" and so fell from favor but, as is so often the case, has recently enjoyed a recrudescence (see, for example, notes at the reference[102]).

Group selection is most clearly applicable to social animals like ants and bees, but one may argue that nearly all animals are social, at least when it comes to the mating game. The ordthodox or classical stance is that natural selection operates on the individual phenotype, improving its eyesight, muscles, teeth, etc., which improvements then diffuse through the group genotype as a result of its greater reproductive success.

12. CLIMATE. The idea that climate or climate change causes speciation is an old one. In anthropology, the notion that climate – or more broadly, geography – is the major determinant of human races and every detail of human culture dates back to the seventeeth century or earlier[103]. This idea is still very much alive today, at least among evolutionary biologists:

> "A number of hypotheses propose that climate-driven environmental changes during the past 7 millions years were responsible for hominin [human-like] speciation and the morphological shift to bipedality, enlarged cranial capacity, behavioral adaptability, cultural innovations, and intercontinental immigration events." [104]

Blame it all on the weather[105]. But most anthropologists, although recognizing the obvious importance of adapting to the environment, no longer take this seriously, for there are too many objections, such as the great diversity of cultures in places like New Guinea or parts of Africa despite identical physical environments. Biologists, however, continue to tender climate changes, such as glaciations, as causes of speciation. However, a test of this hypothesis by studying songbird DNA failed to show any impact of a period of glaciation on rates of change[106]. Rosenzweig has shown that when viewed over the long term, speciation rates are nearly constant, or more accurately, are fractal in time[19, 107]. So, it seems safe to conclude that climate by itself is not a serious contender for the explanation of speciation. For if all the features in the above quotation ("bipedality, enlarged cranial capacity ...") were really caused by the weather, then why didn't all the other animals respond similarly, growing bigger brains and rising up on their hind legs as we did? Conversely, if the weather was so important, why were so many species totally unaffected, plodding along unchanged through tens of millions of years of change?

13. THE ABSENT MIND. Other theories are considered later. Meanwhile, enough has been said to show that the theorists are casting many wide nets in search of a satisfactory answer, but clearly, the problem of speciation is still unsolved. This contrasts sharply with the picture in the 1970s, when this book was first drafted, at which time it was commonly said that the theory was essentially complete, and that only a few details remained. Those who still try to present a united front, such as Richard Dawkins, minimize the differences among these theories; "tempests in a teapot" or "quibbles" he calls them, but that is a half-truth. It is true if you believe, as he apparently does, that everyone still subscribes to the classical New Synthesis; but in this he is behind the times. The many positions sketched above, and others given later, some of which lay claim to a new New Synthesis or a Final Synthesis, or even a Holy Grail[108], actually differ from each other quite fundamentally, reflecting a real and deep disarray[109].

The other half of the truth is that all of these diverse theories are indeed alike in one very important way: they are all *materialistic and mechanistic.* That is to say, they all deny a role of *mentality* in speciation and evolution. On the other hand, they commonly contradict themselves on this, freely speaking of the importance of "mate choice," "preference," and "behavioral innovation" in speciation, apparently without realizing what a fundamental inconsistency this represents.

This is especially true now that sympatric speciation by sexual selection is openly acknowledged. Little wonder that these topics were so long forbidden, for as soon as such matters are opened to discussion, the barn doors of *mentality* are also blown wide open.

The more one reads in the recent literature, the more obvious it becomes that the old Newtonian tradition still reigns: living things are seen as passive responders to the great new god named Environment, and its handmaiden, Genes. Living things are allowed no role in determining their own destinies or qualities. The following typifies the attitude:

> "Sympatric speciation differs most significantly from geographical [allopatric] speciation in that it is triggered by ecological interactions [ref]. It targets ecological opportunities and produces new species with a high probability of survival. Evidence from fossils now indicates that sympatric speciation was crucial [ref] ... new mathematical theory [ref] has removed any remaining genetic qualms about it. ..." [19]

Note that such statements are entirely in terms of *external influences* on the biota. But is it not possible that innate mental properties, of the kind indicated in the previous volume, manifesting as choices and inventions, may play an equal or even predominant role? Generations of theorists have felt that "internal factors" (read *mental* factors) are essential to understanding the natural world, but such proposals could never be taken seriously because they conflict with the reigning epistemology.

No theory yet exists within the framework of science to deal with the psychology of free will. Accordingly, everything is framed in terms of genes and environment, though it is plain as day that the choices made by an organism can lead it to a new lifestyle and new species, even while fulfilling all of the objectivist tenets of orthodoxy. This will be seen in the next chapter.

6.

Maggot Points the Way

1. THE APPLE MAGGOT. Another mode of speciation, once popular among evolutionists but now rarely taken seriously, is that of pure choice – or more accurately, free will, in the sense defined and proved in the previous volume. It is rarely taken seriously because it doesn't fit in with the reigning epistemology. But we hold that this modality is *the* fundamental drive of speciation and evolution. On the other hand, this modality overlaps heavily with another which is indeed taken seriously, namely, *sexual* selection (next chapter). Let us begin with a well-known example.

Most fruit fly species are very choosy about the particular fruits they eat and lay their eggs (future maggots) in, such as the native American species, *Rhagoletis*, specific to the fruit of the hawthorn tree, hence commonly known as the *hawthorn maggot*. However, it was noticed in 1862, in the Hudson River Valley in New York, soon after the introduction of European apple trees, that some of these flies were infesting the apples, becoming known as *apple maggots*. Now, the interesting part of the story, said to have been developed by Guy Bush and now a modern classic[1], was the observation that the apple and hawthorn flies were *not interbreeding,* having adopted distinctive courtship rituals.

In other words, they were on their way to splitting into two distinct species, under clearly sympatric conditions. This case is now often discussed in survey articles[2], but not, in my estimation, with much illumination. (A similar case of an incipiently new fruit fly species due to a shift to a new host plant has more recently come to light[3] but is said to involve hybridization.)

What's going on here? Well, the intuitively obvious explanation is that some enterprising male fly discovered that apples taste pretty good and could just as well serve to incubate eggs; and so, to announce the novelty of his discovery, he performed an unusual dance, which just happened to appeal to a similarly non-conformist female. Anthropomorphic? Definitely, and without apology. Why should we presume that our own psychodynamics are basically different from the rest of the living world?

The anticipated objection to this kind of explanation is that it is heuristically sterile, it doesn't lead anywhere; but that is only because there is not yet any theoretical context for that kind of thinking. Going beyond the counter objection that orthodox explanations are even more arid, we shall indicate otherwise. That there are common psychodynamics at work in the living world opens up a very fruitful line of inquiry – no pun on apples intended – leading to many interesting conclusions; and that very basic laws of universal psychology govern these events and are capable of resolving nearly all of the conundrums in the theory today.

2. RELATIVE PROGRESS. Consider a hypothetical group of bear-like mammals who, once upon a time, on the fog-bound coast of Maine, discovered that they could catch fish in the ocean surf, even in the dead of winter, when other food was scarce. This practice, if continued for many generations, would lead to increasing commitment and adaptedness to that lifestyle (niche) since both natural and sexual selection would conspire to favor the relevant abilities (aquatic prowess, cold resistance, etc.), leading eventually to a marine-adapted mammal, perhaps first a kind of otter, then a sort of seal, or whale.

The origin of the special features of beavers is another good example, since it is self-evident that their customs and adaptations, too, must have arisen from an initial idea, a novel and clever coping strategy, to wit: the invention of constructing dams in which they could cozily hide and store food in through the long and perilous winter. Following the initial innovation, both sexual and natural selection would collude in succeeding generations to improve their adaptedness to this new niche, again giving rise to a speciational trajectory leading to the perfection of beaverdom.

Neither of these outcomes, otters or beavers, can plausibly be viewed as mere passive adaptation to environmental exigencies, much less attributed to random mutations. Instead, they were plainly driven by new ideas, i.e. inventions, discoveries. Yet, the trajectory of their speciation is wholly consistent with the precepts of orthodoxy, since all of the requisite genetic changes could easily have arisen from the operation of selection (natural and sexual) of mutations and polymorphisms. The rapidity of these adjustments would clearly be expedited by the agency of sexual selection, granting that in such a community embarking on a new lifestyle, mates will be preferred on the basis of their relevant talents, e.g., at swimming, or skill and diligence at damming creeks. Indeed, the only feature in which this account differs from orthodoxy is its insistence on *mentality* as the cause and motive of the trajectory.

These examples are of mammals because nobody can dispute the plausibility of these scenarios. On the other hand, as explained in the previous volume and hinted in previous chapters, we cannot draw a line between the accepted mentality of mammals and that of lower forms: birds, spiders, snails, insects, worms, yeasts whose habits often appear to be the result of equally intelligent invention, yet are attributed entirely to "instincts" arising from random genetic mutations. Thus, we must be prepared to accept similar modalities of speciation in lower forms, all the way down, as exhibited and formulated in later chapters.

Now, the important thing to notice here is that from the viewpoint of the animals involved, they are *progressing* towards a kind of perfection of their newly adopted lifestyle. We may call this *relative* progress because it is not necessarily true overall progress in the sense earlier defined (Ch. 2). However, we may postulate that the *motivation* for speciation is always conceived *as if* it were a true and absolute progress, in the same sense that human sociopolitical systems (Marxism, Nazism, Quakerism) are always purported to be not only revolutionary but evolutionary. But, to us dispassionate outside observers, all such efforts appear as nothing more than random probing. Indeed, most end up as mere change, not evolution. Narrow is the way.

3. THREE-STAGES. Accordingly, we see a three-stage scenario in such a trajectory. First, a novel invention or strategy by an individual, appreciated by at least a few peers (and mates), initiating the trajectory. Next is the "approach to perfection" of that trajectory by natural and sexual selection acting on the progeny of the original community over many generations. Last is the virtual perfection of the trajectory, after which no further change occurs. The species may thereafter persist indefinitely, even *actively resisting change*, having achieved its end, even as others of its kind depart on variant trajectories; witness the many species of birds, bats, ants, spiders, etc.

This scenario is consistent with the fossil record, notably the "punctuated equilibrium" concept of Eldredge and Gould, as earlier remarked. It is also consistent with certain outlines of human cultural evolution, since the same three stages are often pointed out for human empires: (i) innovation, defining a trajectory; (ii) development towards perfection of the implicit endpoint; (iii) long-term stability, that is, cessation of further significant change.

Furthermore, we suggest that similar *ideological* principles operate in both cases, animal and human cultures. That is to say, as clarified in the next volume, the lifestyle of each human culture is redolent with *philosophy,* a.k.a. religion, through which lens of logic it views itself as having attained a state of virtual perfection, superior to its neighbors and vital to the proper functioning of the cosmos. The only major difference between the human and animal cases is that human philosophies (and religions) are clumsy efforts to express by *language* the justification of the group's existence and activities.

4. ADAPTIVE CLASSIFICATIONS. In view of the above, it becomes necessary to reappraise the concept of adaptations. The present paradigm attempts to cram together into a single category *all* instances of change under the dogma of "adaptation by natural selection of random mutants." Although the historical reason for this was its commitment to the philosophy of Newtonian mechanism (supported by the discovery of genes), the following are obviously sharply and qualitatively distinct:

(i) *Passive and necessary adaptations*. This is the simple or classical case, easily seen in the laboratory, such as by selecting bacteria for various traits. It is *passive* because the organisms presumably take no active role in their adaptive response (granting that random mutation are the source of variation), and is *necessary* because they will die unless they adapt, and most do die. The changing color of the moth *Bistron* or the human sickle-cell gene fall in this category, as do most adaptations to climate and disease.

(ii) *Active and necessary adaptations.* Here we again assume that the organism must adapt (or die) but now we imagine that alternative coping strategies are possible, whence the response may be *active* – creative, inventive. For example, possible ways for a group of mammals to cope with a harsh winter include (a) deep slumber in a den, resulting in hibernation habits; (b) exploitation of some all-weather food supply, such as catching fish in the surf, or constructing an all-weather den, e.g., beavers; (c) seasonal migration to more hospitable environs; (d) caching food reserves for later use, as do many foxes, birds, rodents, etc.

(iii) *Active but arbitrary adaptation.* Here the organism finds or creates some new way of life, not by any demonstrable *necessity*, e.g., the rise of beavers. That is, the original beaver ancestor must have had ways of coping, hence did not *require* dam building to survive; but the offspring of the founders evidently saw merits in beaverdom, carrying it to perfection, with real survival advantages to boot (as measured by large population). Perhaps in this case we should place the word "adaptation" in quotes because this class of modification is not forced by natural selection; "innovation" is a better term here.

(iv) *Passive but arbitrary adaptations.* This was added for symmetry with the three cases above, yet is actually seen in, for example, sexual selection, whereby a bird's plumage may change in style or color over the generations owing to mating preferences: the male appears to be passively, helplessly, unwittingly adapting to the female's arbitrary whimsies – to her "eugenic agenda". It has been said that she dresses him up to please her fancy. Again, "adaptation" should be placed in quotes here because these features are not adaptive in the usual sense; indeed, they may be downright maladaptive (next chapter).

It is clear that these four kinds of adaptation form a spectrum between extremes and cannot plausibly be lumped together under the same explanatory principle. By analogy, human entrepreneurial enterprises are also subject to "natural selection" by market forces; but to assert that any particular such enterprise is "explained by economics" is absurd.

The existing theory fails to make any such distinctions, lumping them all together under "natural selection." Why? Because it is forbidden to invoke *mentality* as a component of evolutionary theory. Although novel inventions and strategies are widely reported and are sometimes mentioned as factors in evolution, they are actually heretical, i.e. are in conflict with the presumed dispassionate mechanisms, and so are marginalized in accepted treatments. There is no place for them in the logical fabric (epistemology) of the theory.

Hence, the theory is in need of radical revision. Why has not such an obviously needed revision been undertaken before? The answer to that is it would conflict with the assumptions of *physics and chemistry*, saying that mind plays no role. Accordingly, the revision we have in mind inevitably implies a revision also of how the entire material cosmos really works.

That is to say, we immediately face the awkward problem of phylogenetic level. While there would be little objection to recognizing the importance of novel inventions, strategies, and cultural copying in monkeys, other mammals, or even birds, and lower forms such as snakes, insects, worms and yeasts are presumed to be incapable of intelligent invention, devoid of mentality as if they were mechanisms akin to a wrist watch. But there are no grounds for that dogma (book I).

A word is in order about the role of natural selection in the operation of novel inventions: not all of them will succeed. Bad choices will be weeded out by natural selection. Consider a parallel from anthropology. It is well known that neighboring societies, as in New Guinea or Melanesia, commonly had sharply contrasting dietary habits, e.g., one group loves turtle meat while its neighbor is disgusted by it, making it taboo, and so on with endless details. Foods that were "good" *vs.* "bad" were bound up with the religious philosophy; and yet, all were in good health when first encountered, enjoying near-perfect diets, for the obvious reason that if they did not have adequate dietary habits, they would have disappeared. We shall return later to the essential reasons why neighboring groups tend to differentiate themselves so sharply, for therein lies the essential motivation for speciation.

The parallel with species of Darwin's finches, fruit flies, etc., is plain. Of course, some will reject this parallel on grounds that humans are utterly different from lower animals. But that is a myth for reasons given in book I.

5. SEXUALITY. It is a fact of living existence that most speciational innovations in the animal world can be propagated only by sexual reproduction; in other words, requires the complicity of both sexes. (There is mounting evidence that even lower organisms, clear down to viruses[4], engage in something akin to sexual intercourse, though some, such as yeasts, appear to have more than two genders, and a great many can switch between sexual and asexual reproduction.) That brings us to the next chapter.

7.

Sexual Selection

1. PERSPECTIVE. As earlier noted, two major paradigm shifts since the 1970s have been the grudging acceptance of *sympatric* speciation, and that *mate choice*, i.e. *sexual* selection (as opposed to *natural* selection), is a major agency of change. In other words, Darwin's original ideas are slowly edging back into vogue. Here are a few vignettes on the fall and recent rise of Darwin's take[1] on sexual selection:

> "Without knowledge of the genetic mechanisms involved, however, Darwin was unable to provide a convincing hypothesis for the origin of female choice. As a result, sexual selection theory was discredited *in toto*, and for decades was largely ignored. Paradoxically, it is precisely this female choice loophole which has recently been the focus of revived interest and heated controversy." [2].

> "These criticisms persisted, contributing to a 100 year dormancy of the theory. In the last two decades, however, assessment of sexual selection has grown to be one of the most active areas ..."[3].

How did this revival happen? For as we have seen, sexual selection is no more compatible with orthodox precepts today than it was in the 1970s.

Part of the answer is a relaxation of intellectual standards, to hell with epistemology. Another part of the answer is that sexual selection has been redefined to bring it under the umbrella of *natural* selection, thus evading the incongruity of it. For example, the Browns:

> "These scholars [earlier theorists] continue to draw a sharp distinction between natural selection and sexual selection, and they continue to view the genetic logic behind sexual signals as somewhat odd: males have them because females like them ... Over the past ten years, though, a new perspective has emerged, from which *sexual selection is seen as in integral part of natural selection.*" [4] Italics added.

Another tactic of the revisionists has been the tenet that mate choice cannot *initiate* speciation, but is a mere secondary consequence, helping the process of divergence, whence it is called *reinforcement*[5]. But a step in the right direction is the conclusion of Hoskin *et al,* not in the title or abstract of their paper (lest it draw fire?), but near the end:

> "Reinforcement has been viewed to have a role only in the final stages of speciation between already-differentiated lineages whereas the present results show the potential for reinforcement *to be the sole cause of speciation*"[6]. Italic added.

This is the first and only time I have seen that radical conclusion in a respected journal. Meanwhile, most work on sexual selection has portrayed it in utilitarian, materialistic terms, such as that it functions to advertise "good genes." Never mind that such advertisements are by no means universal. Thousands of species show so little sexual dimorphism that only experts can distinguish males from females. But others show such extreme dimorphism as to have been mistaken for separate species, and may actually be adapted to different resources, as is the case with males and females of a certain hummingbird[7]; and female spiders are generally much larger than males and differ markedly in habits, e.g., only females spin webs.

That is to say, sexual selection is now "explained" in terms of genetic mechanisms, material benefits, ecological fitness theory, mathematical optima, offspring sex ratios, life history traits, selfish genes, kin selection, honesty principle, handicap theory, development, and so forth *ad infinitum*, depending on who you read, but never in terms of *psychology*. This ocean of verbiage reminds us of the octopus' defense strategy: it emits a cloud of ink when threatened.

We are proposing the contrary, that all of these materialistic optima are *consequences* of choices, not *causes*. The reason for all this confusion, we submit, is a stubborn refusal, or inability, to confront the reality of *psychological* factors as the motor of evolution, and that these factors are essentially the same in *all* organisms. It was precisely to legitimize the psychological approach that we wrote the previous volume, which may be viewed as a prolegomenon to this one, for it supplies the epistemological justification, if you will, for the reality of psychology in the larger world. Nor do we stand alone with Lotka in defending such a view. The following remark by physicist Richard C. Henry adds to the chorus cited in the previous volume:

> "[After] the 1925 discovery of quantum mechanics ... bright physicists were again led to believe the unbelievable, this time, that the universe is mental. [He then quotes or alludes to Sir James Jeans, Sir Arthur Eddington, Niels Bohr, John H. Marburger III, Murray Gell-Mann, and concludes:] The Universe is entirely mental. ... The world is quantum mechanical: we must learn to perceive it as such. One benefit of switching humanity to a correct perception of the world is the resulting joy of discovering the mental nature of the Universe." [8]

By way of historical perspective, it is instructive to observe how earlier evolutionists wrestled with the problem of sexual selection, often contradicting themselves between the covers of a single book. Thus, Julian Huxley writes the following on pg. 91 of his influential 1948 book, clearly affirming the centrality of mental factors *vis a vis* sexual selection:

> "I can confidently assert that Darwin's theory of sexual selection, though wrong in many details, yet was essentially right: that there is no other explanation for the bulk of characters concerned with display, whether antics, song [etc.] .. then that they have evolved in relation to the mind of the opposite sex; that the *mind* has thus been the sieve through which various courtship characters must pass if they are to survive." [9].

But on pg. 107 of the same book, he takes the opposite tack, discounting mentality:

> "Indeed, if you stop to think about it [the cuckoo] cannot know what he is doing... No, the whole train of actions is the outcome of a marvelous piece of machinery with which he is endowed by heredity, just as he is endowed by the equally marvelous adaptive mechanism of his feathers... The acts are purely instinctive ... Like coughing; it has been brought into being by the long unconscious process of natural selection ...." [9].

Again, Jacques Monod, the Nobel laureate in biochemistry, wrote the following on pg. 126 of his equally influential book of 1972, affirming a key role for choices and desires:

> "It is evident as well that the *initial choice* of this or that kind of behavior can often have very long-term consequences, affecting not only the species but all its descendants. ... One is therefore correct in saying that the sexual drive – or better still, *desire* – created the conditions under which some magnificent plumages were created." [10] Italics added.

But on pg. 112, he conveys the exact opposite:

> "And since random mutations constitute the only possible source of modification in the genetic text ... it necessarily follows that chance alone is at the source of every innovation, of all creation in the biosphere. Pure chance, absolutely free but blind, is no longer one among other possible or even conceivable hypotheses. It is today the sole conceivable hypothesis." [10]

It appears that such authors are themselves torn between the common-sense view that mental motives are important *vs.* the reigning dogma that mind is irrelevant, or is limited to higher mammals. To me, it is almost incomprehensible how men of science, ostensibly in pursuit of the truth, can be so purblind. We say "almost" because this is also an excellent lesson on the blinding power and tenacious persistence of the underlying implicit cultural epistemology.

2. MATING DIVERSITY. One might expect that organisms which descended from a common ancestor would have similar qualities. That is, birds have feathers, and fish have scales, because each group has a common phylogenesis. But this is not the case when it comes to behavior, *sexual* behavior in particular. For what we actually see is a bewildering diversity of mating practices even among closely related species, or perhaps *especially* among closely related species, and often within the same geographical region (reminding us again of human societies, e.g., in New Guinea). For example, American blackbirds consist of:

> "... a family [of 90-odd species] ... one of the most diverse among birds in terms of social behavior: some are monogamous, some polygamous, some promiscuous; some care for their own young while others are brood parasites; some defend dispersed territories, while others are colonial ... wide array of patterns of plumage and size dimorphism ... communication ... sex roles ..."[11] .

Even lowly crickets have a huge variety of mating customs[12, 13], as do starfish[14], crustaceans[15], lepidopterans (butterflies, moths)[16], fireflies, and beetles generally. Likewise for dozens or hundreds of other families of species, if not all[17, 18]. The sexual and parenting habits of amphibia, most of which lay eggs, was once thought to be rather monolithic; but careful study has again revealed great diversity, e.g., among frogs[19, 20] and crocodilians[21]. (A comparative study of frog mating calls among closely related species was somewhat ambiguous but again suggests great variety[22].)

Though more difficult to study, it is now suspected that the hundreds of species of bats, many of which appear very similar and share a common range, employ their distinctive signaling capabilities in the mating game[23]. Likewise, electric fish[24, 25] and perhaps even bacteria[26]. One could go on, such as the immense variety of ants, or to the variegated courtship dances and songs of fruit flies, such as the two species which appear almost identical, yet one of which appeals to visual cues, the other to smell (pheromones)[27].

Genetic methods have brought to light a new puzzle for the theorists, namely, the existence of "cryptic" species, meaning those that appear identical in every respect except that they do not cross breed[28, 29]. I like to think of these as the equivalent of ethnic groups. They are the ultimate extreme of sympatric speciation, since they even live together and exploit the same resources. Incidentally, this violates the hallowed "law of limiting similarity," which holds that two species which exploit the same resources cannot coexist since one must displace the other (a recipe for ethnic strife, too).

What to make of all this? Well, starting with the orthodox dogma, the basic idea is that each such mating practice arose in consequence of natural selection of random genetic variants by strictly utilitarian considerations (that is, *secondary* to the species' mode of resource exploitation) because only that kind of explanation is consonant with the premises. But we take the opposite tack, that the invention of a novel mating strategy was often or usually the *primary* event, initiating the speciation, and that the economic optimalities subsequently observed[30] were aftermaths, in the manner earlier indicated; i.e. natural selection demands that any behavioral innovation will be materially practicable, otherwise the lineage will die out.

This need not exclude, of course, the possibility that exploitation of a novel resource could be used as part of the mating game. For instance, referring to the previous chapter (apple maggot), the first male fruit fly to sit upon a big fat apple might be quite appealing to a female accustomed to the ho-hum hawthorn fruit and her usual boorish suitors. Variation of female tastes, if not sheer whimsy, has been demonstrated in several species of spiders, as well as the satin bowerbird[31], and doubtless many others known to expert naturalists.

3. HONESTY AND HANDICAPS. The recent acknowledgment of sympatric speciation and sexual selection, belated though it was, is certainly a welcome sea change from the 1970s. But it came with a heavy price, namely, the requirement of explaining these things without doing violence to orthodox precepts, a tall order. In fact, it is impossibly tall, meaning that the reality of psychological choices can no longer be evaded, except by appeal to psychology cast in terms of genetics (sociobiology, evolutionary psychology), which we shall later see to be largely a failure, or more precisely, a wrongheaded interpretation of the facts.

The guiding precept in such explanations is that the features in question must be shown to offer some kind of *material advantage* to the animal, or more exactly, must improve the *reproductive success* of the organism, this being the bottom-line criterion of "fitness". Classic examples include the peacock's fan and the antlers of elk and deer. In extreme cases, one hears of "runaway sexual selection," a classic case being the outlandishly huge antlers of the Irish elk, said to have gone extinct for that reason (but more recently attributed to paleolithic hunters).

The very existence of such features was an embarrassment to the New Synthesis, rooted as it was in the Protestant work ethic, since such features had no obvious practical function. On the contrary, they appeared to be anti-utilitarian, wasteful of precious resources (antlers and feathers are made largely of precious protein), needless baggage. Therefore, such appurtenances were ignored for as long as possible. But nowadays, such things are "explained" by the claim that such features actually do have a practical function, to wit, they exhibit genetic quality (fitness) to a potential mate. To set the stage, a perspective on a paper of this kind opens as follows:

> "Peacocks would not have such large and elaborate tails were it not for the fact that peahens prefer their partners to be so ornate. But why are females so choosy? There are two standard answers. Fisher suggested that males have elaborate traits purely because females find them attractive [ref] ... Others have conjectured that showy traits indicate the quality of the male [ref Williams]. If only peacocks with 'good genes' can afford to bear bigger tails, peahens should mate with them to ensure that their offspring are viable and healthy." [32]

But these "two standard answers" are not mutually exclusive. They differ only in that the second and more recent one (attributed to Williams) offers a *material reason* why the female finds such ornaments attractive. In that particular study, the problem was why the females of a certain species of stalk-eyed fly prefer males with eyes mounted on extremely long stalks and concludes that such eyes are indeed a marker of superior fitness[32]. This has been called the "honesty principle" since such features cannot be faked by mere strutting. Other interesting but sometimes less clear examples include the horns of beetles[33] and of lizards[34], the latter involving protection against predation by birds and coloration of bird beaks[35]. Similarly, it has been shown by Gustaffson *et al* that the size of the white patch above the beak of the male collared flycatcher, which is a criterion of female mating preference, is also an honest indicator. The perspective on that paper explains as follows:

"An important theory in sexual selection views display traits as reliable signals of quality – if the traits are costly to maintain, only individuals in good condition can afford to maintain the more elaborate displays and hence may be better mates or more formidable opponents. Such costs lie at the heart of an independent theory of life-history evolution which predicts a trade-off between current reproductive effort and the ability to invest in future reproduction. It is these two theories that Gustaffson *et al.,* have connected." [36].

This is all well and good (though it boils down to the same kind of trade-off decisions we all face in everyday life). But, can it be pure coincidence that the peacock's display is also beautiful? At the same time, we must wonder why, granting the cogency of such accounts, such fancy displays are not universal. This reminds us that similar accounts are often applied to explaining human sexual selection, to the effect that extravagant displays of wealth, for example, serve a similar purpose. However, this ignores the vagaries of taste. Among at least some modern Americans, displays like fancy cars, big houses, and gold watches are viewed as honest indicators of lowbrow stupidity, repugnant rather than appealing. Lavish funerals were once a mark of wealth and status in Europe – until, that is, the "lower classes" could afford them[37].

Herein lies another clue to the real causes of speciation: the differentiation of a population into groups owing to contrasting rationalities of what is *righteous and true,* i.e. alternative values: some human females go for gold watches, others for big muscles, others for political power, and still others prefer artistic types, poets, religious or ideological commitment, musical or literary talent, sense of humor, *etc.* Granting a similar if narrower range of tastes in the animal world, then it is easy to see how variant groups could split off by such criteria.

In that vein, we see in the animal kingdom that extravagant displays are actually the exception rather than the rule, and only occasionally may be explainable in part by natural selection. Thus, ground-nesting birds cannot afford to attract attention, although on the other hand, this was not necessarily by natural selection, it may have been a practical choice mediated by sexual selection, and would obviously be related to the presence *vs.* absence of predators.

Coming back to ornamental features, mention may be made of the "handicap principle" advanced by Zahavi in 1975 and later at book length[38], said by Kodric-Brown[4] to have been the forerunner of the "good gene" theory developed by them and by Malte Anderson (1982) and others they cite. In a nutshell, this principle holds that elaborate features on males (the peacock's fan, the elk's antlers) are actually *handicaps* in the game of survival, effectively hobbling the animal, so that those that can survive despite the excess baggage must be very fit indeed, hence are preferred by females! However, if this was all there was to it then males having various other handicaps – a limp, say, or a missing tooth or eye – should be practically irresistible to the ladies. So the logic is less than compelling. Further arousing our suspicion is the observation that removal of the dominant male of a certain species of fish, the one with the brightest colors and thus most visible to predators, results within minutes in lesser males showing the same brilliant colors, indicating the same "good genes"?[39]. In another kind of fish, removal of the dominant male, the largest and strongest, immediately causes lesser males to grow equally large and strong[40], the term "tortured logic" is apropos here.

What galls us is the aplomb with which such mating displays have been assimilated into orthodox theory, as if all is explained in terms of materialistic utility, having arisen by natural selection of random variants, devoid of the mental dimension that is the real hallmark of life, and holds the key to its understanding. Thus, we predict with some confidence that today's evolutionism will go the way of behaviorism and a thousand other once popular -isms now in the dustbin of history.

4. COLORS, ETC. Few sights are more glorious than the flash of colors of parrots in flight in their jungle homes (not cages), usually in pairs. Nothing more glorious, that is unless you happen to be fonder of some of the hundreds of other birds (it is said that bird-watching is second only to gardening as America's favorite hobby, a hopeful sign) or butterflies, beetles, reef fishes, squid, octopi, chameleons, and so on. As often as not, these colors are extremely sophisticated, biochemically speaking, or else rely upon extraordinary optical effects, as in butterflies[41-43] and squid[44], the envy of modern nano-engineers. In many cases, the chemicals for the colors are actively sought out and acquired from sources in the environment (as are many chemical defenses). This reminds us of certain Amazon bees, which are essential for pollinating Brazil nut trees, whose females demand that their suitors wear a perfume acquired from a specific orchid, thus explaining why those trees fail to reproduce in monoculture: the need for the orchids was overlooked.

So, what is the orthodox explanation here? The first echelon of explanations stems from the obvious importance of colors in the mating game. Thus, they serve the "practical function" of being honest indicators of fitness. Duh. This was shown, for example, by the brightness of some orange bird beaks[35]. But is this really a general explanation for why such splendid colors exist? Can it be pure coincidence that we humans so admire them? Yet, among ourselves, we are often ambivalent in our tastes, as in berating someone for being too gaudy, too flashy.

A second explanation in terms of practical function, commonly given in those excruciatingly boring TV shows pontificating on evolution, is that they warn predators that they are toxic to eat. This appears to be widely true, and may also explain why so many tasty creatures have come to mimic those which are toxic: predators avoid the mimics as well as the genuine article. But, to play devil's advocate, consider the possibility that many insects might innately *prefer* beautiful colors, eliciting their rise, but only those with poison defenses survived. That is to say, one may postulate a psychological motive for the origins of such features, insofar as orthodoxy gives no explanation for how or why the bright colors of toxic forms were acquired in the first place.

In a like vein, consider the fact that a great many creatures are socially gregarious, at least periodically, as seen in huge schools of fish, flocks of birds, marching crabs, lobsters, squid, and so on. The orthodox explanation, as everyone has heard on those TV shows, is that "the reason for this" is safety against predators. But common sense tells us that this is absurd, since a huge school of fish is exactly what big predators prefer, for they can gobble up thousands at a single pass. Indeed, a favored feeding strategy of certain cetaceans is to herd fish into tight clusters by bubble walls, the more easily to eat them *en masse.* The only effort I have seen to actually test the standard explanation was an experiment with a type of marching cricket, which confirmed the theory, but was so poorly designed that I am not convinced[45]. However, another and more convincing experiment found the very opposite, that clustered prey was eaten much faster than scattered prey[46]. One could go on making fun of these *Just So Stories* but we need to move on.

5. JEALOUSY, ETC. One observes in the recent literature a growing tendency to portray animal behavior in human terms. This was previously forbidden as "anthropomorphic" (although in reality it was always implicit) and continues to be studiously avoided. Apart from recognition of sexual selection, with its human-like implications, and thousands of articles on the role of social status in lower forms, one may note growing appreciation that human economic systems are closely comparable to ecosystems[47], as Lotka and others had earlier indicated very clearly.

Closer to the bone is a series of studies showing that jealousy is a real factor in animal mating. Thus, L. Dugatkin and colleagues have shown that the preference of female guppies for males according to brightness of his orange color is modified if another female is present and showing interest in a drabber male[48], confirming and extending prior work[49, 50]. As will emerge more clearly in another book of this series, there are endless such parallels between animal species and human cultures. To mention just one, among lek-mating birds, junior males must often wait a long time before they can acquire a mate, a situation comparable to the social arrangement of the Tiwi people of Australia, among whom only the most elderly males are able to acquire wives[51]. Evolutionary biologists are constantly at pains to find rational explanations for such things, e.g. lek mating, in terms of what might be termed the *Practical Pig* model, i.e. that it must improve "fitness." Once you get tuned into this habit of thought, it becomes quite comical.

This yet-nascent paradigm shift, the new willingness to openly draw parallels between humans and animals, is open to two interpretations, First, the orthodox one, that we are like animals in being instinct driven (as in sociobiology, a.k.a. evolutionary psychology); or the converse, which is the view we are expounding, that animals are like us in having a modicum of intelligence and are driven by the same fundamental ideological motives.

6. MIDSUMMER'S NIGHT DREAM. In the mind of every living being, the cosmos exists only there, in his mind, for him alone; and he is at the center of it, with no direct awareness of the consciousness of others. It is inevitable that he (or she) will regard himself as the pinnacle of all creation, the best and finest living thing in the world and accordingly, will seek to reproduce himself. Leaving for later the mystery of the origin and meaning of the two sexes, the harsh reality is that he must find a mate. How best to do that?

One way might be trying a novel display strategy, which might include exploitation of some new resource (as with the apple maggot), perhaps thereby attracting a female who is herself searching for something higher (and are we not all constantly searching?). Presto! A new species is born. To clarify, let us revert to the Kiplingesque mode imagining the speciation of fireflies, of which some 2,000 species are known, many of which are distinguished chiefly or entirely by the flash codes by which they attract mates. The codes of about 125 species are known, and you can attract a specific species simply by flashing a pen light in the right pattern. Some have even become predatory on fellow species, flashing phony decoy signals to lure prey fireflies to their jaws, sometimes for the purpose of stealing their chemicals[52-55].

Now imagine yourself as a firefly, flashing your heart out in amorous signaling. But no females respond. Why not? "Can they be blind to my gifts, my glossy chiten, and my lovely luminescence? Alas, so it seems. But let them see my virtuosity: dot-dot-dit-dit-dot-*dittah!* Yes, by god, here she comes! Oh, what a night!" A new species is born. We leave for later the supposed "underlying genetics" responsible for the subsequent reproductive isolation. But as a kind of clue or prelude, it is already well established that females of many species are able to eject or neutralize sperm of undesired males, apparently at will.

It might be objected that the above scenario is heuristically sterile, fails to open new doors of empirical investigation, insofar as the criteria for differentiation (and thus speciation) seem rather capricious, willy-nilly, and random. However, this is no worse than the existing theory, which is all down to random mutations. On the other hand, it is possible to develop a very systematic understanding of species diversity by appeal to the diversity of *human* societies. The systematics of human social diversity (languages in particular) is the subject of another book of this series, and an effort was made to encapsulate relevant parts here under the subtitle *Levi-Strauss and Willy-Nilly*, but it was cut from this manuscript because an adequate explanation became too lengthy and entailed lengthy digression.

7. WHO'S DRIVING? It is apropos here to reply to the oft-repeated argument that females drive evolution insofar as males are helplessly obliged to respond to her whimsical tastes. Thus it has been said that she "dresses them up" in whatever feathers or colors please her fancy. While it may be usually (but certainly not always) true that males are slavishly obedient to her wishes (to the norm) that fails to account for the rise of novelty. However, when we recognizes that some fraction of males are constantly experimenting with novel strategies, some of which are occasionally successful in piquing her interest, then we see that males tend to be responsible for innovations, the females then "reinforcing" those which she likes. So it appears that yes, she does the picking and choosing, but it is he who supplies the pool of behavioral variants.

8. SPECIOSITY. The same train of thought may illuminate another mystery: why it is that certain groups have such a large diversity of species, even within the same region, while others do not? This is akin to asking why twentieth century America became such a hotbed of technological innovation. A famous example is the cichlid fishes in the lakes of the African rift valley, such as the 165 species in Lake Tanganyika alone, all of which sprang from the same ancestors:

> "Lake Malawi, for example, has more known fish species than any other lake in the world; indeed it has more known fish species than inhabit all the freshwaters of North America. ... Even more intriguing is that most fish species [in the rift lakes] are unique to a lake. Every Tanganyikan cichlid, for instance, occurs only in Lake Tanganyika, and all but a few cichlids of Lake Malawi occur only there. Although these lakes are only 200 miles apart, they have not a single species of fish in common." [56].

There is no doubt that this was sympatric[57, 58], yet there is still no convincing general explanation for the variety, despite dozens or hundreds of papers on these fish (some very insightful[59]), but all falling short of a solution. The same question applies, of course, to bats, fireflies, the whole biosphere.

Once again, it strikes us that the most plausible solution is psychological (or cultural), and boils down to the proposition that speciation is *contagious and self-catalyzing*, in the sense that the presence of at least one alternative modality tends to suggests the possibility of other alternatives. Variety breeds variety. Thus, firefly A, observing the success of firefly B's artfully modified flash-dance, is inspired to go it one better. This scenario is supported by the increasing acceptance of imitation and "cultural learning" in the animal world[48, 49, 60].

As remarked above, further exposition of this principle is best left for the next volume, where it is phrased in terms of human cultural diversification, the dynamics of which are more transparent; but the general idea is that the potential for departures depends heavily on the visible horizons. For example, the aboriginal Australians, although having considerable diversity of languages and other cultural elements, were nevertheless quite monolithic when compared to, say, those of New Guinea or North America. This leads back to the Levi-Straussian idea of variety generated by a series of systematic contrasts, oppositions, and inversions of the beliefs and practices of the parent group or neighboring groups; and which are inherently limited by the initial supply of differentiable features, for the same reason that the number of words that can be composed from a 10-letter alphabet is much less than from a 20-letter alphabet. One cannot conceive of varying some custom or habit which does not yet exist.

9. SUMMING UP. The bottom line is that sexual selection, speciation, and evolution generally can be formulated in terms of psychological principles just as well as, or even much better than, materialistic ones. In other words, the same facts are open to an entirely different interpretation. This idea is well expressed by the following in which a book reviewer sees an entirely different interpretation for the facts of the matter under discussion, which happens to be crime and violence:

> "This alternative account of crime ... is the one I would defend. But the point of introducing it is simply to note that it is broadly consistent with the [same] empirical conclusions. ... Yet it is a theory radically different from theirs – positing different psychological mechanisms and different dimensions of cognition and personality – and the choice between the two will, for now, have to be made on philosophical grounds." [61]

Let us stress, however, that we do not propose the psychological formulation of evolutionary theory as a total *alternative* to orthodox materialism, but rather as *complementary* to it, the subjective side of the objective facts, as essayed in the previous volume and clarified later.

8.

Altruism

1. ALTRUISM? According to Webster, altruism is a "selfless regard or concern for the well-being of others." As one might guess, such a notion had no place in the New Synthesis and so was dismissed as a romantic illusion, unique to human moralistic canons, disallowed from serious theoretical discourse on evolution, along with sexual selection and sympatric speciation. Until, that is, a way was found to explain it in terms consistent with the traditional paradigms of bastardized Darwinism.

When 'the explanation' finally emerged, found in a long-overlooked paper by superhero W.D. Hamilton, the topic of altruism was catapulted from obscurity to a hot topic of the 1980s and yet today, albeit now muted. The following is a typical appraisal of Hamilton's stature; this by Mark Ridley opening his review of Hamilton's collected papers.

> "W.D. Hamilton is a giant of evolutionary biology who has brought about two major revolutions. The first concerned altruism ... and in 1964 [he] showed how natural selection can favor it. ... The second revolution was sexual ... the existence of sex is the outstanding puzzle in evolutionary biology. ... The papers in the book are already classics ..."[1]

Although the originality of some of his key papers has been questioned[2], few doubted that he opened a new frontier, and similar praises are widely echoed, as in many laudatory obituaries[3]. Hamilton was clearly a main inspiration behind Richard Dawkins' hugely popular and influential *The Selfish Gene*[4] and his others[5, 6]. For an eye-opening profile of R. Dawkins, with comments on his squabbles with his American counterpart, Stephen J. Gould, see Parker's article in the *New Yorker*[7].

2. THE BIG IDEA. Hamilton's breakthrough was partly a factual discovery but also rested upon a hypothesis, a particular interpretation of the facts. The assumption is that the paramount concern of every organism is to project its genes into future generations. However, alternative hypotheses which better explain the same or more facts readily come to mind, such as by replacing the word *genes* with the word *ideas* or *ideals*. The factual part was the observation that in many eusocial insects such as ants and bees, the degree of altruism correlated with genetic relatedness. In other words, it was shown that altruistic behavior actually functioned to promote the genetic representation of the actors in the next generation, if only indirectly.

> "Altruism may in fact be selfish if it aids one's closest genetic relatives. W.D. Hamilton showed in 1964 that the genetic benefits of nepotism [kin-helping] can exceed its costs, and he predicted that organisms would often recognize and respond differentially to their kin." [8]. (Waldman notes that Hamilton's key paper[9] was ignored for 15 years, until about 1980.)

The idea is that by helping close kin, which share many of one's own genes, one is thereby helping to project one's own genes. That is to say, if a male cannot gain direct copulatory access to a female, then helping his nearest kin to gain such access is the next best strategy. This was first documented in eusocial insects such as bees, wasps, and ants (hymenopterans).

> "Hamilton showed that 'hymenopteran workers are more closely related to their sisters than to their own offspring and that they increase the propagation of their genes by aiding the reproduction of their mother queen.'" [10].

This idea was worked out in mathematical detail (see later on game theory) showing that the extent of helping behavior should be proportional to the fraction of one's own genes thereby projected, as recounted at tedious length in R. Dawkins' books. The predictions were often borne out, unleashing a flood of literature on altruistic behavior. A book review shows a graph depicting the rapid rise in numbers of publications citing Hamilton[11], and articles regularly appeared in leading general journals like *Science, American Scientist,* and *Nature*[12-16]. All aspects of the biology of behavior were impacted. One leading expert on ants revised his whole approach to them in light of Hamilton[10], and another major work on ants is completely based on kin selection theory[17]. A sub-category of these studies emerged to focus on higher primates[18-20], with all kinds of moralistic spin-offs, some briefly considered later on.

3. NEW MENTALISM. By the way, the enthusiasm with which Hamilton's thesis was greeted is a further indication of the afore-mentioned 'creeping mentalism' infiltrating evolutionary theory. That is to say, the notion of *striving* had long been banished from the literature in the course of disposing of Lamarckism, but here it is again, quite unavoidably, attached to the idea of *striving* to project one's genes into the next generation. Some writers, notably Richard Dawkins, are at pains to deny and disparage this connotation, repeatedly insisting that everything happens by blind compulsion; but not very successfully, however, in view of the picturesque language he necessarily employs.

But the trend was growing. For example, in connection with some altruistic models discussed later, one authority writing in 1991 applauds a "shift ... from the gene as a unit of selection to a social psychological mechanism as the important unit of analysis"[21]. Others opposed the emerging new paradigm, sometimes bitterly. We already mentioned Gould, and may add is no less an authority than Eldredge[22]. Bernard Davis railed against the new paradigms at book length, defending the 'orthodoxy' of his day[23].

4. CALCULATING ANIMALS. Hamilton's theory is often supported by rather complicated calculations purporting to demonstrate how various types of observed mating strategies are perfectly optimized to project the maximum fraction of ego's genes to the next generation. The implication of such calculations is that the creatures themselves have knowledge of their degree of relatedness and are able to perform those calculations themselves; unless, of course, it's "just instinct", whatever that means. But it cannot be just instinct because in case after case, both in the wild and in the laboratory, such strategies show fine-tuned adjustments to immediate circumstances. So it seems that they do indeed 'know what they are doing.'

Human language may be viewed as a stuttering effort to reformulate the realm of pre-verbal or 'instinctive' knowledge into verbal (intellectual) terms. Science and mathematics constitute one special compartment of that enterprise. To illustrate, Alfred J. Lotka, inveterate math buff that he was, demonstrated that when a cheetah runs down a gazelle, say, it follows a specific path, known as the *curve of pursuit*, which exactly intersects the path of the victim in the least possible time (or distance)[24, 25]. Now, does the cheetah actually calculate this path? Well, yes and no, depending on your definition of calculate. It somehow gets the right answer. Likewise when we catch a ball, or when a falcon dives to its prey, etc.[26, 27] or, coming back to genes, when some creature computes the optimal altruistic strategy for projecting its genes. By the way, when we read that certain nerve cells in the brain "model the laws of physics"[28], the reality may be the other way around: it is our laws of physics that are the models, like toy trains.

Thus, all of the game-theoretical math models are a double-edged sword for orthodoxy, revealing on the one hand that animals actually do often conform to them, but at the same time raising the not inconsiderable question of how the animals manage to know their degrees of kinship and then calculate these complex ratios. To illustrate, the dunnock, a common bird in Europe, is observed to practice a great variety of mating customs, from monogamy to polygamy to polyandry and various combinations or intermediates thereof, enough to stump the theorists who struggle to account for any one such custom[29]. It turns out that all of this variety is a matter of flexible expediency, determined by such factors as how far the bird must range for food, the size of a territory a male is able to defend, and other trade-offs and conflicts of the kind any hard-pressed human parent is familiar with. Clearly, animals are not stupid. Yet we persist in our notions that their behavior is all hard-wired, as if we alone are free – and the sociobiologists doubt even that, insisting that we, too, are hard-wired. Indeed, there is an opinion that only *civilized* humans are capable of scientific-like reasoning, since, for example, it has been seriously doubted whether certain peoples living in the Amazon jungle have knowledge of basic concepts of geometry[30].

5. HAMILTON COMPROMISED. Like nearly all discoveries in biology, after the first blush of excitement over some new principle, exceptions and complications start coming to light. This has been true of Hamilton's formulation, with the result that it is no longer seen as the Rosetta Stone that it seemed in the 1980s. Among the first cases studied which appeared to depart from Hamilton's rule was that of the naked mole rat, the first vertebrate found to exhibit true eusociality, e.g., colonies of some 100 or so of these animals have only one queen, a bunch of drones, and so forth[31]. When Braude and Lacey wrote up their findings from field studies[32], as opposed to laboratory observations, they credited a "landmark paper" by Richard Alexander as best explaining this social structure, not Hamilton's principle. This prompted a letter to the editor by Hamilton himself[33], complaining that the naked mole rats were in fact strongly inbred and therefore were consistent with his hypothesis. We have no space for technical details, but one source of confusion was the misunderstanding that Hamilton's principle applied only to haplodiploid eusocial insects. The exchange was cordial, at least on the surface, but in their reply to his letter, Braude and Lacey stuck by their guns:

> "Mr. Hamilton's famous equation regarding the evolution of altruism incorporates both the degree of relatedness and the reproductive costs and benefits associated with altruistic behavior. ... Our reasons for skepticism parallel the arguments used to reject haplodiploidy as necessary to eusociaity: *not all inbred vertebrates are social, and conversely, not all eusocial vertebrates are highly inbred.* [Our findings] indicate that females do not mate with either siblings or offspring [and field data] show that the breeding male and female come from different colonies and therefore outbreeding takes place in the wild at least some of the time." [33]. Italics added.

This was only the opening salvo against the sweeping generality of Hamilton's law. Another came from study of a type of mongoose (suricate) known as meerkats, which also live in highly social colonies, and which feature the regular posting of sentinels to warn the group of danger during foraging, when they are off-guard while digging for food. This had long been viewed as a mystifying example of altruism since the sentinels are highly visible and so might tempt attacks by hawks and other predators. But this was exposed as a myth by Clutton-Brock and colleagues[34], whose meticulous observations revealed that the sentinels are in less danger, not more, because they are more alert and watchful and stand near a bolt-hole. Furthermore, with specific reference to Hamilton's principle, they found that unrelated immigrants did as much sentinel duty as the others. They also show that much of the sentinel and other behavior depends on individual decisions and varies from group to group, often reflecting local circumstances as opposed to being instinctive (hard-wired), concluding that "we should consequently not be surprised if the distribution of guarding behavior across individuals fails to mirror the precise predictions of genetic models" – meaning Hamilton's. Experts will think of many other examples, e.g., a lek mating bird in which the junior males help the alpha males to mate even though they are unrelated[35].

The reality is that in the majority of real cases animals preferentially mate with distant, not close, relatives; or more accurately, that there exists a delicate balance in preference between nearness and distance of genetic relatedness. Thus, mice prefer mates whose immune type is different from their own[36]. Many studies have shown that nearly all animals, humans included, find strangers more appealing as sexual partners than their own comrades. In sum, then, if you look for evidence of preferential inbreeding, it is easy to find; but if you look for evidence of preferential outbreeding, that too is easy to find. Obviously, therefore, the reality is a balance between these two poles, and this must be recognized as a fundamental principle.

This takes us back to the question of inbreeding, touched on in chapter five. General theory holds that inbreeding is harmful because it results in offspring having a narrower genetic repertoire of potentially useful alleles, and also concentrates harmful ones[37, 38]. But as we earlier remarked, such views may be overstated:

> "He concludes saying 'As we broaden our knowledge of inbreeding and outbreeding in various species, it seems the subject becomes more complicated [not less], and fewer generalities are possible [not more]. Or, as Hamilton puts it ... 'it seems that notwithstanding all the facts and theories ... we hardly begin to know answers to any of these questions.'" [39].

To illustrate, it was thought that the decline of cheetahs was due to inbreeding and low genetic polymorphism but field study showed that the decline had nothing to do with heterozygosity[40]. From the opening of that article:

> "It has yet to be shown that inbreeding depression has caused any wild population to decline. Similarly, although loss of heterozygosity has detrimental effects on individual fitness, no population has gone extinct as a result. ..."[40]

In a 2002 review of the literature, West and colleagues concluded that competition among relatives can reduce and even negate the genetic benefits of kin selection[41]; and in a 2004 review, Clutton-Brock concludes that kin selection theory is too simplistic, the evidence "less compelling" than it first seemed and that "the benefits of cooperation in vertebrate societies may consequently show parallels with those in human societies"[42]. We agree, except that "parallels" is too weak a word, and we see no reason to restrict the conclusion to vertebrates.

6. THE SELF IN SELFISH. The "problem of altruism" is predicated on the assumption that every organism is purely selfish. This in turn assumes that we have a clear definition of *self* and *selfish*. Suppose, for example, that an animal believes that it owns the whole world. There is some reason to suppose that this is indeed the case. The people of many preliterate human societies believed that their religious ceremonies were absolutely essential to literally keep the world running – for the sun to rise, for the rains to come, for the seasons to change – and even today many believe that the world was made just for him, or for those sharing his faith.

Granting such a view, then what is the meaning of selfish? It is observed, for example, that reef fish are careful stewards of the reefs they live upon, even fanning them with their fins while sleeping to ensure good oxygenation[43]. Shades of Gaia! Another take on the meaning of selfish: suppose that we humans all know, way deep down, that we are essentially identical – that all are created equal – and that after we die, there will be others exactly like us, going though all the same travails and searchings, in the light of which we may well labor "selflessly" for the benefit of others but actually are aiming to improve our own prospects the next time around. So no matter how you look at it, "altruism" is always selfish, and is therefore a meaningless word. A mother will do anything to protect her young, for the selfish reason of projecting a copy of herself. A stallion with protect his harem, not because of altruism, etc.

A purely selfish human being is hard to imagine. We do not greedily accumulate wealth in order to live alone on some deserted island. Instead, as most psychologists will agree, the concept of "self" embraces the entire dominion upon which the ego depends, including not only family and kin but community, country, religion, ethnicity, property, territory, and increasingly, all humanity, the whole earth. Accordingly, the notion of "selfish behavior" becomes quite fuzzy. It gets even fuzzier if you start thinking about the immunology of endogenous retroviruses, say, which get right into the genome and stay there; are they alien invaders, or part of "me"?

7. ON GAMES. Not even a brief survey of altruism could be complete without giving some space to game theory, for this is the theoretical basis of the subject. Game theory is usually dated from a 1945 book by the mathematical genius, John von Neuman, with key contributions five years later from another genius, John Nash, of *A Beautiful Mind* fame. The former was recently reissued[44]. Their theorems have been widely applied in many fields but their application to altruism *vis a vis* evolution apparently took off from papers by Trivers beginning in 1971, which inspired Richard Dawkins' now-famed book of 1976, and proved useful also to W.D. Hamilton, as in his oft-cited paper of 1981[45]. Maynard Smith's paper of 1973[46] is also often cited in this connection. It is said that these works were long ignored but they became mainstream in the 1980s[47] and by now, literally hundreds of derivative papers have appeared, many of them in high-impact journals like *Science* and *Nature*. A recent book by Trivers, in collaboration with a geneticist is said to offer the latest (as of 2006) on games about selfish genes[48]. But we feel compelled to remark that no matter how interesting and well-supported the selfish gene hypothesis may be, the fact that all of our genes evidently work together harmoniously to fashion the likes of you and I cannot be forever ignored, which fact is slowly being appreciated, as we shall see.

The premises and rules of the various games on altruism are simple but, like most math-based topics with simple foundations, are hard to appreciate until you set pencil to paper for yourself. Once you do that, of course, you are likely to become hooked, captivated, and seduced by them, because, after all, math is fun. The name of the game is mainly Prisoner's Dilemma, but Snowdrift, Hawk/Dove, and numerous variants such as Blizzard[49] and Continuous Snowdrift[50] are also popular. The basic idea in all of them is that players may either cooperate or not; that is, they may defect, or cheat, stealing rather than cooperating. For example, Hawk/Dove imagines individuals of two kinds: Hawks, who always want to fight and steal, and Doves, who always capitulate. Nothing happens when dove meets dove; when hawk meets hawk, one may pay a serious cost (injury or death); and when Hawk meets Dove the game takes off. The math comes in by assigning numerical values to symbols like cost "c" and benefit "b" of the endeavor, and dividing the totals among the players – usually two, for simplicity – and assigning (or inferring) probabilities that a given player will defect, depending on benefit **b** and cost **c** to him; or splitting the population into some fraction of congenital or habitual cooperators, (altruists) *vs.* defectors or cheats *vs.* losers; or including some form of penalty or punishment for non-cooperators; or allowing for the players to be able to remember previous encounters so that individuals can build trust or mistrust ("reputation theory"); or postulating and comparing various strategies such as tit-for-tat[51], and so forth. The possible variations are endless as seen from the number of publications.

Over the decades, these games have become more sophisticated, beyond simple algebraic inequalities (they are often co-authored with mathematicians), and so are usually played by computer through many rounds of encounters (iterations) among the players; the aim being to find out which numerical values under a particular set of assumptions or rules results in an 'evolutionarily stable strategy' (ESS), one that persists robustly against cheats or other perturbations. For those who wish to get into the details, Richard Dawkins' books provide a good basis, although I personally find them distasteful for their condescending style. As usual, the founders of such fields, like Hamilton and Maynard Smith – and indeed, Darwin – tend to be more modest and cautious about their conclusions than their epigoni.

Since hundreds of such game papers have appeared, we may trust that the subject has evolved, so to speak, insofar as each author is presumably striving to improve upon prior publications in light of the critiques that invariably follow each; but whether this is really true or not is debatable. It might be more like a kaleidoscope, endlessly varying the numerical parameters and assumptions or rules of the games according to shifting fashions, without much real progress. That is to say, one finds upon reading these papers that they always begin with the same refrain, whether from the 1980s, 1990s, or 2000s, to the effect that "altruism poses a major challenge ... prior work has failed ..."

To convey the spirit of the subject by a few examples, Hauert and Doebeli created a Snowdrift game which features "spatial structure," meaning that the players are given neighbors by setting up the game on a checkerboard-like grid, so that each player has four neighbors, or eight, if squares adjacent to corners are included, or three, if the array is made of triangles[52]. The results are quite interesting, showing how isolated clusters of cooperators emerge for certain numerical assumptions, shifting regionally with time (iterations), and prominently includes the novel finding stated in the title: that the structure imposed can actually inhibit emergence of cooperation. Contrasting results are shown for the Hawk/Dove game. By the way, Nowak and May had included spatial structure in a 1992 paper[53], where they remark that the ever-changing fractal patterns of cooperating clusters are "extraordinarily beautiful ... extreme richness and beauty," and "have interesting mathematical properties." That much we do not doubt; but is it really relevant to understanding evolution and altruism? As it happens, another article by Nowak and colleagues[54] follows that by Hauert and Doebell.

The latest word on the subject (as of 2006) is a paper entitled *Altruism through beard chromodynamics*, a title sure to puzzle the novitiate[55]. The idea comes from the Greenbeard hypothesis of fitness, that any arbitrary quality, such as a green beard, might become associated with superior fitness, and therefore would become a criterion for sexual selection. (This idea was supplanted by the honest-indicator hypothesis earlier discussed, e.g., a green beard is not necessarily an honest indicator, since being arbitrary it might be acquired by a less fit individual.) The "chromodynamics" appears to be an allusion to particle physics. The idea of the paper seems to be that individuals having variously colored beards might happen to correspond to their willingness to cooperate, so that we (or animals generally) might learn to recognize features indicating a cooperative comrade. The explanation is bound up with interacting genes and features some high-powered math. The result of the game is shown in a checkerboard figure: the various beard colors have segregated into regions of like colors. It is possible that this paper represents a major advance in modeling the emergence of population structure, i.e., speciation, not limited to altruism within a species. Then again, a great many of its predecessors were each greeted as major breakthroughs. Only time will tell which games will survive as true ESS in the Ivory Towers.

On the other hand, the paper on colored beards is by no means the first to show that the population becomes segregated into groups or enclaves as the game progresses. Nearly all of them either arrive at this end or assume from the outset the existence of different kinds of individuals, cooperators and defectors, cops and robbers, whatever. For example:

> "Here we analyze the continuous snowdrift game, in which cooperative investments are costly but yield benefits to others as well as to the cooperator. Adaptive dynamics of investment levels often result in evolutionary diversification from initially uniform populations to a stable state in which cooperators making large investments coexist with defectors who invest very little." [50]

Indeed, this is a general feature of nearly all the games, and in our view, may point to some quite fundamental principles, which might also be deduced in other ways. We are speaking about the importance of *individual differences,* or subpopulations in every "species", even bacteria and viruses, and inborn differences in temperaments, human or other, as well as divergences which arise from oppositional interactions (deliberate contrasting of self *vs.* other). A promising recent game model incorporates heterogeneity of responses[56]. But further pursuit of that is best left as grist for another mill.

8. TAKING STOCK. There are a number of problems with these games which cause us to question their real importance to understanding altruism and evolution. The first is a critical weakness of mathematical modeling in general. To illustrate, there are some luminaries in my own field who have specialized in math modeling of the clotting of blood, based on the known kinetics of many of the key enzymes involved, and I have dabbled in this myself (unpublished but deserving). Heroic as those labors have been, however, they have exerted little practical impact for the simple reason that real *in vivo* blood clotting is just too enormously complex – like the weather, the climate, the economy – for useful modeling. But at least those efforts are based on real well-defined physical facts.

The altruism games are not based on well-defined physical facts. Or more precisely, we insist that *psychological* facts must be included in the notion of costs c and benefits b; both of which, by the way, can mean almost anything and are often left undefined. For example, humans are commonly observed to sacrifice everything, even their very lives, not so much for others or for any discernible benefit other than mere *principles*: religion, oath, nation, dignity, truth, justice, etc.; and, we submit, animals are not so very different in that regard, insofar as many refuse to be domesticated, for example, no matter how well-fed, preferring death to the humiliation of captivity, subjugation.

That brings us again to the question of fitness, which is the bottom-line definition of "success" in all of these games, conventionally defined in terms of reproductive fecundity. But we have argued that the real determinates of reproductive success are essentially the *tastes* (the preferences) of the mating partners. Natural selection acts only to circumscribe an outer bound of viability.

Another serious problem is that of ignorance, meaning that the factual basis of the models may be incomplete. For example, a major obstacle in modeling the clotting of blood is the ongoing discovery of new factors affecting the process. Likewise in climate modeling: certain effects of clouds were recently discovered which invalidate previous efforts. Again, since many of the game models of altruism are predicated on assumptions about how the genes work, they could all become wrong in light of discoveries yet to be made (or some already made, e.g., epigenetics). Indeed, we shall later indicate that our present knowledge of how the genes work is yet very nascent.

9. INDIRECT RECIPROCITY. Simple altruism is often summed up by the homily, "if you scratch my back, I'll scratch yours." This is called *direct* reciprocity (or cooperation). The experts have decided, however, that the human case is more complicated, more challenging to model in terms of games. The following is fairly typical of many papers of this genre.

> "The existence of such [human] cooperation-enhancing institutions is very puzzling from an evolutionary viewpoint, however, because no other species seems to have succeeded in establishing large-scale

cooperation among genetically unrelated strangers. ... Collective action in large groups whose members are genetically unrelated is a distinguishing feature of the human species." E. Fehr[57, 58].

These statements are open to debate. For example, we must question the meaning of "genetically unrelated", for it is now known that all humans are more closely related genetically than are the baboons or chimpanzees of a single troupe, apparently because of a population bottleneck in our lineage roughly 100,000 years ago. Conversely, it is also well known, to anthropologists at least, that the statement about non-family cooperation is not generally true, since in most preliterate societies, as in aboriginal Australia or pre-Columbian America, which tended to be strongly inbred, outsiders were viewed as subhuman, and a traveler unable to establish a kinship link was often summarily killed.

At any rate, it appeared to some that game theory (as briefly reviewed above) ran into difficulties, especially for the human case, as in "the free-rider problem"[57]. To resolve this, the gamers, most prolifically those of Nowak and colleagues [59-62], developed the notion of *indirect* reciprocity, said to have originated with R. Alexander[63]; the idea being that we humans have a social understanding that if you help me, then eventually, somebody else will help you – long chains or networks of reciprocal acts of helping, charitable acts, honorable acts, similar to the popular notion of karma. Central to this rationality, or at least one view of it, is the idea of one's *reputation:* if you establish a reputation for generosity, then presumably others will be generous to you, when the need arises[59, 61]. Some have tried to work this into an explanation for the rise of humanity, as in the following excerpt from a piece entitled *The evolution of the golden rule,* being a 2004 survey of ostensible advances:

> "An innate willingness to help those who would help you back, he argued, would have been beneficial in early human societies. ... Evidence for the benefits of reciprocity is starting to show up in studies of non-human primates. ..."[64]

Resisting the temptation to write *duh!*, the piece goes on to inform us about reputation theory. Games about indirect reciprocity have become widely influential, such as in economic theory[63, 65-67,] and sociology[68], and have commanded attention from the popular press, as if the social importance of reputation was some new scientific breakthrough. Efforts have been made to test some of these models in animals, such as blue jays[69], and of course, in primates[70, 71], and psychologists have been heavily invested in these ideas for some time[72]. But others are unconvinced and so have resorted to psychology-based theories, such as that of Simon[73], which he calls the "docility model," which seems similar to the formulation of Caporeal and Dawes, who applaud Simon and mention the merits of their own "group-identification" model[21].

It is this writer's impression that the more closely such models or theories approach common-sense reality, the fuzzier and less "scientific" they become, ending up as statements of things we all know from everyday experience or good novels[74]. But it is the cynical assumptions of the whole enterprise that most bothers me. For example, the Golden Rule is rationalized as a genetic outcome of natural selection driven by materialistic advantage. But might it not have originated instead from a profound insight on the nature of living existence that we are all the same?

10. THE ROUGHGARDEN DEPARTURE. It is perhaps a sign of the times that a 2004 book by Joan Roughgarden[75], followed by a 2006 paper[76], appears to have discarded essentially all of the above traditional assumptions surrounding "the paradox of altruism" in the face of presumed ubiquitous sexual antagonism and axiomatic selfishness, as well as the usual game theories – but she has some new game theories. Briefly, her thesis seems to be that cooperation, as much or more than competition, is an innate and central feature of living existence, not limited to the human species. It is remarkable that her rather iconoclastic and offbeat thesis was even allowed to appear in a high-prestige journal. Then again, she dresses it up in some very respectable math (yes, more game theory, Nash equilibria, etc.), which is to say, you can "mathematically prove" just about anything by supplying suitable rules and definitions to the symbols. It may be relevant to note that according to the book reviewer, Roughgarden is herself a transexual[75]. To say more about her thesis here would be to impinge on topics taken up later.

11. SUMMING UP. As usual, the word count says it's time to close this chapter, even though my files collected for it seem as full as they were when it began – the files grow faster than I can use them. Some of those nuggets will find a niche in a later chapter, on genes and behavior, and in the next chapter, an extension of this one. Meanwhile, we may first summarize, and then indicate a few of the more egregious liberties taken with the "science" surveyed in this chapter.

The critical summary is easy: (i) the existing theories of altruism are mostly predicated on doubtful premises concerning self and selfishness; (ii) the game theories by which various conclusions are ostensibly validated fail to adequately define key quantities like costs, benefits, and fitness; (iii) despite decades of efforts, no clear progression towards consensus is yet evident; and (iv) even within the community of theorists active in this area, there is much competition and not much agreement, as evidenced, for example, by the selective absence of certain authors in reference lists, and the radically different analytical approach recently taken by Roughgarden, yet which seems to provide at least as good a set of assumptions (and results) as any of the other approaches. In other words and in sum, the game theories, while certainly intriguing, are like any other algebraic system: they can mean anything depending on how you define your 'x's and 'y's.

This brings us to the question of interpretation. A nice illustration comes to mind. Many animals perform courtship and mating or spawning in large numbers, often synchronously, e.g., fiddler crabs[77], fireflies[78], reef algae and coral polyps[79, 80]. Now, a paper appeared in 1990 which analyzed by detailed mathematical modeling the acoustic patterns of courting male katydids, which tend to sing in synchronous bursts, concluding that "a simple mechanistic explanation ... can account for the versatile synchronization"[81]. The paper was widely cited and a whole book was devoted to a similar but wider analysis of such phenomena in the natural world[82]. However, in 1993 another analysis of the same mating stridulations of the same katydid offered a completely different explanation, to wit, that the observed synchrony is simply the result of each male attempting to anticipate the calls of the others, so as to lead the chorus, since the females of this species strongly favor the leading caller[83]. The 1990 paper is like music theory (frequencies, harmonics, scales, etc.) while the 1993 paper reveals the music itself, why it happens as it does, in terms of *motives*. Same facts, two interpretations. Which is better? They are simply two sides of the same coin, one in terms of objective facts, the other being the subjective motives cause.

Robert May, long active in such matters, is quoted thus:

"The most important unanswered question in evolutionary biology, and more generally in the social sciences, is how cooperative behavior evolved ..."[63]

This is a very strong statement. One might have thought that the rise of language, say, would rank at least as high on the list. But at least he admits that it remains unanswered and notes that no real progress has been made since David Hume writing in 1777. However, a very simple explanation comes to mind, namely, that the human invention of weapons, which made it easy for even a weakling to murder an obnoxious dominant male, doomed the alpha male to extinction, resulting in a kind of detente. Presto! Equality, cooperation. Just do the math.

The popular press, and even some of the scientific press, is hugely impressed by the much-trumpeted findings of game theory, *vis a vis* altruism, which is taken to imply, of course, an explanation for morality:

"Evolutionary biologists have begun to expose the origins, purpose, and biological underpinnings of morality. There is now general agreement that moral practices were somehow evolved." [84]

"Somehow evolved," no doubt. Scientific understanding? Hardly.

9.

Sex and Altruism

1. WHY SEX? It is clear that Hamilton's new frontier, altruism, centers on sex. It therefore behooves us to briefly review another central mystery of evolutionary theory, namely, the question of why sex exists. Many readers will be startled to hear that this is a mystery at all, but so it is. The number of theories about it (more than 20 according to Kondrashov (1995) and more since then) is exceeded only by those on speciation. As noted in the opening quotation of the previous chapter: "the existence of sex is the outstanding puzzle in evolutionary biology." Hamilton himself was aware of the problem and was not as confident as others were that he had solved it. It is a puzzle because it does not fit in with the premises of the theory of evolution.

To see why, one need only reflect that the very act of mating involves the sacrifice of half of one's genes (chromosomes), the other half pairing up with those in the mating partner, a seeming contradiction of the selfish-gene principle (which implies that self-cloning is the ideal). There are many other "costs of sex," such as the great energy often expended in it, at variance with presumed frugality of nature, i.e., much of evolutionary theory consists of cost-benefit analyses.

We cannot here review all the theories, arcane as they often are, but we can provide a rough sampler, sufficient to orient the reader, and to show that this is another huge and basic area of controversy within the theory. The most obvious theory is that the benefit of sexual conjugation owes to the reshuffling of genes inherent in that process, getting rid of bad ones, acquiring useful new ones. However, that theory comes in many nuanced flavors. For openers, let us consider what may be called the "adaptive need" hypothesis that sex evolved and is important only in circumstances where rapid adaptive change is necessary. Evidence has been sought and, of course, found to support it. Thus, according to C. Lively, a New Zealand freshwater snail reproduces sexually more often where it is challenged by parasites, asexually where they are rare[1]. Parasites "can play the same game but in reverse" since according to A. Read, a certain nematode (worm) preferentially reproduces sexually when faced with a host presenting an immune challenge, but asexually in a naive host. On the other hand, at the same symposium, Bell claimed that sexual reproduction in a green alga serves both for mutation clearance and mutation fixing, in the complete absence of adaptive challenges[1]. In a like vein, it has been observed that certain bacteria in the gut of an aphid have persisted unchanged without benefit of sex for 100 million years [1a], but in view of that stable environment, one might argue that they had no adaptive need.

Turning to plants, Stebbins opens with many interesting cases, such as the dandelion, the many that can switch between cross-pollinating and parthenogenic habits, and all sorts of intermediates, concluding finally from his study of pussytoes (*Antennaria*) that selfing forms tend to predominate in stable and homogeneous regions while sexual forms predominate where considerable adaptation is called for; ergo, the costs of sex are compensated only when needed for adapting to novel conditions; and he concludes by venturing that this also explains the origin of sex[2]. By the way, plants with two sexes go back about 400 million years[3].

However, Peck and colleagues assert that it is well established that asexual organisms become more common as one moves north in latitude or higher in altitude and in marginal or impoverished habitats, all of which would seem to demand adaptation and thus sex, in conflict with the adaptive need hypothesis[4]. On the other hand, Marschall remarks that aphids "reproduce sexually only in times of draught or inclement weather"[5], presumably in support of adaptive need?

At the same conference[1], the theory of Kondrashov was much favored, said to be appealing partly because it is "falsifiable" – meaning amenable to experimental test – though it was also noted that such an experiment would be difficult. Rising to the challenge, Elena and Lenski devised an ingenious strategy to test the theory, noting in their introduction that among the 20-some hypotheses reviewed by Kondrashov,

> "Two postulate that sex is an adaptation for purging deleterious mutations from the genome. One of these two, Muller's ratchet, depends on random genetic drift and thus provides an advantage only in small populations. In contrast, the mutational deterministic hypothesis [of Kondrashov] is effective in large populations but requires synergistic epistasis." [6]

The term "synergistic epistasis" is explained in Otto's perspective[7] but roughly speaking, concerns how two (or more) genetic mutations may interact, favorably or unfavorably. The bottom line is that their experiment with bacteria (*E. coli*) engineered with various pairs of mutations failed to support the theory:

> "The mutational deterministic hypothesis postulates that sex is an adaptation that allows deleterious mutations to be purged from the genome. It requires synergistic interactions, which means that two mutations will be more harmful together than expected from their separate effects. ... [Our results] do not support the mutational deterministic hypothesis for the evolution of sex." [6]

Paland and Lynch demonstrated that in 12 clonal sets of water fleas (*Daphnia pulex*) maintained for 200 generations, asexual reproduction (parthenogenic) accumulated more mutations (presumably harmful) than sexual[8]. By the way, they note that their results are at variance with the widely accepted neutral theory of Motoo Kimura[9] since the ratio of synonymous mutations (meaning those resulting in the same amino acids and so without effect) to those resulting in different amino acids departed from expectation under neutral theory. Also, by the way, a somewhat similar experiment (12 sets reproducing for 200 generations but using a purely sexual species, fruit flies) demonstrated a large number of mutations; but the authors were puzzled at the absence of overt change, i.e. no hint of speciation was seen, leading them to conclude that "natural variation is constrained" and that "gene expression does not evolve according to strictly neutral models"[10]. This conclusion is practically self-evident in view of the multi-million year stability of so many species, as shown not only in fossil specimens but in those whose details are perfectly preserved in amber[11]. At any rate, the debate about this continues, such as a theoretical analysis in 2006 [12].

2. MORE THEORIES. Michod gives reasons why all of the above are wrong, arguing on the basis of then-recent discoveries that the key benefit of sex has instead to do with mechanisms of DNA repair[13]. His article supplies good background on these debates, including the several "costs of sex," being the reason for wondering why sex exists. Michod edited a book-length study of the subject in which the DNA repair theory was one of three major positions discussed[14]. (It may be added that the mechanisms of DNA repair have become far more complicated since 1989.)

Among the more interesting explanations is that which appeared in a pair of papers, cited together since they are in the same journal issue[15], arguing that *sexual selection* is the very reason why dual sexes are advantageous. Agrawal, citing experiments showing that offspring from mice whose mothers were allowed to *freely choose* their mates were more fit than those from random pairings, concludes as follows:

> "I have shown that sexual selection on males ... can help to purge deleterious alleles from the entire sexual population. In this way males can reduce and even eliminate the two-fold costs of sex." [15]

Miller reaches a similar conclusion:

> "Consequently, sexual selection reduces mutational constraints on the size of genomes and the number of germline mitotic divisions. The fact that sexual organisms began to choose their mates may have had far reaching consequences ..."[15]

Another stellar biologist, Lynne Margulis, to whose oeuvre I am most partial, has declared flatly in two books[16, 17] that all of these theories are wrong and that sex has no clear biological purpose at all, being instead a consequence of thermodynamics, or an anachronistic vestige of DNA repair mechanisms from the dawn of life. I agree, to the limited extent that the answer must be sought outside of biology, in universal principles of psychology (consciousness), which are reflected in thermodynamics (see book I). She ridicules the existing theories as amounting to little more than "bio-economic 'just-so' stories," according to Michod in his book review.

There are many other theories for the origin and purpose of sex, a recent one of considerable interest involving mitochondria[18], another arguing that it was needed to level the playing field in the ongoing war against parasites and other pathogens[5] (which reproduce much faster than higher organisms), and still another based on observing asexual ploidy cycles[19]. Another, by Doncaster et al., was found to reduce to the Lotka-Volterra model [20]. But enough has been said to indicate the issues. The following quotation from a summary of highlights from a 1997 symposium probably still holds:

> "The issue is further complicated by disagreement about the mutational process in general. For example, is the deleterious mutation rate roughly constant across the genome? ... The measurement of this rate ... is similarly contentious. ... estimates ranged from 0.06 to 10.0 per genome per generation ... On the other side of the coin (mutations that are advantageous ...), the Red Queen hypothesis postulates that host-parasite coevolution is the critical force. ... [In sum] it seems reasonable that sex might be maintained in different populations for different reasons, and that the answer to the problem of sex may be relative. 'Maybe' is the operative phrase, for we don't yet know nearly enough." [1]

3. IS SEX NEW? Nearly all theories of sex make the assumption that it is a recent innovation, unique to higher plants and animals, as distinct from the simpler fissioning of lower forms, presumed to produce only clones of themselves (except for occasional random mutations). Obvious support for that view is the existence of numerous lower plants and animals which reproduce by means more or less intermediate between simple division and true sex, such as those which can alternate between sexual and asexual, or the many worms, fish, plants and other organisms which are hermaphroditic (having both male and female parts), capable of self-fertilization and thus parthenogenic reproduction; though most fish with this capability prefer to take on a gender role and mate sexually. Even insects, although most are sexual, have a great variety of reproductive modes. Gadagkar:

"The complex and diverse life cycles and social organization of the feminine monarchies [bees] are matched by their equally complex and diverse strategies for sexual and asexual reproduction." [21]

He goes on to explain some of their less widely known modalities, such as "thelytoky, which permits the production of diploid daughters without the need for a paternal genome." However, evidence is mounting to blur any sharp distinction between sexual and asexual reproduction, or more accurately, that features of sex-like reproduction extend all the way down to the simplest microbes and even viruses.

Of course, this depends on how you define "real sex," usually taken to involve nuclear fusion, meiosis, and syngamy (gamete fusion). But if true sex was really such a boon, one must wonder at the innumerable lower organisms which managed to diversify, survive, and adapt – or survive and not bother to adapt – through some three billion years without it. The answer, it now appears, is that they enjoy alternative modes of sex, a variety of practices resulting in the same end: genetic exchange. Whether this is "real sex" remains unclear, as in the following quotation from a paper about certain fungi:

"It remains to be seen whether [our finding] is solely mitotic recombination within an individual or whether it also involves genetic exchanges and cryptic sexual recombination"[22].

It strikes this writer that the balance of evidence is sliding in the direction of sexual exchange all the way down, particularly as summarized in a 2004 survey of this matter[23]. For example, the sexuality of the classical laboratory yeast *S. cerevisiae* is now firmly established[24], as is that of fungi[25], the early parasites *C. gattii*[26] and *Giardia*[27], and a primitive placozoan[28]. In connection with *Guardia,* it is said to "provide evidence that sexual reproduction started as soon as life forms that have nuclei and organelles within their cells branched off from their structurally simpler ancestors."

Others go further yet, citing genetic evidence that sex goes clear back to bacteria. Elena and Lenski[6] supply a reference arguing that "bacterial transformation is a form of non-mendelian sex," and a more recent and franker reference[29] is cited in Licensing's survey[23]. If so, then explaining the origin of sex must deal with fundamental properties of life, even with the origin of life itself, as proposed in the theory of Niles Lehman[23], to the effect that it stems from a means for expanding the genome in the primeval "RNA world" (a hypothesis for the origin of life). It is thought that some of the chemical communications of bacteria, which are getting more complicated the more closely they are studied, have pheromone-like signaling functions[30-32].

That even viruses "do sex" is evident from ingenious experiments by Chao *et al.*[33] The aim of their experiment was to test the theory that the advantage of sex is to slough off deleterious mutations. The viruses used normally exchanged snippets of their genomes, a proxy for 'sex,' which could be inhibited by allowing only very few of them to infect each cell, thereby retarding their ability to 'have sex', relative to those where many were present. The remarkable mutability, let's say *adaptability*, of viruses is best reflected in their multifarious strategies for invading cells and evading host defenses[34, 35]. (There are hundreds of fascinating papers on mechanisms by which pathogens evade the host's immune defenses.) One paper speaks of viruses as "not simply a collection of diverse mutants but a group of interactive variants"[36]. Similar views about bacteria, viruses and other plankton in the open ocean[37-39] and in soil[40, 41] is gaining ground.

We will return to this topic in a later chapter dealing with lateral gene transfers, the point for now being that it is beginning to look as if sex-like genetic exchange is a fundamental feature of all life, though only recently (geologically speaking) has it taken the form of just two genders. Indeed, Peck and colleagues cite two references which conclude that most asexual organisms arose from sexual ones, not the other way around![4]

4. SEXUAL ANTAGONISM. A major difficulty in assembling this and the prior chapter has been its Hydra-like character, meaning not only that all of its many topics are intertwined, but also that each occurs in many layers or echelons – Hydras growing out of Hydras – and all are connected to all the other chapters; and so it is with sexual conflict. That is to say, sexual conflict ostensibly occurs not only at the audio-visual level of cats screaming in the back alley, but also at the level of individual eggs and sperms, and even at the level of his-and-her genes. Here is a quotation to set the stage, from a book review:

"Sexual conflict occurs because of the different evolutionary interests of males and females in reproductive decisions. ... There has, however, been little agreement about what sexual conflict is, or what might constitute unambiguous evidence for it. In addition, the relative importance of sexual conflict in driving evolutionary change, and the extent to which it could contribute to reproductive isolation and speciation, is unknown." [42]

This topic, by the way, is also new compared to the state of evolutionary theory in the 1970s. Among the big themes here is the "arms race" between the sexes, a.k.a. co-evolution. For instance, a comparative study of several species of water striders purports to show that as the male's equipment for grasping and clinging to the female grows stronger, so does she acquire better means for repelling his attempts at what might amount to rape[43]. Another study found that mating success in male crickets correlates positively with the mating success of his sons but negatively with that of his daughters[44]. Sexual conflict has been portrayed even in hermaphrodites, specifically, a certain worm which engages in "penis fencing," the two lovers each rearing up and trying to stab the other with a load of sperm while trying to avoid being stabbed[45]. Evidently, each thinks he/she has the right stuff.

One survey of the subject focused on selfish gene theory, meaning sexual conflict at the level of genes, eggs, and sperm, featuring the ideas of L. Hurst, D. Craig, and others, and bearing section titles like *Intracellular Warfare* and *The Battle of the Sexes*[46]. We lack the space to do justice to the many interesting points made in that article, or group of articles, but they include more theories for the origin and nature of sex. We also certainly lack space for such pop-culture inanities as "men are from Mars, women from Venus." But it is important to mention that sexual conflict is the basis of another whole group of speciation theories[47, 48].

On the same theme, with perspective on that issue[50], a paper by Rice in *Nature*[49] garnered much attention, discussed also in *Science*[51]. The working assumption there was that males are forever trying to advance their own interests, while females are defending against such efforts and trying to advance their own. To test this, the authors arranged to block female defensive responses in fruit flies, and sure enough, after some generations, evidence was found that his interests were advanced at her expense. However, a letter in response to the commentary in *Science*[51] took umbrage with the statement that "a basic tenet of sociobiology is that the reproductive interests of males and females are essentially at odds," pointing out that "highly convergent interests [of the sexes] are common in many other species, indicating no fixed pattern of relationships between males and females across species"[52]. Borgia goes on to object to the proposal that "intersexual biochemical competition may be a widespread cause of speciation" by supplying reasons to the contrary, "making it unlikely that sexual antagonism has driven speciation" in several very speciose groups, namely, ants, passerine birds, rodents, and parasitic hymenopterans. This writer's impression from multiple sources is that Borgia speaks for many.

Although I do believe that sexual antagonism (or at least tension) does widely exist, the reasons for it will require deeper thought. In particular, there seems to be a profound contradiction between the reason for sex (which is presumably so that organisms can help each other by sharing genetic information) and the evident antagonism or anxiety that so often surrounds that sharing. One is reminded of how jealously humans are observed to guard their ideas, principles, intellectual systems, ideals – *personal identities*. Therein lays a clue to the solution: sex is a *unifying* act, and the essence of the tension is the ambivalence between *unification vs. apartness*.

5. SEX RATIOS AND HELPER-BIRDIES.

Why are there equal numbers of boys and girls? In 1958, Fisher gave a persuasive explanation, and it goes as follows, abbreviated from Hamilton's encapsulation of 1967[53]. Suppose there are fewer males than females. Then, if a mating pair produced an excess of males, they would have a better chance of mating than otherwise, hence genes for producing excess males would spread. But this advantage would fade away as the number of males thereby increased. The same argument applies by beginning with the assumption of fewer females. Ergo, the ratio is 50:50.

Note that this explanation is in terms of natural selection. Therefore, if correct, the 50:50 ratio is not some fixed law of nature but is instead a result of the dynamics of natural selection, and might be shifted under some circumstances. Predictions of such shifts were the theme of Hamilton's 1967 paper[53], taken up by Trivers and Willard in the 1970s, to the effect that the male-to-female ratio of offspring, as of birds, should not always be 50:50 but should vary according to circumstances which determine if the species would thrive better with a skewed ratio. As usual, details are hopelessly complicated, involving such matters as whether male or female offspring remain in the home territory, their role as helper birds in rearing the next clutch, social status of the parent (which is often passed on to offspring of that sex), abundance of food, and so on, to which we might add the ratio of mean clutch size to number of females inseminated per male.

It is interesting that numerous efforts to test the predictions failed or were frustrated for various reasons, not least of which was the difficulty of determining the sex of baby birds. Therefore, it was considered something of a triumph for the theory when at last a clearly skewed sex ratio was found using blood samples to determine gender of the chicks in Seychelles warblers, a species in which daughters but not sons remain home to serve as (altruistic) helper birds[54]. (More than 200 bird species are known where more than two individuals help raise the young, and they are widely studied in regard to altruism theory[55].) These warblers had the largest skew ever reported: birds on high-quality territories produced 87% daughters but on poor territories, only 23% daughters. The rationality given is that in resource-poor territories, it is better to produce sons in the hope that they will fly away to happier places. (A study of certain parrots reported later that year had even more dramatic results but was suspect since many of the birds were not in their natural state[56].) The Seychelles study was said to be "destined to become an instant classic" and to "remove the tarnish" on "Fisher's principle," as Hamilton called it. Other examples of skewed sex ratios are well known[57, 58].

Now, this piques our interest for two related reasons: (i) the how of it and (ii) the why of it. The first is answered by the *assumption* that it is the result of natural selection; but there are reasons to doubt that, beginning with the fact that centuries of efforts by animal breeders, who would love to breed cows and other animals that produce mainly females, were unsuccessful in getting anything other than 50:50 ratios. This is consistent with the observation that no amount of animal breeding (of dogs, for instance, or fruit flies, or bacteria) has ever succeeded in producing a new species, yet we have noted that in the wild, new or incipient species are constantly emerging, even within apparently homogeneous populations. The explanation, we submit, is that enduring changes of this kind come about *only* with complicity of free will in the natural social setting; selective breeding always fails.

The birds are evidently able to skew the sex ratios of their offspring at will, so to speak, so this is another example where the animals are evidently able to *compute for themselves* the theoretically optimal sex ratio of their offspring to match the prevailing conditions.

Meanwhile, the warbler findings were trumpeted as confirming the theory that "the effect of helper birds on the availability of food for their parents is a crucial selection pressure"[56]. That brings us to the second question, the *why* of the skewed ratio: selection pressure? The obvious answer is almost, but not quite, as given by the theory, i.e. making the best of tough circumstances, exactly as human parents in much of India or China, where sons are favored for obvious reasons (except that those human parents seem unable to determine the sex of their babies at will, resorting instead to killing or selling their daughters). We say "almost but not quite" because the orthodox explanation is in terms of selection pressures rather than allowing that the birds might possess what we ordinarily call common sense. That is to say, what might be taken for natural selection may in reality be underlain by deliberate, reasoned choices, a.k.a. intelligence.

In this connection, it is instructive to consider another bird, the oyster-catcher, whose behavior "varies from monogamy, polygyny, polyandry, double clutching, lekking, and serial monogamy to sex role reversal, and many mixed mating systems"[59]. One is reminded of the dunnock earlier mentioned[60], or for that matter, the great variety of human marrying systems prevalent in bygone eras. Those authors chose to study one variant, polygynous trios, in fascinating detail, remarking that this bird

> "... continues to be a misfit because studies of sexual behavior in birds are dominated by ideas derived from sexual selection and sperm competition. ... Even for a simple society such as that of the oystercatcher, there is no theory that unifies the three main career decisions for an individual – when, where, and with whom to reproduce. Instead, these questions are dealt with by three largely separate fields of investigation: life-history theory, distribution theory and mate-choice theory, respectively." [59]

The paper concludes with an entirely plausible account of the *rationality* of this bird's behavior, not unlike that given for the Seychelles warbler; and that is our main point, that these animals are not essentially different from ourselves and that efforts to give "scientific explanations" are no more illuminating than are such explanations, usually ludicrous, of our own life adventures and choices. The real motivational dynamics are those of free will, acting according to certain laws indicated in the previous volume and clarified in the next, in line with the material demands of living in this world.

6. SELF CONTROL. The point we are trying to make, and which is really the theme of this whole book, is that *psychological* factors underlie all of these interactions, at all levels of life, from genotype to phenotype. In regard to fertility between mating pairs, for example, it has been shown that certain fowl are capable of deliberately ejecting sperm put into them by males they did not wish to mate with[61]. But how are they able to do such things? See later. Meanwhile, it is likely that this ability is widespread; but since the experiments are difficult, that is not easy to show. Again, it is common knowledge that numerous female animals, after receiving sperm, may carry it about for a long time until it pleases them to use it, e.g., squid[62]. "The interaction between male and female often does not end with copulation ... the male must persuade the female to use his sperm rather than someone else's"[63]. In a like vein, bees:

> "Queens have perfect control over the sex of their offspring. To produce daughters, a queen lets sperm flow from her spermatheca [where it is stored from her nuptial flight] into her oviduct and then lays fertilized diploid eggs. ... To produce sons, however, a queen prevents the flow of sperm into the oviduct and lays unfertilized haploid eggs. Such parthenogenetic development of males – known as arrhenotoky – is a universal and well known feature of the hymenoptera." [21]

Following the remarkable discovery that the sex of alligators and others is decided by the temperature at which the eggs are incubated, called temperature sex determination (TSD), it came to be known that these animals carefully regulate this temperature by adjusting the amount of rotting compost piled on the eggs. (In the lab, temperature below 88°F gives all females, and above 91°F, all males.) More recently, this has been extended to a viviparous (live-born young) reptile, namely, several species of skink[64]. The author remarks that TSD is "offering the mother the chance to select the sex of her offspring and a mechanism to help to balance sex ratios" in the wild. So it looks like natural selection has nothing to do with deciding sex ratios. The animals figure it out for themselves and control it according to prevailing conditions.

Many fish, wrasses in particular but also some cichlids, can change their sex at will, so to speak. In one study, when a dominant male is removed, a female becomes a male, replete with brilliant colors, larger size, and personality to match[65]. This is clearly a result of purely socially perceived cues, i.e. is psychology-driven. The traditional view of animal sexuality is that it is all controlled by pre-programmed sex hormones[58], but it is increasingly clear that hormone levels can equally or more often be *responses to* social cues. Now, can we accept that hormones are sometimes responses, sometimes causes? Must there not exist a unifying principle or framework by which such phenomena can be understood without contradictions and exceptions?

Yes, such a framework must exist, and it must be in terms of *will* (psychology) as the prime mover. (Bear in mind that we are viewing the genetic repertoire in terms of long-term memory, the most recently acquired being most amenable to behavioral flexibility and heritable modification.) A major reason we have resisted such ideas is that we humans are unable to do such things. We cannot switch our genders, for example. Why not? That's another whole story, but the general idea is that something happened to us in our recent evolution, popularly known as the Fall of Man, which impaired the limited abilities of this kind that we formerly had. It had to do with the invention of language.

7. SPERM MEETS EGG. Let us venture a step further, to the fertilization of egg by sperm. A great deal of first-rate biochemical sleuthing has recently improved our understanding of this process. In one note on work about the basis of rejection (incompatibility) by the egg of sperm from a dissimilar species, the authors are quoted as wondering if this rejection is "the forerunner of a speciation event, or does divergence take place after speciation has occurred?" [66]. In either case, they may be mistaken in viewing the incompatibility as the result of an accidental genetic quirk. For it strikes us that they may be one and the same event, that it is within the female's power to allow or forbid fertilization to occur, by means similar to those by which she selectively utilizes some but not other packets of sperm that she carries about – or, for that matter, by which she selects one mate but not another, for whatever reasons.

Indeed, it seems to be quite well established that females can deliberately select which of the sperm cells within her, from multiple matings, she allows to fertilize her eggs, e.g., sand lizards[67], much as noted above for chickens[61, 68]. Yet, we are not doubting the 'physical' (biochemical) basis of egg-sperm compatibility, which crucially depends on a protein called bindin[69], among several others, or that such incompatibility is critical for fixing a speciation event, i.e. reproductive isolation.

8. THE IMMUNE SELF. Those ruminations bring us to the doorstep of another great field of modern biology: immunology; for it bears on the matters of speciation and sexual antagonism, or more generally, the mutual antagonism between groups of any particular identity. Now, it goes without saying that all organisms must have means of defending themselves against attack from without (against predators) and from within, speaking of intruders such as parasites, bacteria, viruses, fungi. The latter defenses are called immunologic. But the lines are often blurred. For concreteness, let us consider a few examples.

It has been found that colonies of a type of sea anemone regularly make war on each other. It seems that each colony consists of clones (genetically identical) which fight with neighboring groups having a slightly different genotype, making use of special tentacles equipped with weapons, even employing juveniles as "scouts" in the no-man's land between colonies[70]. It appears that each group has a special "warrior caste." When clones are grown on floating tennis balls in the lab, the balls become sorted according to clonal identity. In a like vein, groups of a primitive marine tunicate, *B. schlosseri,* a type of sea squirt, will

> "... reach out and begin an interaction [with neighbors] resulting in either fusion of the two ampullae to form a single chimaeric colony sharing a common blood supply, or a rejection reaction during which the interacting ampullae are destroyed, thus preventing vascular fusion." [71]

Whether fusion or rejection occurs depends on certain combinations of MHC markers. (MHC abbreviates major histocompatibility complex, used in tissue matching.) Stem cells from one colony sometimes invade and take over another, reminiscent of invasive cancer cells[72] and of the transplant rejection response often seen in human recipients of organ donations. But our point here is to suggest that the immune phenomenon is not merely a means of defense against pathogens, but more fundamentally reflects self and group identity, in the sense of species and speciation – and in the sense of inter-group antagonisms.

These antagonisms seem to exist at all levels. We earlier mentioned the recent discovery by genetic techniques of numerous cryptic species, that is, reproductively isolated groups not previously recognized as distinct, e.g., bats[73], salamanders[74], fruit flies [75], a bacterial species[38], and others76. It appears that the closer we look, the more such groups come to light, as with the sea anemones mentioned above, which look alike; and by the way, these examples must erase all lingering doubts about the reality of sympatric speciation, although it is still discussed as if it were still controversial[77]!

10.

More Sex, and Some DNA

This chapter fissioned off the previous one in the course of final editing because it had gotten too long and was wandering. Also, the author was advised that a bit of introduction to the next chapter – genetics in evolution theory – was needed. However, the chapters which follow do not require any high-level knowledge of molecular biology. High-school biology is enough.

1. THE SAGA OF Y. In connection with the question of sex, it is instructive to briefly review the saga of the Y chromosome (another of the Hydra's heads) being that which makes half of us males, XY, the rest females, XX (but the opposite in birds). By about 1990, the sages of the genome had concluded that the Y was degenerate and fading fast, doomed to disappear "in about ten million years"[1].

> "'It's a pattern you see over and over and it's always been a puzzler: Why the hell does the [Y] chromosome end up as a little blob with hardly any genetic activity on it?' asks Jim Bull, an evolutionary geneticist ..."[2]

Among the evidences for this conclusion was the fact that over phylogenetic time, meaning from the earliest sexual organisms to later ones such as mammals, the Y tends to shrink. It was believed that this was mainly the result of accumulated deleterious mutations or "junk DNA."

> "These properties include the failure of X and Y to recombine through much or all of their length, the genetic inertness of much of the Y chromosome, dosage compensation of the activity of X chromosomal loci, and the accumulation of repeated DNA sequences on the Y chromosome." [3]

Experiments by W.R. Rice indicated that the decay was due mainly to the lack of recombination of the Y[2]. A book about our impending loss of maleness, *Adam's Curse,* achieved best-seller status[4]. It is said that the Y chromosome of the papaya tree is youthful and recent, for it is quite large[5].

However (as another lesson about premature announcements based on incomplete knowledge), around 2003 a flurry of papers appeared suggesting that these conclusions were wrong, and that Y is perfectly healthy and here to stay[6, 7]. Furthermore, far from having "hardly any genetic activity," the latest finding (as of 2006) indicates that it is changing 10,000 times faster than average[8]. Among other curious and rather mysterious findings is that Y contains long stretches of palindromic DNA, meaning that it could code for the same protein regardless of which direction it is read! This may have something to do with its stability. Regarding its shrunken size: bear in mind that the genome of our mitochondrion, the "powerhouse of the cell", is also highly depleted, simply because it does not need all the genes it once did as a free-living organism, whence many of its genes now reside in the nuclear DNA; but it has, of course, retained those essential to its identity and function. Likewise, perhaps, the Y.

Incidentally, *vis a vis* sexuality, the control of expression of the X chromosome is also of current interest. That is to say, the female genome has two of them, XX, and it was discovered in the 1960s that one of them becomes silenced (inactivated). But this, too, is a complicated story still under active research[9, 10], and it will probably remain under active research for a long time (see later).

2. ASIDE ON NEUTRAL THEORY.
We earlier remarked on the "neutral theory" of Motoo Kimura[11-13], and it warrants further comment here, because it seems to be spreading as a mode of explaining things. Considered a radical

abomination in the 1970s, Kimuran neutrality is now a pillar of evolutionary theory, especially in molecular studies, and holds in brief that the great majority of mutations in the genotype, and hence their effects in the phenotype, are adaptively *neutral,* neither favorable nor unfavorable to fitness. On the one hand, this conclusion should have been practically a corollary of the assumption that mutations are purely random, except that it conflicted sharply with another assumption of the theory that all features, all changes in an organism, must be adaptive and result from natural selection. As pointed out by Kondrashov, this marked the beginning of the end of classical evolutionary theory.

> "Once upon a time, the world seemed simple when viewed through the eyes of evolutionary biologists. All genomes were tightly controlled by various forms of natural selection. DNA encoded functional genes, and most mutations that occurred were rejected through negative selection. Those exceptional genes that were beneficial substituted for the original gene variant (allele) and spread through the evolving population ... This idyllic world began to crumble in 1968, when Kimura made his modest proposal that most allele substitutions and polymorphisms do not substantially effect an organism's fitness ..."[14]

In a later chapter, we shall see that Kimura's neutral theory often fails (with certain snake venoms, for example) but here we indicate how the same concept has been extended into theoretical ecology, Lotka's old bailiwick, to wit, the neutral theory of *ecology.* Now, some experts may object (even "emotionally" according to one authority) to including ecology in the context of evolutionary theory; but the reality is that ecology, too, is steadily being sucked into the black hole of evolution theory – along with psychology, sociology, anthropology, and Christianity. Indeed, it seems to be posturing as a "theory of everything."

To illustrate, another subdiscipline called *macroecology* has sprung up, toadstool fashion, embracing the study and theory of biodiversity in space and time, a recalcitrant old chestnut, early examined by Lotka. The idea of it is to try to figure out how and why certain assemblages of species, plants, say, occur in particular makeups or patterns, differing from region to region. Indeed, the simple fact that numerous species of flowering plants, for instance, do in fact generally occur together, all of them feeding from the same soil, is by itself problematical for the theory. Since "the law of limiting similarity" implies that one of them is bound to out compete the others and thus take over. But that doesn't usually happen (except in occasional cases of invasive species) [15]. Therefore, great effort has gone into trying to account for biodiversity, resulting in lively debates among rival theorists, and forbiddingly dense books and papers on this and related matters[16, 17].

Enter the neutral theory. Two of the founders of this movement, Graham Bell and Stephen Hubbell[18], argue that "you don't need to invoke adaptation to explain biodiversity," because their formulation

> "... predicts the individual species abundance of the more than 800 tree species in a Malaysian rainforest with just three numbers. ... If correct, neutral theory means that the species in a habitat have been thrown together – and will come and go – at random." [19]

Unfortunately, however, it may be just a bit too simplistic since, although successful in many applications, it failed many other tests, such as for birds in South America[20], coral reefs[21], and a more rigorous study of tropical forests[22]. One of many reasons why it fails for forests might be the extraordinary degree of connectedness of tree roots between various species by networks or filaments of a type of fungi called mycorrhizas (literally *fungus roots*).

"In an undisturbed forest ecosystem almost all of the trees" are interconnected by "mycelial systems" and "groups of tree species joined together in this way have been recognized as functional guilds"[23]. Using radioactive carbon dioxide, Simard et al., showed that nutrients can move through this network in both directions between trees, with up to a dozen trees connected by a single fungal network. For example, nutrients flowed copiously between birch and conifers but not adjacent cedars, which were colonized by a different type of fungal root symbiont. Some trees "share up to ten mutually compatible fungal symbionts." Of special interest is the observation that disadvantaged trees, such as young sprouts in gloomy shade or others lacking chlorophyll, were nourished preferentially ("subsidized") by surrounding trees *via* the ectomycorrhizal networks. Read concludes by invoking "Clements' concept of the community as a 'superorganism' ... that all of the components of a stable ecosystem are interdependent." Similar statements have been made about other ecosystems, ocean bacteria, for example. "In the 1980's, Sonea and Panniset argued that the discoveries of molecular biology render the bacterial species concept obsolete and that bacteria are indeed a superorganism with a common gene pool"[24]. More on that later.

Numerous fascinating papers about these soil networks have since appeared[25-27], further supporting and extending the findings. Moreover, plants can make war on one another by underground chemicals secreted from their roots, and the defenses of some can also be routed to protect their neighbors[25, 28]. Add to this the obvious importance of supporting networks of pollinators, the effects of fauna on soil properties, etc., and one comes to appreciate what the ecologists well knew: that such systems are immensely complex. Thus, Hubbell's neutral theory, while doubtless a useful approximation for some rough purposes, is not really very illuminating of ulterior dynamics, tending to obscure the underlying realities in much the way that statistical approaches to economics or crime fail to penetrate to the motivational roots.

3. A WORD ON GENES. Ever since the New Synthesis of the 1940s, the foundation of evolutionary theory has rested upon genetic science. In the next chapter, we show that this foundation is built on ever-shifting terrain – a castle built in air – and is essentially irrelevant to the theory of evolution for one simple reason: *ignorance.* Not enough was known, or is yet known, to draw any clear conclusions about anything beyond obvious facts known already to Darwin.

Knowledge has been improving, of course, decade by decade, each requiring major revisions of the prior concepts. Indeed, genetic science today is probably the most vibrantly active and exciting area of research in all of science. But this must not obscure the fact that recent advances, brilliant as they are, have not so much clarified our questions about inheritance as they have opened new doors to immense vistas of mystery, in which almost anything is possible. How true it is that the more we learn, the more we know how little we know.

We have definitely learned a lot. The rise of molecular biology since Watson and Crick must be counted among the great achievements of the twentieth century, along with quantum mechanics and relativity theory. However, while the latter two were almost completely formulated by about 1930, molecular biology is by no means yet complete. Contrary to sensational headlines, such as those which accompanied the sequencing of the human genome, it is very much a work in progress with no end yet in sight. In an earlier chapter, we quoted an authority who estimated that *centuries* would be needed to fully understand DNA.

For the benefit of those who missed Biology 101, the human genome consists of an extremely long chain of DNA, actually, a pair of them wound together in a double helix, made of a sequence of four bases (A, G, C, T), triplets of which code for the manufacture of proteins. The proteins, mostly enzymes, are responsible for just about everything that happens in the organism. Proteins are chains of a few hundred amino acids (aa), of which there are about 20 different ones. For example, triplet ATC specifies the aa isoleucine; GCA codes for the aa glycine, and so on. Many triplets are synonyms for the same aa, such as CGG = CGC = CGT, all coding for the aa argenine. There are also codons for "stop" and "start." Nearly all of the three-letter codons are universal for all living things, with a few exceptions, e.g., the start and stop codons may be different in bacteria or mitochondria compared to more complex organisms (eukaryotes). The actual translation from DNA code to protein is accomplished by messenger RNA, which reads the code and delivers the aa to the site of assembly of the proteins.

In this way, the thousands of enzymes and other proteins in our bodies are produced, each with astonishing specificity. There are also mechanisms for modifying the proteins after they have been translated, such as by the addition of chains of various sugar molecules, or by "alternative splicing." Granted, the details are very complicated, yet the basic concept is astoundingly simple. The word *elegant* comes to mind. The notion of a "gene" corresponds roughly to a stretch of the DNA which codes for a particular protein.

In the original draft, this section was much longer. It was decided to cut it, for two reasons. First, nowadays, everybody under the age of about 50 knows the basics from high school biology. Second, it is not possible to summarize the basics in a few pages. Enough said. The following chapters are largely self-explanatory.

In parting, we may quote from a book review which indicates how rapidly the field is changing and how complicated it has become. It is said to be "a big book, full of information" about RNA, but apparently was already out of date by the time it was published. Some of the topics mentioned come up in the next chapter.

> "However, there is no chapter on introns themselves, despite the fact that they comprise over 95% of pre-mRNA transcripts in humans, nor any discussion of the functions, if any, of the excised intronic RNAs, especially in view of the fact that many miRNAs and snoRNAs are encoded within them. In addition, there is neither a discussion of the thousands of non-protein coding RNAs (ncRNAs) that have been recently identified in cDNA ... nor any treatment of the now significant subset that has been functionally studied in various contexts. Admittedly, this field has exploded in the past year or two ..."[29]

> "Also disappointingly, the book does not have a section devoted to the expanding field of RNA editing ... [New work on miRNAs] raises the spectre that there may well be many more miRNAs, as well as sno-like RNAs and other classes of yet-to-be-discovered regulatory RNAs derived from non-coding transcripts, which span at least 70% of the mammalian genome. These vast tracts [of sequence] have been dismissed as evolutionary relics and transcriptional noise ... because the alternative explanation [that they may be functional] challenges the protein-centric orthodoxy of molecular biology, despite the fact that there are more conserved non-coding than coding sequences in the mammalian genome, and that chromatin modification (genetic memory) and alternative splicing, the hallmarks of differentiation and development in complex organisms, have no satisfactory explanation in terms of protein-based sequence recognition and control."[29]

11.

Mysteries of the Genome

1. EMBARRASSMENT OF RICHES. It goes without saying that the foundation of the New Synthesis was genetics. The aim of this chapter is to show how recent advances have destroyed nearly all of the simplistic assumptions in that old paradigm. In other words, present knowledge of how genetics works is still too primitive to draw decisive conclusions *vis a vis* evolution. Konner's guess, quoted in Ch 1.6, that "centuries" will be needed to achieve full understanding sounds about right, especially in view of the fact that the human mind – *consciousness* – in all its glory and richness, must somehow be a part of any ultimate understanding of how the genes work. Accordingly, the sequencing of the human genome marked only the beginning of figuring out how it works, as pointed out in an article aptly titled *End of the Beginning* [1, 2].

Paradoxically, open admission of our ignorance has come only lately, with huge advances in knowledge revealing layer after layer of awesome complexity, rolling back the horizon of the unknown. That is to say, for some 50 years following the New Synthesis, a simplistic genetic theory was invoked to explain every detail of evolution (and to squelch the skeptics), but it is now clear that the High Priests were talking through their hats, spewing out a smokescreen of jargon and half truths. Indeed, geneticist

P.A. Lawrence argued in 2001 that high-flown jargon is still used as a cover for ignorance and compares the language of modern genetics to that of the medieval alchemists[3].

This chapter acquaints the reader with some of the more remarkable developments casting doubt on the traditional mechanistic scenario, or at least, greatly expanding the range of plausible alternative scenarios. Underscoring the pace of advance in the field is the fact that in the short decade since beginning this update of the manuscript (1997) and now getting around to finishing it (2006), additional truly revolutionary principles have come to light.

2. HORIZONTAL TRANSMISSION. Everyone knows that the genome is normally passed from parent to offspring, called *vertical* transmission. In the 1980s, as more genomes were sequenced, it was noticed that several bacteria contained genes and other sequences that didn't belong and must have come from other species, called *horizontal* transmission (or *lateral transfer*). As more and more of these examples turned up, it became clear that this phenomenon was quite common, even threatening to "shake the tree of life"[4]. It shook the tree because by that time, everybody was using sequence data to establish phylogenetic relations, the "family tree" of organisms and time of branchings; but if bits of DNA could be transferred between unrelated organisms, that approach was in serious trouble[5].

It was initially thought that such transfers were restricted to lower forms. Indeed, the mechanisms of such transfer are an active area of research[6, 7]. But Mother Nature had more surprises in store. It was reported in 1991 that a gene from one species of fruit fly had naturally spliced into the genome of another in the 1970s (some say it was the 1950s), apparently mediated by a parasitic mite[8]. Like many other major developments in the field, think of Barbara McClintock's jumping genes[9], this was initially shrugged off as a rare and freakish event, limited to closely related species. But by around 1993, it became undeniable that lateral transfer was commonplace. The significance of this to evolutionary theory is still only dawning. For example, a 1997 commentary on the sequence of the genome of the ulcer-causing bacterium, *H. Pylori,* remarked as follows on the probable transfer of DNA from bacteria to plants:

"Of the many details reported, the generalizations about molecular evolution are the most interesting ... Although the significance of these anomalies is not clear, the possibility of rampant horizontal gene transfer is unnerving to a community that hopes to reconstruct the history of life on the basis of amino acid sequences. To test this point I examined a number of bacterial urease sequences [and] found it unsetting that the plant [jackbean] urease was almost as similar to each of the bacterial ureases ... as most of them were to one another." [10]

In other words, the gene for this enzyme may have been passed back and forth between plants and bacteria. Similar comments were made regarding another bacterial genome[11], and assessment of lateral transfers is now a routine part of genome sequence analysis. It has been said that "between 1% and 8% but possibly as much as 60%" of genes in bacteria arose by lateral transfer[12]. New examples of such transfers between higher organisms have come to light regularly, e.g., between plants[13, 14] and from a plant to an animal[15]. Lateral transfer is big news, the subject of meetings[16] and books[17].

Of course, it had long been known that bacteria of a given species routinely exchange such information by "parasexual" means, the study of which helped lay the foundations of genetic science [18-20]. But lateral transfer *across species* is a horse of another color. From a practical perspective, it is believed to underlie the rapid spread of antibiotic resistance[21-23] (but many microbes naturally possess such resistance[24]). Antibiotic resistance is mediated by certain proteins that literally pump toxic agents out of the cell and is closely related to the development of human resistance to anti-cancer drug therapy[25]. Some of these pumps are remarkably similar between human and bacteria[26], suggesting lateral transfer. The protein structure of one of these transporters was recently solved[27]. Vancomycin has been the antibiotic of last resort but even that is now meeting resistance, once again most probably by lateral transfer[28]. Lateral transfer has been a cause for concern in genetically engineered food crops, the worry being that the inserted genes will escape to other species[29]; but since that happens naturally anyway, the worry is probably unfounded.

It was also known that certain viral DNA can insert itself into host genomes, more or less permanently (see below). Indeed, the human genome is rife with viral material, known as retro-transposons. Methods for the deliberate laboratory transfer of genes exploits similar mechanisms, e.g. laboratory transfer of a fruit fly gene affecting behavior from one species to another[30], now routine in biotechnology. What was new was that such transfers occur naturally and commonly, including to and from higher species.

> "For years ... DNA viruses were thought to contain numerous 'non-essential' genes. These genes are now found to code proteins used by the virus to counteract the host immune response. ... These viral genes are thought to have been captured from host cells during viral evolution ..."[31]

Thus, lateral transfer must have played major roles in evolution, independent of natural selection, with implications undreamed of in the New Synthesis, where each adaptation was supposed to have been painstakingly acquired by natural selection of random mutations.

> "Recently, however, a wealth of evidence has accumulated indicating that horizontal [transfer] may be more broadly practiced in nature. It may be so widespread, in fact, that it has affected and continues to affect the evolution of all living beings." [32]

3. GOING FURTHER. In sum, it is now clear that significantly long segments of genetic code (DNA) can be transferred from any species to any another, literally overnight, vectored by viruses, bacterial plasmids, fungi or parasites, alone or in combination, resulting in heritable changes. Of course, any newly inserted piece of functional DNA would presumably be subject to natural selection, so in most cases might be weeded out as detrimental, excised as 'foreign' by mechanisms of DNA repair (see below). Hence, the long-term persistence of such elements could imply either that the host finds the new gene useful, or that the invasive new element has succeeded in subverting or manipulating the host for its own gain, as viruses seem to do. An intermediate view is symbiotic coexistence.

It has been found that certain genes for the formation of eyes are remarkably similar in many distantly related genera, previously believed to have independently evolved but now suspected to have resulted from horizontal transfer[33-35]; likewise for pigment melanophores in the skin of vertebrates[36]. An early form of the visual pigment, rhodopsin, is apparently regularly traded among marine plankton[37]. By the way, advanced eyes have been described in a jellyfish, although it has no brain[38]. Lateral transfer may even have a bearing on the "melange" of characters seen in certain fossil forms[39].

In a like vein, fully half of the 80-some orders of animals include species that glow in the dark (worms, squid, sponges, insects, protists, jellyfish, snails, octopi, etc., as well as bacteria, yeasts, mushrooms, and more[40-43]) many of which rely on the enzyme, luciferase, to produce light of various colors with high efficiency. Since it is hard to imagine that each of these forms independently invented this beautiful and complex system, it seems more likely that the genes for luciferase and its substrates were laterally transferred. The original genes probably arose only once, or rarely, and thereafter became available to anybody who wanted to pick them up from some passing vector, and modify them for sundry purposes.

In many cases, the luminous display is supplied by symbiotic bacteria, which may or may not pass directly to the offspring. Luminosity is exploited for many purposes including attracting mates or prey, or frightening predators. That is to say, acquiring the genes for producing light (or toxins, venoms, aromas, etc.) is only part of the story, the rest being their *utilization*. Many or most luminous organisms confine this enzyme system to a specific organ and control the luminosity for various purposes; likewise with toxins and venoms. In other words, in order to be useful, such genes must be smoothly incorporated into purposeful strategic equipment serving the larger behavioral repertoire. (Compare this idea to the invention of the electric motor: it was immediately applied to washing machines, vacuum cleaners, golf-carts, etc.)

Another example which comes to mind are certain enzymes of carnivorous plants[44, 45] which, one may suspect, were acquired by lateral transfer since they are so similar to those in animals. Like luminous animals, such plants occur in many orders only remotely related, traditionally implying that they arose independently on multiple occasions[46]. But here again, lateral transfer of genes for digestive enzymes cannot by itself account for carnivorous plants, since in order to utilize those enzymes, it is necessary to come up with structural equipment for the luring and trapping of prey.

Hence, we may imagine that the potential host of a new gene wafting its way along might glimpse it as exploitable, leading it to facilitate incorporation of the new gadget into its behavioral repertoire. Parallels between human inventions and those of organisms are popular in children's books (like *Nature Thought of it First*) but are not taken seriously by the theorists. Not yet, anyway. Our case will get stronger. But we are reminded of what anthropologists call *diffusion*: being the spread of new inventions from culture to culture, sometimes piecemeal, or else resisted, rejected (e.g., certain sub-Saharan Africans resisted iron cookware because of a religious taboo against iron). Organisms seem to behave similarly, as in picking up various toxins or perfumes from their neighbors or prey for their own use[47], as further taken up in a later chapter.

At any rate, it is clear that lateral transfer supplies a whole new dimension to the acquisition of traits, entirely apart from natural selection of random variants. It cannot have acted willy-nilly because, in the first place, numerous organisms have *not* been infected with such new properties, remaining unchanged for eons; and in the second place, many organisms which have made such acquisitions have harmoniously incorporated them into their repertoire. This includes some of the endogenous retroviral elements in our own genomes, discussed below.

In parting, it may be added that some authorities attempt to find alternative explanations for some cases of apparent lateral transfer. Thus Roelofs and colleagues, commenting on the finding of 113 bacterial genes in the human genome, propose instead that those genes were shared in a common ancestor but were subsequently lost by non-vertebrates; however, they admit that this seems "far-fetched"[48, 49]. A similar argument was given for the many genes we have in common with sponges[50], and with sea anemones[51], as discussed in more detail[52]. But whatever the final disposition, such theories, which are indeed far-fetched, do not diminish the reality and importance of lateral gene transfer.

4. JUNK DNA AND TRANSPOSONS. Another bombshell in evolutionary genetics has been the still-dawning realization that most so-called "junk DNA" isn't junk. To back up a bit, when large-scale sequencing studies started taking off in the 1980s, everybody was looking for genes (and finding them by the thousands), whether or not their functions were known. However, it soon became clear that the genes themselves, defined as protein-coding sequences of DNA, represented only a tiny fraction of the total genome, around 2% for most of us mammals. The rest consisted of "gene deserts" and was termed *junk DNA* on the rather arrogant assumption that it had no purpose.

All kinds of high-flown theoretical explanations were given for why and how the genome had accumulated so much junk. Whoever invented that term *junk DNA* doubtless rues the day, for it is emerging that much or most of the junk (if not all of it) actually has important functions. It is now more wisely spoken of as *noncoding* DNA (ncDNA)[53, 54].

Some of these functions are considered in the next chapter. Meanwhile, let us focus on *transposons*, the jumping genes discovered by Barbara McClintock[55-57]. These were viewed, and still are, as parasitic or selfish elements whose only (known) function is to replicate themselves and jump around the genome, often wreaking havoc by messing up any genes they happen to land near or on. There are lots of them, up to 50% of the human genome, or more by various estimates. Of this 50%, most consists of fragmentary remnants of viruses, but a significant fraction is fully coded RNA viruses (retroviruses, retrotransposons). From the abstract of a paper titled, *Altruistic functions for selfish DNA*:

> "Mammalian genomes are comprised of 30-50% transposed elements (TEs). The vast majority of these TEs are truncated and mutated fragments of retrotransposons that are no longer capable of transposition. ... [They are] commonly perceived as neutrally evolving and non-functional ... In a major development, recent work has strongly contradicted this 'selfish DNA' or 'junk DNA' dogma by demonstrating that TEs ... generate RNA on a massive scale across most eukaryotic cells. This transcription frequently functions to control the expression of protein-coding genes via alternative promoters, cis regulatory non protein-coding RNAs, and the formation of double-stranded short RNAs. ... In sum, these findings challenge [the dogmas]."[58]

(Note that this reference is 2009. Up to this point, the author has not altered the 2006 manuscript except for minor editing; but the topic at hand has been evolving so rapidly that a few updated references, as of December 2010, are needed.)

Relatedly, if these "gene deserts" in the DNA sequence were of no use they would show a lot of random variation, but in fact they are often highly conserved, indicating they are important. Thus, in a commentary entitled, *Fruit fly Genome is not Junk*[59], it is revealed that up to 65% of noncoding DNA shows signs of being conserved; that is, of being subject to selective constraints, adaptation, or *exaptation*, to use the neologism. As noted above, transposons are now seen to play major roles in the regulation of gene expression[60] and, indeed, in speciation[61, 62]. They have been shown to play an important role in acquiring resistance to insecticides[63] and antibiotics[28], and to play other roles[64], including in the human immune system[65]. Another report gives them a role in embryonic development[66].

Of special interest here is the finding that junk DNA is involved in *social behavior,* specifically, in reflecting the preference of some groups of voles (small rodents) to practice monogamy while others are promiscuous; the difference being attributed to the activity of a receptor for the hormone, vasopressin[67]. The vole story has been floating around for some years[68-71]. But is this hormone the "cause" of the preference? Perhaps it is now, but this should not be taken to imply that its origin was an accident. To argue that this was the *cause* of the behavior is like arguing that acquiring our vocal cords was the *cause* of human language; yet nobody doubts that our vocal cords depend on certain genes.

By way of some updated references, it is now clear that "junk DNA" was responsible for some *major and fundamental* evolutionary events. One is the rise of the placental mammals [72, 73], another is the acquisition of the higher adaptive immune system, meaning our ability to generate antibodies [74, 75]; for general review, see Cordaux and Batzer [76]. Who can say where further studies will lead? But they are definitely not leading in the direction of orthodox classical assumptions.

5. DNA REPAIR. Another bombshell in evolutionary genetics was the discovery of DNA repair: that mutations often or usually disappear, thanks to machinery in the cell nucleus which repairs faulty, alien, mismatched, or broken bits of the double helix. One might have guessed this just by noting the long-term stability of so many species over the eons despite being constantly bombarded by radiation, free radicals, climate change, and new selection pressures (and presumably also by attempted horizontal gene transfers).

The first hints of this complication emerged from classified studies during World War II on the health effects of radiation. It was found that different organisms responded very differently to the same radiation dose, some being more resistant than others to injury or death by mutagenic agents. Still more puzzling was that the mutations induced in various animals often disappeared after a few generations. The explanation remained a mystery for some 40 years. To make a long story short, the observations were at last explained by the discovery of DNA self-repair machinery[77]. A DNA repair enzyme was named "Molecule of the Year" in 1994 in a special issue of *Science* devoted to the topic[78, 79]. Books on the subject were quickly outdated by the rapid pace of advance[80]. Over the years, review articles have revealed increasing complexities[81, 82], the latest take (2006) being an effort to comprehend more than 5,000 interactions among some 30 proteins thus far known to be involved in genetic repair[83]. That is to say, a whole family of repair systems exists, each with specialized duties.

Some of the repair engines chug along the double helix like tiny cog railway trains, at speeds around 1,000 base pairs (bp) per second, "interrogating" or "surveilling" as they go, then pausing to make repairs as needed[84]. Some seek out tiny or subtle imperfections while others correct various kinds of major breaks or other damage, sometimes calling a halt to cell division until repairs can be made, or, in the case of irreparable damage, inducing cell suicide by a complex process called *apoptosis*[85].

Perhaps most challenging is to imagine how double-strand breaks, likened to "a three-alarm fire"[86], get repaired. But repaired they are, by a whole array of fancy equipment, comparable indeed to that in our fire stations[87, 88]. For example, the exceptionally radiation-resistant bacterium, *Radiodurans,* "can within hours stitch back together the hundreds of fragments of broken DNA," including double-strand breaks[89]. A young friend reading this page over my shoulder remarked, "Hey, that reminds me of lizards regrowing their tails!" There may well be some underlying commonality. It remains a mystery why *Radiodurans*, the Humpty-Dumpty bug, came to possess such outstanding repair equipment, and why others lack it. Astronomer Hoyle took this resistance to radiation as evidence that life came from outer space[90].

That brings to mind an equally or more amazing example, named "Mutant-of-the-Month" in one account[91], concerning a laboratory plant caused to have a key mutation, but the mutation was erased or repaired in later generations[92]. This finding "has the potential to push the laws of Mendelian inheritance to the breaking point."[91] The authors are mystified as to how the genome can somehow 'remember' the original genetic sequence and restore it. Some sort of repair mechanism must be at work here, but of a kind not yet known. I am reminded of the slow return of memories and speech to some victims of stroke or brain trauma, long after those parts of the brain were destroyed.

Faulty DNA repair systems have been implicated in several inherited diseases. For example, study of the unfortunate disorder, ataxia telangiectasia, has led to important new insights[93, 94]. Defective DNA repair is also linked to a far more prevalent disease: cancer[95-99]. DNA repair has even been offered as an explanation for speciation, for "reproductive isolation," based on the idea of genetic mismatch between sperm and egg[100].

It is believed that mechanisms of DNA repair originated as a means of defense against viruses, acting to block lateral transfer of alien DNA, or to excise it if it gets in[101]. That may well be true, but it raises again the overlooked question of why or how the genome knows, or cares, if it is being invaded. How are we to imagine a genome, conceived as a mere chemical entity, 'defending itself'? Or, if these mechanisms arose to maintain fidelity of replication (genomic identity) as some have proposed, the same question arises: how does the genome 'know' to resist change? The more one learns about DNA, the more likely it seems that the genome is itself a living entity. In later chapters, this concept emerges as well-nigh inescapable.

6. MUTATIONS AND IMMUNITY. Another crumbling pillar of the New Synthesis is the idea that mutations, supposed to be the ultimate source of all genetic variation, the 'raw material' of natural selection, are purely random and uncontrolled. We have already seen that mutations are often or usually self-repaired, but that is only the beginning of *The Muddle about Mutations,* to borrow the title of an article[102]. The following quotation sets the stage:

> "These papers [cited] transcend the assumption that mutations are spread more or less evenly through a population: this assumption was only reasonable when mutation was considered to be a direct result of random insults from outside the organism. Mutations are

now known to be due to processing of the internal consequences of such damage, as well as to endogenous processes. ...." [103]

What is a mutation? There are many kinds with many effects, but for present purposes it is sufficient to liken it to a typographical error in copying the DNA sequence; that is, one of the four bases gets replaced by a different one, under the radar of the repair machinery. Known causes include radiation, chemicals, and copying errors; but the end result is a slightly modified protein, or no change at all for synonymous substitutions. Dozens of inherited blood diseases alone are known, each of which can be caused by a hundred or more known different mutations, with effects ranging from sub-clinical to mild to life threatening. But most mutations have little or no apparent effects, i.e., are adaptively neutral.

This has led to the idea of the genome as a 'molecular clock,' since if we know the average rate of mutations, then by counting the differences in the DNA sequences (mutations) in sister species, we can estimate how long ago they diverged, and thereby construct family trees (phylogenetic diagrams) of relations among related species or groups of species[104, 105]. This program has seen a rocky road, however, punctuated by clashes with traditionalists employing fossil-based methods, e.g., "cladistics"[105-107]. The new methods rely heavily on computers[108, 109]. But the molecular approach has been improving as workers have learned to select certain limited regions of the genome where the clock hypothesis appears to hold fairly well, so there by calibrating the method against known times of divergence using fossil or amber-preserved specimens[110-115]. Still, the field remains unsettled, with new findings and criticisms appearing regularly, such as mutation rates ten-fold more than previously thought[116], and another report finding that mitochondrial DNA, which has been the basis of hundreds of phylogenetic studies, is unreliable[117]. To say more about this would entail major digression.

More relevant to our thesis is the fact that different regions of any one genome mutate at grossly different rates, and likewise the genomes of different species. Perhaps the most dramatic example is the huge acceleration of mutation rates employed for immune defenses. To explain, one of our crucial defenses is the production of antibodies (immunoglobulins, Ig) such as the class called IgG. These are Y-shaped protein molecules where the tops of the Ys are like little hands, perfectly adapted to cling to pathogenic microbes such as bacteria, flagging them for destruction either by phagocytes that eat them like Pac-Man, or by the complement system which shoots holes in them, or by neutrophils which spray poison at them, and/or by other means. But there are thousands of different bacteria, so the tiny hands of the IgG must be modified for each new kind that comes along, without accidentally darting one of the body's own cells (which causes autoimmune disease).

This modification is accomplished by a process called *somatic hypermutation*: the generation of numerous IgG molecules, each a little different, by rapidly mutating the proteins in the little hands of the Ys; and the IgG molecules quickly 'mature,' getting better and better at locking onto the invader, within a few days[118]. A specific mutator enzyme that does this was recently found[119] and is now (2011) better understood[75].

Lower organisms, which of course must also resist disease, employ similar techniques – for example, the lamprey, a primitive fish[120]. Others achieve similar ends – if not for their own immunity, to evade or confuse the host's immune system – by mixing and switching between certain genes, as in the fruitfly[121], the malaria parasite[122], the parasites of Chagas disease and African sleeping sickness[123], the virus of whooping-cough[124], the bacterium of gonorrhea[125], a sea urchin[126], and dozens of others. The effects of mutations are sometimes confused with recombinations of genes that are already present[127-129]. The sea urchin varies its immune defenses by a trick called *alternative splicing;* see next article. Infectious (pathogenic) organisms often mutate their coats to overwhelm immunity at the same time as they are blunting the host's immune response by highly specific and sophisticated strategies, e.g., malaria[130] and a parasitic worm[131]. It is known that environmental stresses can foster increased rates of mutation[103], and that DNA repair mechanisms are involved with those of immunity[132].

The point here is that mutations are not necessarily or generally random accidents, as was assumed in the New Synthesis. Instead, it appears they are often tightly controlled, are frequently corrected by repair systems, and are commonly exploited for immune defenses, host invasion, and only God knows what all else. But by exactly *what* upper-echelon system is it doing the controlling? That is taken up in a later chapter.

7. FINE-TUNING. There are other truly major recent discoveries: epigenetics and non-classical RNA, to name the big ones, but they are left for the next chapter. Meanwhile, let us mention some important additional complications that further render classical evolutionary theory obsolete, or more obsolete than it already was in the 1970s. Chief among these are means for modifying proteins aside from mutations. To set the stage, the following is from the concluding section of a review article on protein evolution, being a frank summary of what is known about this, which is not very much:

> "Clearly our discussion on protein evolution, which focused on point mutations, is incomplete. There are at least three other principle genetic mechanisms that contribute to the evolution of new functions: duplication and functional divergence of genes, protein domain shuffling, and horizontal gene transfer across species." [133]

We already discussed horizontal transfer. Turning to gene duplications, it was found that genes often exist in multiple copies, only one of which is normally active. Sometimes the duplicates exist in a degraded or nonsense state (junk DNA again) termed *pseudogenes*. However, it has been shown that at least some pseudogenes, a.k.a. paralogs, have acquired novel and important functions[134, 135]. "After a mutation of a gene, a transcriptional reprogramming mechanism allows the intact paralogue to rescue the organism"[136]. Yet, this 'spare-parts' or 'backup' theory is not the whole story.

Many duplicate genes have acquired new functions, inspiring another theory of how evolution works, stemming from S. Ohno's book of 1970[137] and gaining adherents as ever more genome data pours in. Eichler is quoted in 2003 as saying that "we've known for some time that duplications are the primary force for genes and genomes to evolve over time"[138]. All genomes sequenced to date are rife with evidence of past duplications. A nice example related to my field (hematology) is a snake venom component, which the snake made from a duplicated gene for an ordinary clotting factor made in the liver, which it moved to its venom gland[139].

To explain by analogy, one might begin writing a business letter, say, by duplicating a previous one and then editing it to fit the new purpose. Likewise, many if not most 'new genes' appear to have arisen by duplication of old ones, followed by editorial modifications to suit new functions. One is reminded of the many specialized tools and machines of modern technology, most of which are made by assembling old parts in new ways. But is this 'mere analogy'? Might it not instead be a recapitulation of the selfsame principle on a higher echelon of organization? For example, comparative genomic analysis of four yeast species

> "... reveals that the different yeast lineages have evolved through a marked interplay between several distinct molecular mechanisms including tandem gene repeat formation, segmental duplication, a massive genome duplication, and extensive gene loss"[140]

Indeed, the entire yeast genome underwent an ancient duplication, followed by extensive loss of unneeded parts, and modification of the remainder. "Strikingly, 95% of cases of accelerated evolution [of duplicated genes] *involve only one member of a gene pair,* providing strong support for a specific model of evolution"[141], meaning Ohno's model (italic added).

Depending on interpenetation, this modality could have radical implications for evolutionary theory. Efforts have been made to find alternative explanations for duplications, such as that they provide redundancy (spare parts, back up) to make the genome more robust against errors, as defended by Gu et al.[142]. But there are reasons to doubt that as the primary explanation, for if such redundancy were really important, then one would expect to see it for all genes. Details are complicated, involving the various mechanisms that might give rise to duplications of many kinds and sizes: tandem duplications and crossings-over, segmental duplications, polyploidy, etc. For example, here is Postletwait, as quoted in an article:

> "Instead of a single gene doubling to make two adjacent copies as in tandem duplication, in a segmental duplication you could have tens or hundreds of genes duplicating either tandemly, or going elsewhere on the same chromosome, or elsewhere on a different chromosome ..."[138]

In other words, pretty much anything can happen. To explain the final outcome of the duplication, a new trait, the theorists fall back on the old dogma of random mutations and natural selection because no other hypothesis exists. We do not doubt that genetic change involves a good deal of poking around in the dark, so to speak — random searching, as if for a light switch on the wall — but we cannot shake the conviction that *active searching* underlies the events.

Another way that proteins are modified is called *alternative splicing*. Among the surprises of the human genome sequence was that there were so few genes, about 23,000, whereas more than 100,000 were expected. (Many simple plants and animals have genomes much larger than ours, but most have about the same number of genes.) The question was, how could this possibly be enough to produce, say, the human brain? Alternative splicing! To explain, it is a fact that many genes have several distinct segments which act to produce proteins in pieces, which are then stitched together (by *spliceosomes*); however, they can be spliced together in multiple functional ways. Thus, each such gene can potentially make "tens of thousands" of different protein products, and at least 74% of our genes are amenable to alternative splicing[143]. The caption of a figure announces *The 38,000 Protein Gene*[144]. Related new discoveries continue apace, such as that by Hanada connecting this mechanism to immune function[145]. As far as I can tell, nobody knows exactly how alternative splicing is controlled but hundreds, or possibly thousands of known proteins (especially membrane receptors) exist in variant forms with distinct functions due to alternative splicings of the same gene. With regard to evolution:

> "Alternative splicing is also seen as a method for enhancing evolutionary adaptability of an organism. ' It's like a big storehouse of potential genetic information,' says Douglas Black, ... 'a mechanism that allows testing a new mutation without losing the wild-type version.'" [144]

The implication for traditional orthodoxy hardly needs comment. Another modality for modifying protein function is called *post-translational modification* (PTM), meaning that after the protein is produced, it goes through a final finishing process, adapting it for special functions. For example, many of the proteins involved in blood clotting acquire in this way an extra carboxy group attached to certain amino acids that makes them bind calcium in order for the protein to bind to certain kinds of phospholipids[146]. (This particular PTM requires vitamin K, accounting for the anti-clotting action of drugs like warfarin.)

For another example, many if not most proteins have attached to them some carbohydrates (sugar chains), added during PTM (think of tinsel on a Christmas tree) called *glycosylation,* and this can be critical to their functions. For example, the blood group antigens (A, B, O, and many others less familiar) are mostly of this kind. As a rule, only higher forms of life practice glycosylation, but there are exceptions, such as a nasty virus[147]. Biochemists are trying to learn to duplicate this process[148]. Some PTMs require the embrace of appropriate chaperone molecules, 'chaperonins', some of which can compensate for errors (mutations) in the gene[116, 149]; but faulty chaperones can also cause disease[150]. Another way that faulty genes can be counteracted is by the emergence of other mutations which override, cancel, or circumvent the problem[151, 152]. Clearly, genomes are not passive lumps of clay.

11. BOTTOM LINE. Although more could be said, and will be said in the next chapter, enough has been said to make the point: genetic science today is still in its infancy, no better equipped than it was in the 1970s to resolve basic issues in evolutionary theory. Indeed, rather than clarifying how evolution proceeded, it has made it all the murkier; for so many possible mechanisms of change are now known that practically anything is possible. As someone recently put it, "just when you think that biology cannot possibly become any more complicated, it gets more complicated." If we liken "full understanding" to the summit of Everest, we are still toiling in the foothills.

Even if today's science did have clear implications for evolutionary theory, as opposed to shaky speculations, we would be well advised to hold our tongues, for there is no telling what new discoveries yet lie ahead, waiting to overturn hasty conjectures based on the latest findings. For that reason, we here resist being tempted into fruitless arguments about whether genetic theory can or can't account for this or that fact. Genetics is still so infant a science that it is *practically irrelevant* to such problems; and besides, we have already said enough to show that no clear "theory of evolution" any longer exists to argue about.

12.

Decoding the Code of Codes

1. THE SECOND CODE. The previous chapter gives only a foretaste of the big revolutions underway in genetic science. The two other real biggies are RNA interference and epigenetics. To back up a bit, recall that under the New Synthesis the theory of evolution rested almost entirely on the dogma that random gene mutations are the sole source of variation in the biosphere. It now appears that the genes are of only secondary importance, as in the following excerpt from a piece entitled *Searching for the Genome's Second Code* (the first code being, of course, the sequence of codons in the DNA that spell out the string of amino acids specified by any given gene):

> "Genes ... can't by themselves explain what makes cows cows and corn corn. The same genes have turned up in organisms as different as, say, mice and jellyfish. Instead, new findings from a variety of researchers have made clear that it's the genome's exquisite control of each gene's activity – and not the genes *per se* – that matters most. ... These [regulatory] elements are 'a major part of the [evolution] story that's been overlooked,' Levine says." [1]

That is to say, all organisms have a similar set of basic genes that supply the nuts and bolts for reproduction, growth, metabolism, and so forth. But how the nuts and bolts are assembled determines whether you end up with a Model-T Ford or an F-111 fighter jet, a flea, an oak tree, elephant, or human. Control of gene expression also determines that each cell in the human body, say, will function as a liver cell, brain cell, sperm cell, bone cell, whatever. Of course, this has always been obvious, whence much attention has been devoted to trying to understand how the many gene enhancers and repressors work together to govern the intensity and timing of expression of each gene[2-5].

An early clue that the genes are only part of the story might have been the unexpected discovery that some 98% of the human genome is non-coding ("junk DNA"). But as noted in the previous chapter, more and more of this 'junk' is now known to be functional, as further shown in this chapter.

On the other hand, it has proved possible to remove rather long stretches of non-coding DNA from a bacterium[6] and from a mouse[7], without gross ill effects, although in that same journal issue is another article showing that at least one-third of junk DNA is functionally active[8]. This reminds us of the fact that rather large pieces of the *human brain* can be destroyed or removed, also without gross ill effects. To illustrate, fully half of the human brain has been removed from several hundred patients, many of whom went on to lead normal lives, some even graduating college[9]. Are we to conclude from this that most of the human brain is junk?

Be that as it may, the search is now on for the Second Code, to finish the *Unfinished Symphony*[10]. Comprising that myriad of systems are repressors, enhancers, silencers, restriction factors and other means for controlling and orchestrating, *choreographing* the dance of gene activity in the developing embryo and in the specialized cells and organs of the mature organism, minute by minute. However, it is beginning to appear that there are multiple "second codes."

2. THE RNA WORLD. Nobellist Phillip Sharp has called the discovery of RNA interference (RNAi) "the most important breakthrough of the last decade"[11]. RNAi was deemed "breakthrough of the year" in 1992 and "arguably the most important advance in biology in decades has been the discovery that RNA molecules can regulate the expression of genes"[12]. Like so many other advances, the initial findings were believed to be rare and anomalous, or restricted to lower organisms or plants, but RNAi is now known to be centrally important in all living things, or at least, those composed of more than one cell.

Recall from high school biology that the classical function of RNA, known since the 1960s, was that of 'messenger' (mRNA) and 'transfer' agent (tRNA), acting to copy the gene (transcript) from the master DNA template to the protein assembly machinery of the ribosome for translation into the finished protein. But a number of seemingly unrelated and very puzzling experiments were about to converge on a major revolution.

One of those was an effort to make the color of petunia flowers more intense by adding some extra genes for the pigment. The result, however, was the exact opposite of what was anticipated: instead of darker colors, the flowers were paler, or had white blotches, or were completely albino. It later emerged that this effect, discovered by R.A. Jorgenen and independently by J. Mol in 1990, was caused by the extra dose of RNA from the added pigment gene[13]. This effect was called *co-suppression*, or co-repression, because the inserted gene as well as the original gene were silenced, by a mechanism still obscure.

A few years later, W.G. Dougherty engineered some tobacco plants to contain a gene for a virus coat protein (of tobacco etch virus) and found that it conferred resistance to infection, a kind of built-in vaccine. But how did it work? It turned out that it was the RNA transcript from the added gene that co-repressed that from the invading virus. Baulcombe and colleagues made similar findings with a potato virus that infected tobacco[14, 15].

Another key discovery was made in an animal (worm) while exploring the use of anti-sense RNA as a promising approach to therapy, such as against viruses. It was expected that this backward RNA would bind to the target RNA and disable it. As a control, they added the same amount of the normal RNA for that gene. "But to everyone's bewilderment" the normal RNA was equally or more effective[13, 15]! Yes, it was co-repression again. This set the stage for the 'eureka experiment' of Fire and Mello, who showed in 1998 that the effect was mainly due to double-stranded RNA, naming the effect RNAi[16]. The hunt was on[17, 18].

Incidentally, the terminology is somewhat confusing to us outsiders. It appears that RNAi is the name used by those working with worms, but was known as 'co-repression' to those working with plants, and as 'quelling' by those studying fungi, and as 'squelching' to still others[19]; all of which are now supposed to be called 'post-transcriptional gene silencing,' PTGS[15]; but "RNAi" is the name that stuck. Similarly, one encounters many names for the various types of RNA causing these effects, such as micro-RNA (miRNA), small or short interfering RNA (siRNA), small nucleolar RNA (snoRNA), small temporal RNA (stRNA), short hairpin RNA (shRNA), and several others, some of which seem to be equivalent, depending on authors. Such fine points need not concern us here, though it may be mentioned that miRNA and siRNA are about 21 to 25 nucleotides (nt) in length, while stRNAs are about 70nt in length. Some 95 of the snoRNAs seem to perform a "guide function," hence are called gRNA[20]. A table of seven classes of small RNAs was given in 1997[21] and some 18 classes in 2002[22], but new ones keep popping up, most currently at this writing (2006), piRNA[23]. No attempt will be made at further updates in final editing (December, 2010), except to mention that the field continues to explode.

Thus began a veritable floodtide of discoveries. It is hard to find a single issue of the relevant journals of the last ten years that does not contain reports of new findings about RNAi[24]. The pace of advance may be gauged by comparing the special issue of *Science* on RNA in 2002[25] with that of 2005[26]. Some idea of the complexity of the subject is conveyed by the quotation we gave in chapter ten from a book review[27].

3. IMPACTS. Aside from revolutionizing genetic science, Ran has emerged as a powerful new research tool for the study of gene functions because it is easy to construct saran for any known gene to silence it, allowing one to observe the effects[28]. This is called the 'knock-down' method, as opposed to gene 'knock-out' (the actual deletion or excision of a gene of interest) which is far more tedious and expensive. A nice *tour de force* of this kind was that by Klamath et al., who used RNAi to probe the function of about 85% of the roughly 19,500 genes of a worm[29]. Although most of the knock-downs were not viable (were lethal), enough worms survived that Kamath et al. were able to quickly assess the effects of knock-downs on 1,722 genes, an impossible labor by previous methods. Another such application has refined our understanding of Down's Syndrome[30,31], another was to search for genes involved in cancer[32], and another puzzled out how a virus evades immune defenses[33].

The same method has potentially huge medical applications[24, 34, 35], some relevant to hematology[36]. The general idea is that it is now possible to attenuate any gene that might be causing health problems, either within the body or from pathogenic microbes. For instance, infection by the SARS virus has been cured in monkeys using RNAi[37], and efforts are afoot to control cholesterol levels in this way[38, 39]. Unfortunately, however, as usual with new approaches, rough spots in the road are already coming to light, such as fatal liver disease using shRNA in mice[40]. Adverse outcomes like this are called 'off-target' effects[41-43].

A number of diseases are now attributed to aberrations of RNA control systems, such as heart defects in newborns[44, 45]. Faulty RNA control now appears to be a crucial missing piece of the cancer enigma[46]. RNA silencing has been called *The Genome's Immune System*[47]. It has been shown that RNAi functions in the immune system of the fruitfly[48] and probably likewise in humans. Immune recognition of viral RNA is known to be important in human defenses[49], but in a bizarre twist, it appears that some viruses exploit RNAi as a means of evading the host's immune response[50]. The hepatitis C virus seems to be controlled by miRNA[51]. RNA control is said to be responsible for the unusual meatiness of a kind of sheep[52], suggesting another whole area of practical application and with deeper implications. Of more than passing interest is the finding that a natural miRNA found in the brain regulates dendritic spine development[53], intriguing because these tiny spines on our neurons are important for learning and memory[54]. Human sperm is loaded with small RNAs[55], function unknown; and by the way, it appears that sperm cells sniff their way to the egg[56].

One could go on, for this is a very active field. The point to be made here is that it is an *entirely new and unexpected level or layer of complexity in gene regulation*, having come to light in just the last decade or so. One article sums it up nicely in its title, *RNAs Running the Show*[57]. Though the surface has barely been scratched thus far, it is already clear that it is immensely complicated, hugely important, and is served by an array of specialized molecular machinery. In sum, then, this further warrants our skepticism towards evolutionary theory based on a 1970s concept of genetics.

4. DARK GENES. All of this RNA floating around is being called the "dark matter of the genome," even a "secret society"[26, 58], a kind of Twilight Zone. Mattick has estimated that non-coding RNA may account for up to 98% of all transcriptional output[17]. Some of the dark RNAs have their own special genes, mostly found out in the junkyards of gene deserts. For example, codings for Lin-4 and Lin-7 are conserved from worms and fish, to humans[59]. The two that were known in 2000 became 150 by 2002[17], and yet more by 2003[60], but counting them is difficult because they are not genes in the classical sense, they lack the standard signatures. Thus, the definition of a "gene" has become fuzzier[61]. The old beads-on-a-string picture is fast fading.

> "Over the years geneticists have also documented overlapping genes, genes within genes, and countless other weird arrangements [such as genes without borders. R. Flavell has] documented human immune system genes that seem to be controlled by regulatory genes from another chromosome. ... 'We've come to the realization that the genome is full of overlapping transcripts' [said Kapranov]. The picture these studies paint is one of mind-boggling complexity. ... a whopping 63% of the genome is transcribed [*vs.* just 2% classical genes] ... the crux of the debate now is the extent to which all the extra RNA plays a part. ... A study from last year [by J. Hogenesch et al.] hints that at least some of the mass of RNA is doing something useful. ... 'The degree of complexity we've seen was not anticipated' [says Roderic Guigo]." [62]

Nobody yet has dared to speak of 'junk RNA.' At least two main functions for the dark RNA are clear, most obviously being for immune defense against viruses, at least in plants[63, 64], but the regulatory role is huge, though is yet rather murky. So this is a wild-west frontier of science. The 'dark RNA' doesn't just float around to stick here and there, instead there exists an impressive array of special machinery for refining and processing it, involving enzymes and complexes with names like Dicer, Slicer, Argonaute, Clipper, Drosha and Symplekin[65], some of which assemble into large complexes, e.g. the 'RNA-induced silencing complex,' RISC[11, 14, 65-69]. It is almost certainly central to 'molecular evolution'[70], another whole can of worms. One authority is quoted as saying that our knowledge of dark RNA "is just the tip of the iceberg [of] a vast control system"[59]. Deciphering this code – if it is a code – poses a formidable challenge to generations yet unborn.

5. EPIGENETIC MARKS. Equally or more revolutionary has been the discovery of the epigenetic phenomenon called *imprinting*: heritable changes from either parent, apart from changes in the DNA sequence. There were hints of these effects back in the 1960s, as in experiments showing heritable effects on the offspring of cage-crowded mice and, more obviously, in changes in animals such as pigs after a few generations of domestication, or upon reversion to the feral state. But since there was no explanation for such things back then, they were ignored, swept under the rug. Now, of course, it's a headline topic, right up there with RNAi.

To set the stage, one of the early experiments was by B.M. Cattanach in 1985[71]. As told by Sapienza[72], he took advantage of an unusual strain of mouse to show that if both copies of chromosome 11 came from the same parent (instead of the usual one from each), the pups were radically different. For example, they had stunted growth if both were from the mother, but were gigantic if both from the father. Clearly, the chromosome had acquired some distinctive features, an *imprint*, depending on which parent it came from, and this was revolutionary stuff.

The big discovery that opened this field to systematic research was that of the chemical basis of the imprints: they are distinctive chemical marks affixed to the DNA (and/or histones, see below). These marks are simple chemical groups such as methyl (-CH3), acetyl (-CH2COOH), phosphate, and some others. The act of adding them is thus *methylation, acetylation,* etc., and of removing them is *demethylation, deacetylation,* etc. There exists an array of specific enzymes, often acting in complex with other proteins, for adding and deleting these marks, with names like *methyl transferase, histone deacetylase,* and *histone acetyl transferase*[73]; and experiments have shown that inhibiting these enzymes or silencing their genes disrupts imprinting.

Old beliefs are being overturned on a regular basis, as the new findings are "breaking Mendel's laws" of inheritance[74]. Such work is not always greeted with open arms. Thus, one recent paper, already being called a landmark, showing that the grooming and nursing of pups can trigger lasting changes in gene expression, was initially rejected by *Science* because it was "too radical"[75]. Another shocker has been the demonstration that dietary components, in this case, the vitamin folic acid, can modify epigenetic marks to restore faulty gene expression[76]. It seems possible that a lecture I heard at a recent meeting has a similar explanation[77].

Epigenetic findings are now making daily news, with headlines like, "Men inherit hidden costs of dad's vices"[78]. Methylation can also modulate gene activity by acting on tRNA[79].

Faulty imprinting seems to hold the key to a number of serious inherited neurological diseases[80] including Beckwith-Wiedemann syndrome[81, 82]. A whole new and unexpected dimension of cancer, a missing piece of the puzzle, has opened up with epigenetic research[83-87]. For example, the long suspected relationship between genes and environmental factors in cancer now has a more or less clear physical basis[88]. A number of hormone activities are under epigenetic control[89]. Among the papers to catch my eye was one showing that the gene expression of identical twins steadily diverges as they age[90]. Methylation state appears to regulate whether jumping-genes will jump or not and was apparently involved in the formation of a hybrid of two kangaroo species[91]. One could go on.

But let us come to the bottom line: *Lamarck is back, big time!* So are Darwin's original thoughts along these lines. After generations of ridicule, it is now established that heritable changes acquired in the parent's experience can pass to offspring. (Lamarck's beloved daughter, mourning over his grave, vowed and predicted his vindication.) By the way, *vis a vis* teaching evolution in the schools, I distinctly recall a teacher, in about the 8th grade, pointing out the "disproof" of Lamarck's theory: cutting off the tails of successive generations of mice failed to produce tailless mice.

6. THE HISTONE CODE. It is now known that the DNA itself carries only some of the epigenetic marks, the others being carried on the little spools called *nucleosomes*, composed of histone proteins, around which the DNA is tightly wound in the chromosomes. It was previously thought that these little spools were just that: spools for compactly storing the DNA; but it is now clear that they are active players in the regulation of gene expression. Also, they are loaded with epigenetic marks, initially just the methyl and acetyl marks, but now (2006) numbering some 50 kinds[92]. Each of them is more or less specific for certain target bases and is accompanied by its own enzymatic machinery for the addition or removal of the marks.

How do they work to affect gene expression? Roughly speaking from casual readings (bearing in mind that this writer is not expert in these topics), it seems that segments of DNA cannot be read, cannot be transcribed, if they are too tightly wound; and it now looks like the epigenetic marks can act to loosen the coils so that the transcription machinery can ride along the gene segments to be read[93, 94]. However, the details are complicated[95] and are not yet clear to anybody, much less to me. This might be guessed from the existence of some 50 kinds of modifying marks, only a few of which have yet been investigated. It was thought that some of the epigenetic marks on the histones were irreversible, but it turns out they are not[96], and many or all may be in play at once. Once-secure knowledge is overturned on a regular basis.

> "In the field of transcription there is little room for dogmatism — principles that were once deemed sacrosanct can [now] be violated with impunity. The latest example of this is illustrated on page 187 of this issue ..." [97]

The epigenetic marks do not exert their effects in simple independent fashion, but rather modify each other in complex ways. For example, there is mutual interplay between methylation and acetylation[98], affecting neighboring or distant nucleosomes[99]. Furthermore, the marks on DNA interact or overlap with those on histones[100]. Another added dimension of complexity is that epigenetic mechanisms often involve components of RNAi[101, 102], as well as systems of DNA repair[95, 103]. The enzymes performing these duties are not, generally speaking, simple ones in the classical sense (if any enzyme can be called 'simple'), but are large complexes made up of perhaps a dozen of more proteins termed 'molecular machines' or 'holoenzymes' – "transcription by a committee" of proteins, is how one reporter phrased it[104]. All this stuff is going on in the cell nucleus:

> "Much like a train station at rush hour, the nucleus seems chaotic at first glance — packed with a jumble of chromosomes, RNA molecules, and proteins. Yet, just as careful observation of a train station reveals an orderly flow, research ... has revealed [that these activities] are far more coordinated than had been suspected." [105]

But of course it must all be coordinated! In the same article, Wolffe is quoted as saying the nucleus is "like a Swiss watch" but we prefer the 'train station' metaphor – or better yet, a *city*. And by the way, anybody who thinks that science takes the mystery and wonder out of life is missing the boat. It's total mystery. And if the theorists would only admit that, instead of their usual pretense to knowledge that doesn't exist, then perhaps more creative and visionary young people would be attracted to the field.

7. WHAT KIND OF CODE? Maher mentions "stacks of papers" written about histone methylation alone, and quotes a remark by C. David Allis, who coined the term *histone code*, that "something is writing and something is reading this code"[106]. What's that? As I see it, anything that can read and write deserves to be called *somebody*, not *something*. That is to say, whether it is really a 'code' or not is open to debate, and has been debated[92, 107]. But I see it more like a language than a code, in the sense that the items specified by a true code, such as Morse code or the genetic codons, are quite unambiguous, whereas the things specified by the words of a language – and by the histone marks – can vary depending on context. Hints of a language-like structure have been claimed for sequences of junk DNA, granting the validity of the statistical criterion behind that result[108]. Unfortunately, we must postpone further discussion of this point to a later book of this series because at the present time, there is no consensus on how language works, or what it is, or how it arose in the history of our species. But the parallels between language and the genetic code(s) is so striking that we must be prepared to accept a direct correspondence between the mental workings by which we speak, and those of the genome; which, after all, must embody all that we are.

8. GETTING IT ALL TOGETHER. To sum up, what we have before us is a sumptuous plate of new knowledge but no way to digest it, no way to see how it works overall in producing the likes of you and I, keeping us healthy through thick and thin. The world's computers are bulging at the seams with terabytes of genomic data, a kind of inventory of parts, proteins galore, complete gene sequences of hundreds of organisms, a plethora of RNA species, carbohydrates and lipids by the thousands, motors and cables and turbines and generators of every description, and all on a scale so tiny as to make Lilliput seem Brogdingnagian. Truth (reality), indeed, is stranger than fiction.

So, on the tenuous assumption that we at last have enough data, the problem confronts us, *how does the whole thing work?* One approach favored by a certain cadre is known as *systems biology*[109]. What is systems biology? Good question. Advocate Wiley says it "seeks to understand how the molecular processes of cells are linked to higher biological functions"[110]. Hot stuff. He also confesses that the enterprise is "host to disagreements fueled in part by a lack of uniform definitions," and that skeptics scoff that it is "long on style and short on substance." Amen.

My impression from some dozens of such articles over the years is that 'systems theory' belongs to a family of buzz words including also network theory, complexity theory, and chaos theory, and is thereby related to efforts to make sense of and predict earthquakes, hurricanes, volcanos, the weather, climate change, economic trends, internet crashes, stock prices, crime waves, and power grid failures. Huge computers serve these exercises, with a resounding lack of success.

The editorial with that article avers that a recent clinical disaster (eight volunteers testing a new medicine all collapsed, critically ill) could have been averted if the Systems Approach had been used. The same idea was floated following the first space shuttle disaster, the idea in both cases being the need for more integrated input from related disciplines. But this ignores the very nature of the beast For example, in the case of the space shuttle, the warnings of certain engineers were ignored, much as the warnings of experts on the Middle East were ignored by the Bush administration bent on civilizing Iraq. Similarly, many of us correctly predicted failure of early clinical trials of a certain cancer treatment; but a good friend went to the High Priests of Harvard anyway to receive this much-touted treatment, returning a few month later in a coffin (but not before being relieved of 20 million dollars). Systems Biology to the rescue?

Wiley is more specific in showing how our understanding of a certain hormone, epidermal growth factor (EGF) and its receptor (EGFR), could be improved by a five-step approach in the "systems perspective". Granted, the EGFR signaling system, like most hormone-receptor interactions, is horrendously complex, and some laudable understanding of it has been achieved, but not by application of any five-step method, suspiciously reminiscent of the fictional Scientific Method, not to mention a 12-step method for kicking the bottle.

Closer to my bailiwick is an article he cites as a model of good systems biology, concerning apoptosis, by Janes et al.[111] Now, apoptosis, also known as programmed cell death, or induced cell suicide, was a simple idea in the late 1980s: activation of certain enzyme cascades in the cell either killed the cell or it didn't. Over the years, however, it became increasingly complicated, needing wall-size posters to show all of the dozens of agents and signaling pathways involved, making it more and more difficult to decide if a cell was apoptotic or not, which is how our lab got into the fray[112]. In any case, the article by Janes et al., purports to put all of the details together, identifying a network involving "7,980 intracellular signaling events that directly links measurements to 1,440 response outputs associated with apoptosis"[111]. They are even able to make some correct predictions of the effect of a given input on the outcome. A similar approach was taken in attempting to organize the growing complexity of DNA repair mechanisms into a comprehensible body[113].

To manage and analyze the deluge of data, computer whiz kids in 'bioinformatics' have emerged and are much in demand[114, 115]. All of this is well and good, but it is hard to see how the 'systems approach' differs from ordinary hard work and common sense, the only novel feature being the effort to assimilate vast amounts of data.

9. THE -OMIC APPROACH. Everyone has heard about *genomics*, comprising efforts to understand how the genome works by trying to make sense of all the genes[116], such as by drawing computer-assisted maps and diagrams of gene interactions[117, 118]. The set of all such gene interactions is being called the *interactome*, and the results of such efforts are diagrams of dizzying complexity[119-122], some being almost black with connecting lines and dots marking nodes[123]. We already remarked on efforts to organize knowledge of all the little pieces of RNA floating around, now being called the *transcriptome*.

Upon examining such networks, one notices that that they resemble human social interactions, being similar to, for example, the set of all connections among computers using the internet. One study of interactions of proteins in a worm draws the unsurprising conclusion that the most highly connected hubs are the most important and vulnerable[124]. However, another report indicates that most such biosystems are constantly changing, "rewiring the network" as they go[117], reminiscent of the brain's wiring.

The totality of gene regulation mechanisms has been called the *regulome*[125] and many approaches are taken to try to figure it out, although nearly all are based on computer-intensive mappings of interactions, dense with connecting lines[123, 126]. Several have taken the so-called combinatorial approach [4, 127], i.e. that genes work in various combinations. No big surprise there either. But every time somebody thinks they have it all figured out, another report comes along to set the heads scratching anew. Thus, it had been assumed that similar sets of genes which are conserved in different organisms will interact in much the same way, but a study of such interactions in the malaria parasite upset that applecart:

> "It is generally expected that conserved genes will retain their functions and interactions. From this comparison, a different principle emerges: conservation of specific groups of related genes does not necessarily imply conservation of interaction among their coded proteins." [128]

The output of the genes is a myriad of proteins, many or most of which interact in many ways, whence we have *proteomics* and the *proteome*[115, 129-131]. To explain how this is done, a method called micro-array analysis is used in which one begins with a plastic plate having some hundreds of tiny wells or depressions, each containing a speck of some specific type of protein – or segment of DNA, RNA, antibody, whatever – in such a way that when you add your sample, anything in it that happens to stick to the speck in the well causes a fluorescent signal, or a color change[132, 133]. This color change, or some corresponding number, is fed into a computer where the results are displayed on a big multi-colored checkerboard, to be analyzed in various ways. Other methods rely on mass spectrometry.

The results are presented as a kind of map, an *interactome*, featuring hundreds or thousands of tiny circles, squares, and triangles, all color-coded or numbered, with lines showing how each interacts with the others. The first big effort I saw of this kind caused me to burst out laughing, whence I herewith propose the generic name, -*comics*. Or, since nobody really knows what to make of these maps, perhaps a better generic term would be *Theomics*. For example, one such paper (with 52 authors) gave "a draft map of 7,048 proteins and 20,405 interactions," partly shown in a full-page diagram of some 3,500 tiny color-coded circles and triangles connected by thousands of lines[115]. There are many of this genre[134, 135], and their resemblance to networks of human social interactions has not gone unnoticed, one survey bearing the apt title *A society of proteins*[136], another carrying this remark: "'It's a bit like human society,' suggests Gitte Neubauer, one of the researchers ... 'People come together in different groupings to fulfill particular functions.'" [135]. So they do.

Not to be left out, those who study lipids (fat-like substances, of great variety and protean roles) are busily creating a *lipidome*, towards a science of *lipidomics*[137]. Similarly, those who specialize in carbohydrates, also of vast structural variety and importance, are pursuing the *glycome*[138, 139] and *glycomics* (where *glyco-* refers to sugars, *-comics* to my term). In the same spirit, one hears of the *metabolome*[140, 141], the cancer *kinome*[142], the malarial *secretome*[143, 144], and, not to be outdone, the *epigenome*[145, 146]. Some work hints at a coming *signalome*[147]. We have already alluded to the *RNAome*[25], later called the *Ribognome*[148]. Plans are now afoot also for a brain *connectome*[149]. Even a proposal for a psychology-based *behaviorome* has been floated[150]. Those in my own field, blood clotting, seem to still have their wits about them; although we, too, have some big diagrams, nobody yet has dared to propose a *coagulome*. It is interesting to observe how the original aim of the genomic enterprise, to embrace the whole system, has now fragmented into the multitude of above-named *-omes*.

10. STUMPED. No ridicule is intended. The above named *-omic* approaches are well-intended efforts by people of normal intelligence to try to get insight on a very knotty problem, to wit, how the cell works, how the genome works. The underlying science is great stuff, the money is well spent; good science has always been a hallmark of a great civilization and is one of the few things our society can be proud of. The fact that there is not yet any glimmer of understanding how it all works is nothing to be ashamed of.

The above-cited *-comics* are nothing more or less than a valiant attempt to get our arms around the problem of figuring it out by sheer computational power, as if sufficiently many numerological permutations will suddenly reveal, displayed in full color on the computer screen before us, the genie in the Lamp of Life, the Code Behind the Code, the Wizard of Oz in all of his emerald-green glory. But don't hold your breath.

It's not more data that we need, or bigger and faster computers, or larger international collaborations. What we really need are a few real individuals with old-fashioned brain power, deeper thinking, a brood of young Isaac Newtons. Even as I work on this chapter, startling new wonders are appearing, such as the bewildering structure of one of the chaperonins[151], and breaking news on "gene factories"[152] (yes, their given name) showing that widely separated genes, even on different chromosomes, can somehow literally reach out and touch together. So it's all very mysterious, and getting more so; and the experts know it. They are stumped. It's dumbfounding.

11. MAN AND BRAIN. What we actually see in the genome is a recapitulation on the plane of molecular biology the same types of phenomena seen in the macro spheres of biology: ecosystems, human societies, insect societies, economics; and above all, beneath all, the brain that affords us the limited consciousness that we have (or the equivalent in creatures without brain, such as jellyfish which have eyes but no brain; see book I).

Now, in a very real sense, you don't have to be a scientist to know how the body works. Everyone knows perfectly well: by a simple act of willing, you can stand it up from your chair, walk it, run it, sing with it; and what is the brain but an expression of the genome? So, we circle back to the hypothesis, that the genome is itself a conscious entity, and that the consciousness we all experience in everyday life is a small window on the giant mentality of the total genome. And where did that total genome come from? Well, it's made of carbon, oxygen, nitrogen, hydrogen, sulfur, phosphorous (see book I). Granted, this is only the beginning of an answer, a crude outline. But it's a pretty good beginning, and our case will get stronger. Read on.

13.

The Biochemical Tool-Makers

1. WILDER YET. Previous chapters have compelled us to some relatively radical convictions, if Charles Darwin's convictions can be called "radical." But it so happens that further pursuit of these ideas inevitably leads to conclusions wilder yet, and the reason is this: it is not generally possible to sharply dissociate animal *behavioral* innovations from the physical and biochemical machinery which supports them. Venoms are a good example for opening discussion. The general public is rightly fascinated by the powerful venoms of so many snakes and other creatures, but that fascination greatly increases, the more one learns about their biochemical intricacy, and how perfectly they are tailored to the task of disabling or killing their victims. The big question asked in this chapter is, *how did the snakes get their venoms?*

We have already found reasons to doubt the conventional explanation: natural selection of random mutations, which amounts to hand waving – faith – as opposed to the nitty-gritty of actually confronting specific questions, like venoms, and supplying a detailed, step-by-step scenario that is really persuasive. We begin with some examples.

1. THE HEMATOLOGICAL SNAKE. Anyone who works for long in research in hematology is bound to encounter snake venoms. Several are used in routine clinical testing, notably, Russell's viper venom, which causes plasma to clot, is commonly used to measure the lupus anticoagulant. The southern copperhead, *Agkistrodon contorix*, is used for measuring protein C, a natural blood anti-clotting factor. Ecarin is a component of venom from the saw-scaled viper, *Echis carinatus*, used for its action on prothrombin; and Botrocetin (trade name) is purified from Wied's lance-head snake, *Bothrops neuwiedi*, for its special mode of action in causing platelet aggregation, a diagnostic test for von Willebrand disease. Some have even been used as drugs, such as ancrod[1], trade name Arvin, from the Malayan pit viper, *Ancistrodon rhodostoma*. But many similar agents have been purified from other snake venoms, e.g., from the timber rattlesnake, *Crotaus horridus horridus*, and the lance-headed American pit viper, *Bothrops atrox*, trade named Defibrase and Reptilase, respectively, all of which break down fibrinogen to destabilize clots, causing bleeding. Dozens of other snake venom components are used in special research applications, e.g., the many with thrombin-like activities[2]. Perhaps a thousand hemo-active components from hundreds of snake venoms have been purified and studied[3, 4]. New ingredients are constantly coming to light, including from this institution[5, 6], and often at meetings, published only as abstracts[7].

These venoms are not simple poisons but are complex mixtures or "cocktails" of enzymes (proteins) or peptides (tiny proteins) in definite proportions, each of remarkable specificity. In addition to their hemoactive properties, actions on blood, causing either hemorrhage or clotting (thrombosis) — many or most also contain potent neurotoxins. The human blood clotting system is delicately balanced and extremely complex, consisting of at least 25 proteins in normal blood, and more if you count the many other plasma proteins that modulate clotting (phospholipases, complement system, bioactive lipids, etc.), not to mention platelets. Snake venom components are known which target just about all of these parts, usually in lethal combinations. For instance, several have specific actions on platelets, such as one which targets certain proteins on platelets, called gp 1b/V/IX[8, 9], and another, crotavarin, specifically inhibits platelet aggregation[10], while still another, crotalin, was proposed for medical use[11]. For a list of snake-derived agents which mimic aspects of just the fibrinolytic system (the body's system for dissolving blood clots — see Markland[11]). These venoms have been extremely useful in helping to tease apart the structure and function of the normal blood clotting factors[12], as well as the complement system of innate immunity[13], the latter by use of a cobra venom factor[14]. However, venom of cobras, such as *Naja kaouthia,* has many components, such as the agent called kaouthiagin[15]. Snakes often or usually have specific inhibitors to protect them against their own toxins, as do a few of their prey animals[16].

With that introduction, the question before us is this: how did these venoms arise? Or more accurately, how did such a *system* of envenomation arise, meaning not only the venoms but the complex apparatus for producing them (numerous specialized enzymes), delivering them such as by elaborate fangs, glands where the venom is stored, special muscles for injecting the venom, and so on. On top of all that is the requirement that the snake *must know how to effectively employ* this equipment and must be motivated to do so. Bearing in mind that even a single wrong amino acid among the hundreds in the chain that make up a protein can ruin its action, it is clear that something must be going on beyond the ken of the orthodox theory.

2. THE ANTICOAGULANT LEECH.

While we're at it, mention may be made of the many creatures whose food is blood and therefore have developed (or *devised?*) means for preventing blood clotting that would otherwise interrupt their meal. Best known of these are the leeches, each species producing its own distinctive cocktail. One agent, hirudin, purified from a leech, achieved some commercial success as a so-called "blood thinner"[17, 18]. In fact, it worked too well, and so fell from favor; but at least a dozen other active components from leech saliva have been characterized, e.g. antistasin[19] , decorsin[20] , calin[19], a prostaglandin[21], lefaxin[22], ornatin, hementin, bdellins, ixin, and others, many of them listed by Rigby and colleagues[23]. There is growing interest in using whole live leeches for therapy[23, 24]. Leeches also secrete an anesthetic cocktail, so their feeding goes unnoticed.

The saliva of the kissing bug has its own special anti-coagulants[25], as do deerflies[26], mosquitoes[27, 28], vampire bats[29-31], ticks[32-34], and parasitic worms[35], including hookworms[36]; the latter leading to a patent dispute because of commercial potential [37]. For a partial list of those which specifically affect platelet aggregation, see Chow et al[38]. Many venoms contain agents which attack specific lipids (fat-like substances) involved with blood clotting and other physiologic functions, especially those from bees and wasps but also spiders[39], and of course, snakes[40]. Some creatures appear to deploy hemoactive agents for defensive purposes, as with caterpillar spicules[41], and perhaps sea cucumbers[42, 43]. Again, how did they get this equipment?

3. THE NEUROLOGICAL SNAIL. Although the above examples focus on attacks on the blood, attacks on the nervous systems of prey or predators are even more widespread, such as at the neuromuscular junction[44]. These agents are equally beautifully designed and specific. For example, proteins used by related snakes target either the heart (cardiotoxin) or certain nerves (neurotoxin), yet differ only in very subtle but crucial details[45]. A certain scorpion toxin must be folded just right to be effective, complicating laboratory efforts to study it[46]. Likewise, so-called three-finger neurotoxins from certain snakes all have similar structures, with only 60-74 amino acids, yet include widely diverse specifics:

> "Members of the family include alpha-neurotoxins, which antagonize muscle nicotinic acetylcholine receptors, kappa-bungarotoxins which recognize neuronal nicotinic receptors, muscarinic toxins with selectivity towards distinct types of muscarinic receptors, fasiculins that inhibit acetylcholinesterase, calciseptins that block L-type calcium channels, cardiotoxins ... that exert their toxicity by forming pores in cell membranes, and dendroaspins, which are antagonists of various cell-adhesion processes."[47]

A family of snail species is at least as astounding in their neurotoxins. By way of background, Baldero Olivera returned to his native Philippines after earning a PhD in biochemistry but was dismayed at the low budget and meager equipment available to him. He therefore took on the modest project of studying the locally known "killer snails," the *cone snails,* which hunt by shooting out venomous harpoons to impale and almost instantly immobilize passing fish, which they then reel in to eat. What he discovered was dozens of species, each with a cocktail of potent neurotoxins, collectively the *conotoxins,* now of great importance to research because so many of them target with pin-point accuracy neural receptors previously unknown[48], with potential medical applications[49].

4. HOW DO THEY DO IT? Olivera finds that conotoxin genes exhibit more rapid changes than others, and commentator Caporale reminds us of hypermutation in our own antibody production, stating: "It is clear that genomes have developed the capability to focus genetic exploration."[50] His word is that the *genomes* have acquired this ability, but we prefer to be politically incorrect by saying that the *snails* have this ability "to focus genetic exploration." Similarly, Ohno, a leading researcher of snake venoms, states in his abstract that

> "... the molecular mechanism by which proper base substitutions occurred in particular sites of venom isozyme genes *is a puzzle to be solved in future studies.* It should be noted that *accelerated evolution occurred until the isozymes acquired their particular function* and, since then ... less frequent ... The relationships between mutation and its driving force are speculative and the real mechanism remains a mystery." [51] Italics added.

This quotation and others like it are among the strongest factual supports for our hypothesis, earlier stated, which is that following an initial discovery, such as (in this case) an accidentally acquired weak venom, the animal then mounts a purposeful strategy of improving and perfecting it, as reflected in the accelerated pace of mutations leading to the final product, after which mutations cease. Of course, this raises many questions which might lead the closed-minded skeptic to scoff and throw this book across the room. But he who thinks open-mindedly will find that all of those questions can be answered in self consistent fashion.

At the very least, it is clear from such statements that natural selection of random mutations is out the window, no longer a viable option, if it ever was. Another clue to the solution is found in an advertising blurb, from a company that sells venoms for hematology labs: *"Snakes know everything about human hemostasis. And we know a lot about snakes."* [52]. Again, an article about cone snails refers to them as "the most innovative genetic engineer"[50]. Indeed, it is commonplace to hear bacteria being called "master biochemists." Of course, these are thought to be mere figures of speech, with joking overtones, but we submit that such statements are literally true, that lower organisms do in fact have a clear knowledge of their own biochemistry and how to work it, and by extension, knowledge of the biochemistry of their prey or adversaries. The rationality here follows from our general thesis that all life and its evolution is psychologically driven and from the pyramid model of self-control, that as one ascends the *scala natura,* increasing numbers of bodily functions become "unconscious," having been delegated to autonomic modules. In terms of genetics, then, the most fundamental attributes of life constitute the base of the pyramid of genetic memory.

Thus, we no longer have conscious control our body temperature, heart rate, respiration, etc., much less over the genes and enzymes underlying these functions. But simpler organisms, such as the bacteria, who must do daily battle against a myriad of enemies – including the antibiotics we deploy against them – have no such intermediary autonomic modules. They are all "bare-fists & shields" at the biochemical level. A parallel may be seen in our own habitual skills: learning to ice skate or ride a bicycle takes intense concentration but eventually becomes automatic, taken over by a lower echelon.

This thesis is further supported by the observation that biochemical-based adaptations become less and less common as one ascends the phylogenetic tree; with rare exceptions, higher animals do not produce venoms or toxins, or shoot poison darts from their bodies, or spin webs, or change color, or regenerate lost body parts. It is as if they have forgotten how to do such things; or more accurately, they no longer have easy access to or control over such functions. How does the snake know to continue pursuing perfection of its venom, generation after generation, for thousands of years? Answer: the genome remembers. More on that later.

Similarly, the chemical communications so rampant among bacteria, protists, and insects become progressively less prominent in higher taxa, presumably displaced from the sensory space – i.e., from deliberate conscious access – by higher-echelon modalities, until finally with humans, we have lost even the keen sense of smell of our mammalian compatriots, perhaps displaced by such faculties as language.

This is not quite to suggest that lower forms intellectually plan and then execute such biochemical devices. Without eyes, they are working pretty much in the dark. Many or most of their innovations were likely originally serendipitous. For instance, the ancestor of the archer fish probably first downed a passing insect by fortuitously spitting a drop of water at the right moment. Indeed, it is very likely that our own ancestor discovered, initially by accident, that it could stun or kill small prey by throwing rocks, resulting in the invention of weapons. But once a small advantage was realized, discovered, by the first venomous snake (or snail, or etc.), it is plain that a *motivated trajectory*, a vision or ambition, was set in motion, giving rise to the dozens and hundreds of other species, each of which sought and found new ways to modify or improve their venoms.

Notice here a certain parallel with orthodox theory, which holds that a small initial advantage, such as an accidentally acquired venom, would lead to natural selection to improve it (though natural selection cannot explain the acceleration of mutations). We could almost agree, except that we insist that the small initial advantage was a *discovery,* which motivated its deliberate improvement by shuffling genes that *it knows from its own blood and nerves,* building the accessory organs – the venom glands, the perfect fangs, and so forth, all having the appearance of "intelligent design", because that's exactly what it is!

There is no other credible way to explain, we submit, the sum total of observations (the above-quoted acceleration of "mutations," etc.), except by allowing a central role for the mind, for consciousness. Or, if you insist, "the genome," the DNA, for it contains everything that the animal is, hence must be the ultimate seat of consciousness in living beings. Our ordinary consciousness is but a tiny window on that vast mind; and the brain is among its advanced tools.

5. PLANTAE. Not even the most cursory survey of natural chemicals used for defensive and offensive purposes would be complete without at least passing mention of those from the kingdom of plants. From ancient times to the present, plants have been the source of the great majority of our medicines, and many other products as well, including poisons. A recent survey hints, without quite saying so, that hi-tech approaches to drug discovery have largely failed to live up to the promise of the hype, and that pharmacologists would be well advised to return to the study of natural agents for inspiration[53].

Many of these agents belong to the loosely-defined class of plant chemicals called *alkaloids,* and include such familiar items as quinine, strychnine, morphine, nicotine, belladonna alkaloids, and thousands more. For medical purposes, they are usually classified according to physiological effects such as on blood pressure, the heart, water retention, psychoactivity, and so on, as in the textbooks.

Although these molecules are generally small compared to proteins, many if not most are devilishly complex and difficult to synthesize in the laboratory, posing a continuing challenge to organic chemists. A Nobel prize was awarded for the synthesis of vitamin B12, and more recent challenges have included the anti-cancer drug from the Pacific yew tree, taxol; the malaria drug, artemesin; and the antibiotic, vancomycin[54] (the latter from a bacterium, not a plant). However, accomplishing a difficult laboratory synthesis of some natural chemical, which is often a highly competitive race among top-drawer chemists, is frequently only an academic *tour de force*, since the procedure can be so laborious as to make the synthetic product too costly. Let's face it, laboratory methods are hopelessly clumsy compared to the series of enzymes which produce these agents in nature. Clardy and Walsh note that 20 to 40 enzymatic steps are not uncommon, and 55 steps are known for one important agent, enediyne C-1027. Here are a few sentences on how nature does it:

> "When monomers dedicated to secondary metabolic pathways are required, such as ... methoxymalonyl CoA for some polyketide initiations, they are produced by a 'just-in-time' cellular-inventory strategy. To this end, biosynthetic gene clusters for non-ribosomal peptides ... contain both genes for the assembly-line enzymes and genes for the enzymes to make the monomers needed for the assembly line to run. A third set of clustered genes typically encodes enzymes that tailor the nascent products released from assembly lines, most notably for glycosylation and oxidation ... required to make the product biologically active." [53].

We humans, of course, have no idea how to make such series of enzymes, which so effortlessly manufacture these compounds in nature, let alone how to arrange them to work seamlessly together in assembly-line fashion. Likewise, our efforts to capture solar energy, at best maybe 15% efficient, are pathetic compared to any green leaf. Such compounds are called "secondary metabolites" because they are not vital to the plant, and this fact occasioned much early speculation on why and how the plants produced them, particularly those which have highly specific effects on humans, such as those which are psychoactive or hallucinogenic. However, it is now believed that most of them are produced for defending the plant against attacks by herbivores, fungi, viruses, bacteria, worms and other parasites, and exert their actions on humans because of the increasingly recognized parallels in the biochemistry of all species.

On the other hand, the possibility remains that some or many of these agents have functions known only to the plant. For example, many chemicals involved with the nervous systems of higher animals are now known to occur also in plants, perhaps suggesting that they, too, have some kind of "nervous system." Another large class of plant products, the *lectins,* first identified by the observation that some of them agglutinate human red blood cells, remain largely mysterious in their natural function[55], since only a few of them are significantly toxic to enemies of plants.

Be that as it may, the variety and specificity of the chemical strategies deployed by plants against their enemies is nothing short of astounding, and more so as researchers probe ever deeper with better tools. Among my favorites are the many trees which send out airborne chemical signals when they are attacked by leaf-eaters, announcing "free food!" to birds and other down-wind predators, which then eagerly fly in to eat the offenders[56]. It is now known that many or most plant chemical defenses are dormant until switched on by an attack, a response parallel to the immune response of higher animals[57-64].

6. THE LOWLIES. Bacteria are traditionally seen as the lowliest of the lower organisms. As biochemists, however, they are unrivalled. Most modern antibiotics come from them (and from fungi), and the above quotation on the 50-odd steps of natural biosynthesis was actually in reference to a bacterium. As Crick famously put it, "anything that is true of *E. coli* [a bacterium] is also true of elephants but more so." Of course, he was speaking of DNA whereas we are speaking of psychologically-driven chemical inventions.

In book I, we sketched some of the complex and apparently motivated behaviors of the protists — one-celled animals. Their stinging barbs, toxins, projectiles, etc., are well known [65]. Bacteria do not have such elaborate hardware (though they do seem to possess a penis of sorts, and a nose [66]), but their chemical skills more than compensate. Hundreds of antibiotics have been isolated from them, used by them to assail their competitors, and doubtless thousands more await discovery since only a tiny fraction of their teeming varieties have yet been studied. Only a small few can be used in the human pharmacy, however, the others having too many side effects, but all are valuable research tools because of the many ways by which they disable specific targets in the adversary, such as particular steps or components of DNA replication, or ATP production, etc. The botulism toxin, a.k.a. "Botox," specifically attacks certain proteins called SNARES in our nerve cells that are essential for synaptic transmission, making us wonder what their natural function is in the bacteria that make it[67].

To consider just one class of antibiotics with which I happen to be familiar, the *ionophores* were first isolated by a former colleague, the late Bert Pressman, from various *Streptomyces* and soon included also monensin, valinomycin, monactin, nigericin, lasalocid, A23187, salinomycin, and many others[68-70]. What they have in common is killing enemy cells by inserting an ion leak in the target cell membrane. Each is astonishingly specific in this action, such as by passing only calcium, or only sodium, or only potassium, etc. Their chemical structures are fascinating and beautiful, but also frustrating to chemists who would like to understand and reproduce their effects synthetically, but have thus far produced only such comparatively crude imitations as the crown ethers.

To understand how these ion-leak effects work, it must be explained that the cell membrane resembles a rain forest whose trees are species of proteins, replete with "vines" (carbohydrates) and "birds" (small molecules that flit about their preferred trees, carrying messages), and floating islands on a sea of fats ... and that beneath this forest floor is the interior of the cell, teeming with activity and machinery, densely strung with little pipelines and cables on which ride tiny trucks ferrying cargo this way and that, all in a precisely orchestrated manner. The ionophores fatally disrupt this microcosm by allowing vital salts to flow out, or injurious ones in. (By the way, many other kinds of membrane hole-punchers are used in the natural world, including in our own immune systems, to wit, the *perforins*, also *C5b9* of the complement system.)

What is the point we are driving at here? It is not a new one, but we repeat it anyway and then comment: this biochemical machinery is just too elaborate, too complex, and perfect to have arisen by the orthodox principles. Now, whenever this objection was raised in the past, the theorists shouted it down and the skeptics backed off, with tails between their legs. But in this book, we have shown, and will further show, that the theorists are all wet, skating on thin ice, and never really had a good theory at all. It was just hand waving. It is now obvious that the modern theory of evolution, King of Biology, is no clear theory at all, it is just a lot of empty talk about vague generalities, reminding us of *The King's New Suit of Clothes:* the king is actually stark naked! It was all hucksterism! Beyond that, we are offering a perfectly plausible alternative hypothesis, one which explains all the same facts, but much better.

Getting back to ionophores, what intrigues us is that each species of bacteria which produce them has seen fit to produce its own unique variety, reminding us of the variations-on-a-theme seen in the snake venoms, snail toxins, firefly flash codes, spider web designs, bat echo signals, electric fish languages, etc., strongly implying a common underlying principle of species radiations (diversifications). Though usually called "adaptive" radiations, that term is loaded with the dogma that these are passively acquired and necessary adaptations, whereas we are proposing that they are active innovations, only then subjected to natural selection.

7. TINY SOCIETIES. New appreciation of the complexity of bacteria was counted among the top ten scientific advances of 2004.[71] This was for two main reasons: first, recognition of vastly more "species" than previously suspected (earlier cited); and second, new appreciation of the ubiquitous extent of their social and "sexual" activities[72, 73]. One article opens by quoting Costerton, a leading authority on biofilms, those slimy yet tough protective coverings produced by bacterial colonies:

> "'If you found yourself in a biofilm, you'd be going along a channel full of water, like the canals in Venice, and up from the bottom of the channel, on either side, would be these slime towers. ... The channels would be bringing in oxygen and nutrients, and removing waste. And within each building, so to speak, some of the bacteria would be cooperating with each other, making a compound [molecule] and passing it along to the next. It's at least as complicated as a tissue, and possibly as a city.'"[74]

14.

Evo-Devo and Plastic Man

1. EVO-DEVO? No survey of evolutionary theory, or rather, *theories*, would be complete without evo-devo. This term is a contraction of *evolution and development*, where *development* refers to the process of making a baby from an egg: the development of the embryo, formerly called *embryogenesis*. The field has emerged almost as a rival theory of evolution.

There are several reasons why it arose. First is that the protein coding genes tell us nothing about why we are born as humans, not as otters or fish. The genomes of all animals look pretty much alike and, indeed, are not so different from those of plants. But obviously, we are not just a bag of proteins. We have heads, eyes, bones, faces – not to mention brains – every detail of which is well-nigh perfect. "What a piece of work, is a man." These features, the appearance, or morphology of the animal, are known collectively as its *phenotype*, and presumably arise from its *genotype*. But how? Evo-devo to the rescue!

In a nutshell, evo-devo argues that it is not so much the genes that make the phenotype, but is rather the timing and intensity of expression of certain genes, or sets of them ("modules"), governing development. The idea goes back to Ernst Haeckel (1866), an early Darwin champion who played a role similar to Origen's in spreading Christianity, and to von Baer before him (1828). Haeckel is famous for his euphonious principle, "ontogeny recapitulates phylogeny," the idea being that the developing embryo passes through all prior stages of evolution. This is an appealing idea since, for example, even the human embryo goes through a stage which appears to have gill slits. Upon close inspection, however, that idea seemed to fall apart[1, 2], and interest in development receded to the sidelines of evolutionary theory. (Incidentally, it has been charged that Haeckel falsified some of his drawings, or at least fudged them, to better support his idea[2].) Aside from that, since nobody had a clue as to how development worked, there wasn't much to talk about.

The status of developmental biology was reviewed in a special issue of *Science*[3] ca. 1994. Its increasing connection to evolutionary theory reviewed in 1997[4], when the term *evo-devo* first appears. A book-length treatment in 1996 was said to be one of the most important of the decade[5], which laid the foundations of evo-devo. At least two related books appeared in 1997[6, 7], and later ones are cited below. Raff's book[5] includes a diagram of major events in phylogenesis, showing such key branch points as multicelled organisms, differentiated tissues, anterior-posterior axis, central nervous system, and the advent of Hox genes (see below).

Evo-devo advocates do not mince words about the weakness of the orthodox theory. Conversely, the orthodox theorists are not entirely pleased with the claims of evo-devo, as in the following 2005 quote from the eminent Brian Charlesworth, in a book review:

> "[The authors] argue that the Darwinian explanation is incomplete ... Are they right, or does their effort represent the latest entry in the catalog of failed attempts by developmental biologists to supplement or replace neo-Darwinean evolutionary biology? ... It is very difficult, even now, to determine the potential range of variability in a given character ... Until we have a predictive theory of developmental genetics, our understanding of the molecular basis of development ... sheds little light on what variation is potentially available for selection. As a result, it is currently impossible to evaluate the idea." [8]

2. THE HOX BOX. By all accounts, it was the discovery of the *Hox* gene family in the 1980s that spawned evo-devo[9, 10]. Briefly, it was found that very similar sets of genes in all animals seemed to control the formation of body segments, limbs, wings, eyes, head, and so on. Accordingly, the general thesis of evo-devo is that comparatively minor adjustments to these genes can cause major changes in the phenotype: the addition of more or fewer body segments, more fins or legs, perhaps even convert a fin to a leg or a wing.

Among the more spectacular experiments supporting this notion were those showing that eyes could be made to sprout on the legs of a fruit fly, or even at the tips of its antennae, looking like tiny stalked crab eyes, just by tinkering with these genes – or in this case, with just one, called *Pax6* in vertebrates[11]. Similar studies have targeted the formation of the head[12, 13], body segments, limbs[3, 14], crustacean appendages[15], anterior *vs.* posterior[16], and so on[16]. These discoveries made headlines, even featured on the cover of *Time Magazine*[17].

The fact that a single gene could induce such a complex structure as a fly's eye, composed as it is of some 800 perfect facets (*ommatidia*), numerous component parts, implies that such a gene acts as a "master controller", as in the title of Gehring's book, *Master Control Genes in Development*[9], somehow orchestrating the activities of the hundreds or thousands of subordinate genes needed to make all the parts. But the experts seem chary of this notion, as expressed in a review of that book.

3. HOX AND EVOLUTION. So, what can these genes tell us about evolution? A lot, or not much, depending on point of view. But one thing is certain: these discoveries have thrown into doubt many of the old dogmas, some of them very basic. Among these is the long-accepted belief that homologous structures (eyes, in particular) evolved independently, some 32 times, in the many animals that have them; but it now turns out that they all have the same set of Hox genes. It was for this reason that we earlier cited some evidence that these genes (or controllers of them) may have been passed around horizontally. But most experts doubt this, and strive instead to figure out how they evolved by vertical descent.

For example, observing that vertebrates have two sets of these genes, Hox and ParaHox[18], it was concluded that one of these arose by duplication of the other (even though they are distributed on several different chromosomes); and a later study sought to deduce the original ancestral "ProtoHox" genes by investigating some primitive forms, the hydra and the sea anemone[19].

However, not all agree that Hox genes are a defining feature of all animals. Indeed, another study found no sign of them in several lower forms (including the sea anemone) despite complex body plans[20]. Also puzzling is that development in plants appears to have arisen independently, even though most agree that higher plants and animals must have a common ancestor[21].

Another classical principle now in the dustbin is that evolution could proceed only in tiny steps, by micromutations; but now the "big gene" theory is back, i.e. that small changes to a single master gene can have large (macro) effects[22] – albeit controversially. For example, Davidson and Erwin argued at length that only the big-gene hypothesis of evo-devo could explain the origin of phylum-level differences (meaning major divisions of living things), but that paper was sharply criticized a few issues later by heavyweight Jerry Coyne, asserting that classical "natural selection adequately explains the origin of new phyla"[23]. Really? Even back in the 1970s, master genes with "pleiotropic effects" were often invoked to reply to skeptics, although back then, it was purely hypothetical.

There is now no doubt that single small changes in the genotype can have big effects, even in bacteria. A single mutation altered production of some 50 different proteins, enabling it to adapt to a different environment[24]. The many and far-reaching effects of mutating a certain heat-shock protein, Hsp90, in which effect was transmitted to subsequent generations[25], was earlier mentioned and more recent work on that is cited later. Also worth mentioning in this context are "co-activators," described as:

> "... molecules that can move among hundreds of transcription factors and coordinately control the appropriate expression of subsets of genes to produce a desired end-goal, such as growth. ... [Their] actions have expanded to chromatin modification and remodeling, initiation of transcription, RNA elongation and splicing, and protein degradation." [26]

The example discussed in this quote is "master gene" SRC-3, controlling aspects of certain steroid hormones. Another likely tip of the iceberg is the dawning realization that small interfering RNA (recall Ch. 12) may be crucial to evo-devo. For example, a 2000 paper implicated siRNA in development of a worm[27], and in 2004, David Bartel and colleagues extended this to mammals, showing that an siRNA, called miR-196, acted in developing the hind legs of mice[28]; and in 2006, Giraldez et al. showed that another siRNA family, miR-430, contributes to the switching of control of development from the mother's genes to the embryo's own genes in zebrafish[29]. A single exposure to an RNAi can induce major effects that persist for indefinitely many generations in the lab worm *C. elegans*[30]. There are many other interesting findings on this matter[31], and others are more recent, but here we are attempting only an overview.

That said, let us look at a developing fly wing in which some cells were given an increased amount of a growth regulator: those cells got much bigger but, unexpectedly, surrounding normal cells got much smaller than usual, and the overall final wing was normal[32]!

The bottom line? First, it is clear that these experiments, fascinating as they are, basically amount to tinkering with systems nobody yet understands. It reminds us of work a century ago in which assorted monsters were produced by cutting and pasting bits of amphibian eggs. Second, none of this tinkering has resulted in a new species. Third, these observations imply the potential for major changes in very short times, in conflict with traditional assumptions. It is also in conflict with another fact of observation that evolution is very slow, and that many animals and plants do not change at all for tens of millions of years. That brings up the next topic.

4. CANALIZATION. In the 1970s, C.H. Waddington was still a major figure in evolution theory, noted for his concept of *canalization*. The general idea arises from the fact that the phenotype of any species is very stable: humans all look much alike, as do foxes, ravens, swordfish, etc. Even among the hundreds of human birth defects caused by genetic errors, the resulting baby is always a recognizable human, never a unicorn or a mermaid. Likewise, in all the experiments of gene knock-outs with flies, fish, mice, whatever, the outcome is flies, fish, and mice. Thus, there is a certain *robustness* to the phenotype, as if development moves in a fixed groove, or *canal*, as Waddington put it. The concept fell from favor since no way could be seen to test it; but I happened to notice it come up recently (2010) in an article titled, *"Is canalization more than just a beautiful idea?"* [33], in connection with experiments on the above-mentioned Hsp90. The author demonstrates that absent or defective Hsp90 appears to release a much wider range of phenotypic variation than in the normal offspring. Interesting? Yes. Meaning? Obscure.

5. DEVELOPMENTAL PLASTICITY. Conversely, among the more interesting theses of evo-devo is the phenomenon called *developmental plasticity,* referring to the fact that the final phenotype of an animal depends very much on environmental factors. Few are yet openly saying that this means that environmental factors can directly affect heritable qualities, for that would be Lamarckism. But they do openly assert that it is commonplace in evolution for "phenotype to precede genotype," in the sense that an environmentally modified phenotype ("ecotype") is then subject to modified natural selection, which leads to heritability of the revised phenotype[34]:

> "For more than a century, evolutionary biologists have fretted about how novel phenotypic variation originates. ... Although few biologists question the importance of mutation, debate continues over whether ecophenotypic responses represent simple microevolutionary fine-tuning or an important source of new variations. ... In the unconventional phenotype-precedes-genotype mode, sometimes called genetic assimilation, developmental plasticity creates novel phenotypes before heritable variation that affects their development. This ultimately arises later by means of random mutations." [34] References omitted.

Palmer goes on to cite the example of the asymmetry of crab claws, one being much larger and stronger than the other, an obviously heritable trait — but how did it originate? It must have been *behavioral,* similar to the innovations we spoke of earlier. He then broaches the origins of more prosaic asymmetries, such as the fact that many of our internal organs lie to the right or left of center, noting that left-sided heart asymmetry is conserved in vertebrates but that the frequency of exceptions (with reversed symmetry) has declined over time[34]. He concludes by remarking that "genetic assimilation may be much more widespread than currently believed."

A particularly nice example of this hot-button topic is the ability of some snakes to acquire a larger mouth in response to larger prey, beyond the genetic control of mouth size, in two populations with different prey size[35]. This is not far from Lamarck's idea, so long ridiculed, that the necks of giraffes became lengthened by constantly reaching to higher branches.

Although developmental plasticity is well recognized, there is much dispute about the degree of its importance to evolution: is it a major driving principle? Some seem to think so; see notes at reference[36]. Needless to say, it is not well understood but has potentially radical implications for evolutionary theory. West-Eberhard has taken a particularly strong position on this, envisioning

> "... a synthetic theory of evolution and development in which environmentally induced phenotypic changes give rise to adaptive evolution as readily as, or even more readily than, mutationally induced phenotypic change"[37]

The author is said to propose that this sequence accounts for "nearly all evolutionary novelty, adaptive radiation, speciation, and macroevolution." Although the reviewer of the book is not impressed, his main objection seems to be that this theory is outside the mainstream, and offers no genetic explanation or hypothesis to test. Then again, as we have seen in previous chapters, the mainstream is in no position to thumb its nose.

6. THE FORGOTTEN MIND. West-Eberhard seems to be moving in a productive direction, but we must object to the frigid language. It is generally not "the environment" that causes changes, but is rather the animal's behavior, its innovations (recall Ch. 6.4, types of adaptation). The lobster's claw did not get bigger because of "the environment" but because of the way the animal used it. The same applies to the many clear cases of animal innovation, such as the earlier-mentioned archerfish, or beavers, spiders, birds of all kinds, the whole nine yards. That is to say, the ideas of West-Eberhard et al. will need to ripen to include the role of deliberate innovations and choices made in the minds of living things. They are not passive lumps of clay, as if you could stick a wing on here, a foot there, by pure random accident, and expect a new phylum to arise, like some Venus emerging from a sea shell.

7. NEOTENY. Although now out of fashion, the neoteny hypothesis could be of more than passing interest, *vis a vis* evo-devo. Its best-known advocate was anthropologist Ashley Montague, and his ideas intrigued this writer back in the 1970s. Montague was still at it as of 1983[38]. In a nutshell, the general idea is that humans arose from an ape whose development was arrested in an early juvenile stage. The physical evidence is impressive and includes our large head, hairless body, small teeth, and many other details known to physical anthropologists, all of which seem to imply that we are baby apes. What was new in his 1983 book was an effort to show that we also retain baby-like *psychological* features into adulthood; and that this can explain many of our uniquely human qualities, especially our openness to new ideas (*psychological* plasticity), marked by qualities such as heightened curiosity. He argues that juvenile forms are more likely to initiate novel adaptive strategies, and this seems well-supported by studies of apes and monkeys, where novel innovations are most often accomplished by young-sters. The reviewers find it less convincing in this respect (perhaps because the reviewers were at work on a rival book on the same theme), pointing out that previous thinkers had come to opposite conclusions.

In earlier works, the illustrious Montague has argued for neoteny as a gen-eral evolutionary mechanism, on grounds that juvenile forms tend to be more plastic, more receptive to behavioral modifications, and that only minor genetic changes would be needed to terminate development, thereby bringing about a major evolutionary shift at one fell swoop. Accordingly, one might argue that Montague anticipated some of the key notions of evo-devo today.

His neoteny hypothesis also addresses the many interesting changes observed between wild and domestic animals, e.g., pigs, whose return to the wild phenotype (feral state) upon release or escape is marked by numer-ous modifications within one or a few generations. According to Montague, humans have many phenotypic features in common with domesticated animals.

8. REGENERATION. Some authorities have wondered in print if recent advances are much more significant than those of a century ago in which various parts of frog and other embryos were cut and pasted to other parts, giving bizarre freaks analogous to those now produced by the gene tinker-ers. The following is from a review of a book about developmental studies with focus on the regeneration of tissues and body parts (such as livers or limbs), often seen in certain animals such as salamanders:

> "It is a humbling work to read, in that it appears that remarkably little has been accomplished toward our understanding of what T.H. Morgan called 'vital factors' ... since the 18th century." [39]

Granted, we are now tinkering with more-or-less defined genes rather than gross chunks of tissue, yet tinkerers we still are, fumbling in the dark. One is reminded of the many "spectacular advances" in brain research announced every decade. Indeed, much the same could be said of medical science, since if you believe the headlines, eternal life is just around the corner.

Do not misunderstand. Beautiful research is going on all the time, and this book highlights but one corner of that marvelous work. The objectionable part is the overblown claims, and the evolutionists are the worst offenders in this respect. This is ironic since the theory of evolution is perhaps the least secure of any science, if indeed it deserves to be called "science" at all.

9. MODULES OF THE REGULOME. Coming back to evo-devo *per se*, the whole idea is that development is key to evolution and that altered *regulation* of development is the mechanism. Thus, "gene regulatory networks" (GRNs) was the theme of an article by Davidson and Erwin[23], suggesting another -ome, the *regulome*. A nice example is the work showing that the sharp spine on the stickleback fish, which is present or absent depending on habitat, is due to effects of modifying, not the gene for the spine itself, but regulators of its expression[40]. Similarly, the eyespot of butterfly wings was traced to a specific gene, but details of its location on the wing and other features were attributed to other genes[41]. In later related work, a comparison of the genetics of wing-spot patterns on 77 species of fruit fly found that the specifics resulted from regulatory modifications of a single "pleiotropic" gene[42].

Interesting, yes. How does it work? Nobody knows. One position emphasizes the timing and duration of gene expression, termed *heterochrony*, in determining phenotypic differences; this being a prominent theme of at least two book-length expositions[43, 44].

10. MORE MODULES. The why and the how of evolution is not much clearer in evo-devo theory than it is in the conventional theory (whatever that is), but evo-deco offers many novel ideas. One of these is the notion of *gene modules,* the idea being that groups of genes (and related regulatory elements) affecting various forms or parts, can be "redeployed ... across species, thereby creating novelty and fueling the ... evolution of biological complexity"[45]. The Hox box comes to mind. Clear as this might seem, however, there is much dispute about exactly what a "module" is, and how it works. But enough fairly clear examples are known to suggest that it might be a fruitful notion.

It goes without saying that the actual operation of these modules must be exceedingly complex to produce the likes of you and me, to somehow orchestrate the network of controlling mechanisms, which has been spoken of as that "hairball of proteins such as transcription factors and signaling molecules ... all connected by means of regulatory DNA into multiple feed-back loops"[46].

11. PARTING COMMENT. The growth of a human embryo into a baby has always been viewed as a miracle. It would be rash to suggest that science will never understand it. However, it is the thesis of this book that science *as we know it today* will never succeed in understanding it, and this is for one paramount reason: science today has no place for the *mind* within all living things, motivating all of their actions. If you wish to say that life reflects the hand of God that is entirely reasonable. But the present writer prefers to map out a path towards the rational understanding of the properties of mind, so that some day we may have a glimmer of insight into what's really going on.

15.

Evo-Psycho and Gene Man

1. WHAT? Unlike evo-devo, the term, *evo-psycho* is a contraction of my own devising, better known as *evolutionary psychology,* or EP. Much as embryology was revitalized by the new era of genetics, so, it was hoped, would the flagging field of psychology. That is to say, academic psychology is a mess and has been ever since the unfortunate rise of behaviorism. After behaviorism's fortunate demise, psychology never regained its footing. This opinion is widely shared, as in the following two brief quotations, which could be multiplied:

> "Cognitive science is a mess. That's largely because the shape a theory of cognition takes depends on the theory of concepts it endorses [which is] likewise a mess. In this respect, this collection of papers fairly represents the state of the art." [1]

> "It has been about 30 years since the first rumblings of discontent with the state of academic psychology began to be heard. ... It is a remarkable feature of mainstream academic psychology that, alone among the sciences, it should be almost wholly immune to critical appraisal ... Conceptual muddles long exposed to view are evident in

almost every issue of standard psychology journals. This is a curious state of affairs." [2]

The hope is that it can gain some much-needed scientific credibility by becoming more "biology-based," as one author phrased it. One direction it has taken in this quest is by way of an impressive array of new real-time brain imaging devices, e.g., functional magnetic resonance imaging (fMRI) [3], and new and better methods of brain mapping. Both have afforded rich new insights on neurology, although the actual functioning of the brain overall remains enigmatic. But since this book is about evolution, we focus on EP. A thumbnail history and survey of the field, *ca.* 1994, was given in a special issue on *Genes and Behavior*[4].

By all accounts, the initial inspiration and foundation of the field was E.O. Wilson's *Sociobiology* of 1975, warmed over as befits its new moniker, EP. The essence of EP is based on the premise that we are what we are because of the *instincts* we acquired during our million-year childhood in the wilderness. By definition, instincts are genetic. As we discussed in book I, Wilson's thesis was controversial from the start. In a nutshell, Wilson, an expert on ants, proposed that we are like ants. This can mean two things: either we are like ants in the orthodox sense – helplessly instinct-driven, deterministic – or, it could mean that ants are like us in the ordinary sense: passionate, intelligent, self-determining. Needless to say, the present framework of science demands the former.

2. MR. NATURAL. According to the EP paradigm, there exists a *natural* man bottled up within us, like a cat in a bag, just beneath the proverbial "veneer of civilization." Of course, natural selection is part-and-parcel of this not-so-new gemish. Most readers have doubtless already heard of it, judging by the fact that shopping is nowadays called "foraging." Here is one encapsulation of what EP is all about:

> "Evolutionary psychology is a research paradigm that proposes that the human mind consists of cognitive modules that evolved in response to selection pressures faced by our Pleistocene ancestors." [5]

To illustrate, Berrenby's book, *Us and Them: Understanding your Tribal Mind*, fits comfortably in the EP camp, dealing as it does with our supposedly instinctive allegiances to groups – the "tribal mindset" – although no major new insights are conveyed, at least not by two reviewers of it, although one expressed some doubts about the "contact hypothesis" (that intimate familiarity with "other tribes" tends to break down prejudice[6]).

3. OF GODS & ORGASMS. If one subscribes to the EP doctrine, then one has a lot of explaining to do. For example, religious beliefs are to be understood in terms of evolutionary adaptations, or fitness. No problem! At least three books in 2006 alone, one by Daniel Dennett[7], another by Lewis Wolpert[8], and a third by Tremlin[9], endeavor to explain religion in just such terms. All seem to make the tacit assumption that religious beliefs are so absurd, so fanciful, and so ridiculous that the only possible explanation for them is some sort of wacko instinct. Tell that to the Ayatollah. Sigmund Freud took a similarly dim view of religion, seeing it as primitive wish-fulfilling fantasy, imagining himself, of course, to stand at the very pinnacle of clear-headed scientific enlightenment, miles above such nonsense.

The rationality of religious beliefs will be considered in more depth later in this series, but meanwhile, regardless of its rationality, the passionate intensity with which religions are defended clearly signifies that *nothing* is more important – not even life itself – than one's beliefs, ideals, convictions about the truth of things. Indeed, as we indicated in an earlier chapter, the pursuit of truth is precisely the motor of evolution; and this motor is fueled by communication and is refined in the furnace of inter-group rivalries, the better to thrash out the truth. Also, if one wonders how such foolish beliefs can possibly persist in the great light of Science, the preceding chapters may offer some clues.

Dennett's book made a big splash, being reviewed everywhere, even featured in a *New Yorker* article[10]. Michael Shermer, who puzzled over why so many people lack faith in the theory of evolution[11], in the course of reviewing Dennett's book[7] alludes to his own two books on religion[12, 13] and offers the following summary of the evolutionary reason or purpose of religion:

> "What is the value of religion to evolutionary fitness? In two books [cited], I have outlined at least four such values: (i) mythmaking to explain apparently inexplicable phenomena, (ii) redemption (forgiveness in this life) and resurrection ... (iii) morality (reinforcement of prosocial behavior and punishment of anti-social behavior), and (iv) sociality (encouragement of within-group amity and between group enmity.)" [7]

These are actually time-worn explanations that have been penned hundreds of times in the twentieth century, by all sorts of anthropologists, psychologists, and miscellaneous atheist scriveners. Phrasing these four explanations in terms of evolutionary fitness, however, seems ill-advised, since they all beg the question of exactly why and how such beliefs arose and came to possess such power. *Why* do people feel a need for forgiveness, redemption, moral authority? *Why* should religion be needed to "encourage amity"? Above all, *why* is it so important to people to explain the "apparently inexplicable"? We have already intimated the answer to the latter.

One could put the shoe on the other foot. Any smart Christian must be tempted to wonder at the madness of evolutionary theory and to speculate that it must be the work of Dark Forces of Evil. How else to explain the irrational denial of self-evident truths? How else to explain all the bloodshed in the name of science, the millions dead as the result of Darwin-inspired beliefs of Hitler about racial genetics, which at the time was respectable science (and may still be, if only whispered)? How else to explain the Republican Party, except as a result of belief in the alleged Law of the Jungle, a Mother Nature "red in tooth and claw"? How else to explain the collapse of morality, fidelity, charity?

(In fairness, it must be added that more thoughtful books have appeared seeking to reconcile evolution and religion[14, 15].)

Dozens of other questions have been posed and answered by the EP canon. Take, for instance, the evolutionary purpose of the female orgasm. What might be the adaptive reason for it, *vis a vis* evolution? Elizabeth Lloyd (having decided that this particular ecstasy has no apparent adaptive function) evaluates twenty-one theories and rejects them all, concluding finally that it must be a genetic accident, comparable to the vestigial nipples on males[16]. More of that pesky junk DNA? Konner pondered the same problem some years earlier[17]. But other explanations come to mind, such as a celebratory culmination of union resolution of inner conflict, *par excellence!*

4. WADE'S WORLD. Consider a book of this genre, *Before the Dawn*. According to reviewers Weiss and Buchanan, Wade explores "the lost history of our ancestors" of the last 50,000 years by "focusing on how genes can reveal the hidden past that can't be inferred from fossils." The well-told tales

"... will doubtless captivate many readers. However, what could have been a tempered and timely treatment of an important subject is, in our view, regularly undermined by Wade's determination to find simplistic natural selection behind every trait, and by a lack of attention to [the complexities of genetics]." [18]

These "simplistic" attributions amount to more of the *Just-So* stories that dog the reputation of EP[19], as they do the New Synthesis. The reviewers list some of Wade's notions, e.g., that "Jews were selected for higher intelligence because of the calculational demands of money-lending"; that Chinese excel at ping-pong for genetic reasons; that "Europeans resist mad-cow disease because their ancestors were selected for cannibalism"; and so on. This kind of junk science was popular in the early twentieth century and supplied much of the rationality of Hitler's regime[20]. The reviewers coin the apt term *slopular science* for this kind of book and remind us that it's not only bad science but dangerous[21-23]. Dangerous, yes, but this danger comes directly from the New Synthesis, as mentioned above. Yet, the reviewers are not as scathing as they could be, leading us to suspect they are sympathetic to EP.

Another review of the same book – yes, it somehow got a lot of press – opens by reminding us of our ancestors' life in the frigid ice ages, then remarks that this book has "peaks of superb science writing followed by dramatic collapses into spurious assertions that left me cold"[18]. Mithen, a career archeologist, is not pleased with Wade's dismissive nod towards archeology in favor of genetics as the way to reconstruct our past. But Mithen, too, seems to have mixed feelings about the book. It strikes me that a crucial weakness of many books of this kind is their evident ignorance of anthropology, of the power of *culture*. Like many others, it seems that Wade uses only selected examples of violent societies (the Yanomami) in defending his thesis, and ignores the huge numbers of very different ones. Specifically, Wade thinks that the crux of human progress hinged on new genes to counteract our previously violent instincts. It seems that everyone has forgotten the hard-fought anti-racist campaign of anthropologist Franz Boas in the early twentieth century, which succeeded in toppling the myths of racial genetics so rampant at the time, bringing to a close at last the era of eugenic witch hunts.

5. PRIMATOLOGISTS AND KINDRED. Moving now to another of the central tenets of EP, that much can be learned about ourselves from the study of our animal cousins, and we observe that many ethologists (students of animal behavior) are joining the EP fold, or at least, so it appears from the spin put on many of their books. One from the eminent primatologist, Frans de Waal, claims in its title to answer the immodest question, *Why We Are Who We Are*[24]. It follows his very readable and successful book of a decade earlier, with the slightly less ambitious subtitle, *The Origin of Right and Wrong*[25]. These books are primarily engaging accounts of his studies, only secondarily carrying messages about us. Among his favored themes is the resolution of conflicts among monkeys, and de Waal stresses their moral side: tender, caring for their sick, injured, or bullied fellows, and easily learning of strategies for reconciliation. "Monkeys adjust their behavior with mentally retarded or physically handicapped individuals, suggesting a rudimentary sympathy, and chimpanzees show consolation towards victims of aggression"[25].

In sharp contrast, Wade's book is obsessed with the presumed violence of our ancestors, apparently based on limited readings about chimpanzees. "Wade's view of the past is one of continual violence"[18]. Indeed, chimps can and do exhibit savage behavior. Then again, certain human societies exhibit savage behavior *en masse* from time to time, e.g., Germany and Japan in WW II. By instinct? Of course not. De Waal's books are written with warmth and affection for his animals, leading him to stress their 'good side.' Gee comments that de Waal's *Our Inner Ape* "is perhaps the most humane treatment of the human condition you can read, for all that it is mostly about chimpanzees"[24].

The salient point here is to illustrate how strongly *interpretation* can color impressions from animal studies. This is a well-known problem in the many related books about chimpanzees, or the smaller bonobos, making it difficult or impossible to draw clear conclusions. The same happens with anthropologists reporting on human societies: one may stress the nasty, back-biting, scheming, and violent side; another seeing in the same people their warm, affectionate, loyal, and tender side. It is not uncommon for such different perceptions to lead to acrimonious disputes, exposing the violent genes of anthropologists.

For another example, it is widely assumed that pre-modern human life was "nasty, brutish and short," in constant fear of attack by animals against naked and cowering humans – or of attack by neighboring humans. But there is equal or more evidence to the contrary, that even our earliest ancestors, armed as they certainly were with spears, were already King of the Jungle, so to speak, fearing none and enjoying ample leisure for sensitive interpersonal relations and intellectual discussion. How else could we have acquired our noblest sensibilities? Indeed, there is factual evidence of a precipitous decline in stature and general health associated with the shift from subsistence hunting to sedentary agriculture[26, 27].

It must also be pointed out that not all chimpanzees are alike. For example, there was said to be a sharp contrast between the social lives of the common chimp (*Pan troglodytes*) and the much more civil pygmy chimp, or bonobo (*Pan paniscus*, now recognized as a distinct species). Bonobos are said to attain their more peaceable ways by resolving conflicts through sexual favors (but this distinction has recently been challenged, another example of the vagaries of interpretation). Then again, there are the amicable gorillas (now often murdered by humans for their meat and tourist knick-knacks). Think also of the different habits of baboons, orangutans, and innumerable monkeys, not to mention the "lower primates" (lemurs, tarsiers) [28]. But it is now emerging that local groups of nearly all of these exhibit marked *cultural* differences – incipient species? So the question arises, how is one to draw any clear conclusions about humanity from them?

Primatologist Robert Sapolsky, writing with J. Cape, has published some collected essays under the cute title, *Monkeyluv*[29], apparently also in the spirit of EP. The review is not very illuminating, comparing some boy/girl decisions with those of apes and remarking on brain reward pathways (dopamine) and brain imaging studies to "explain," for instance, why [some] people like to gamble, and why [some] older people are less adventurous than [some] youths.

6. BIRD BRAINS. A number of avian biologists (students of birds, ornithologists) are also of the EP feather, such as Forbes, who, in his *Natural History of Families*,

> "... takes up this theme [EP] with a vengeance to reveal the many fascinating aspects of human family life and reproduction that arise from evolutionary conflicts of interest. ... He describes research showing that morning sickness and even pre-eclampsia ... during

human pregnancy may be the manifestation of an evolutionary struggle between mother and child over placental blood flow ..."[30]

The reviewer remarks that this book "sells itself on the salacious revelation that all is not as nice as it seems within our cozy families." In view of current divorce rates, one wonders how many "cozy families" still exist among humans, which is to say that it is a well known fact that psychological theories tend to parallel the attitudes and ambivalence of the larger society; so perhaps our present epidemic of shattered homes has influenced what bird watchers see in the bird's nest.

Be that as it may, the reviewer goes on to note that "at the end of the day, this foray into popular science perhaps does not do justice to the author's own research on avian family biology," and suggests that Mock's book, on the family life of birds, probably offers "a better compromise between the popular and formal science formats". Indeed, Mock's book has a nifty title, *More than Kin and Less than Kind,* subtitled *The Evolution of Family Conflict*, and is said by reviewer Godfray to have a "lively and engaging style" in presenting topics such as siblicide, widely observed in birds, and aspects of parent-offspring conflict, all in light of the latest theories of EP[31]. Godfray dissents on one point though, affirming his allegiance to the theory of Trivers on parent-offspring conflict, dismissed by Mock as an "interesting theory [which] has failed to produce much insight." Otherwise, "Mock has done a superb job of bringing a large area of contemporary behavioral ecology to both a biological and a general audience"[31]. But is it not so that conflicts of interest (trade-offs, risk/benefit analyses, energy efficiency, etc.) are confronted by all of us every day and by all living things? Vaunted theory seems superfluous.

In closing this section on a brighter note, it is gratifying to find that birds are at last recognized to have *minds,* and to have as much relevance to human life as do monkeys or apes, which is to say that as recently as the 1970s, birds were classified along with reptiles, fish, and wristwatches as insensate mechanisms.

7. THE GROWING EP TENT. It is said that the eminent psychiatrist Nancy Andreason has moved into the EP tent with her book of 2001, *Brave New Brain*[32], said to be a major departure from her 1984 book, *The Broken Brain*[33]. Reviewer Plomin explains the shift from her former perspective, which we might term "classical," to a more "biological model that combines neuroscience and genetics"[32]. But it seems she is not buying into EP whole-heartedly, adamantly resisting one of its central tenets, genetic determinism, stressing instead the ease with which "brains rewire themselves," a pivotal point[34]. Some of the EP people deny a commitment to genetic determinism: that we are what we are and do what we do because of our genes; but without that postulate, the whole thesis of EP disappears. It seems that nobody has carefully thought out exactly how to resolve this either/or dilemma of nature *vs.* nurture.

Another now in the EP tent is philosopher David Buller, said to have experienced an epiphany while watching a TV show about biology, precipitating his conversion to EP and his book, *Adapting Mind: Evolutionary Psychology and the Quest for Human Nature*[5]. However, the conversion was apparently not complete, for he seems to harbor lingering doubts. As explained by reviewer Bolhuis, Buller dissents on the basic EP postulate, that "our modern skulls house stone-age minds." He adds that "the bulk of the book consists of a thorough analysis of key data offered in support" of EP, including three chapters on mate preference, marriage, infidelity and parenting. "Buller is very critical of the paradigm" of EP, not because of any fundamental flaw but because "it makes little mistakes at nearly every theoretical and empirical turn." Yet, Buller winds up "on a hopeful note" for the future of EP, although the reviewer "does not share his optimism." Buller's views are more briefly and candidly set forth in an article which is distinctly anti-EP[35].

Also of interest in this connection is the recent EP by-product, *literary Darwinism*, evidently an attempt to analyze the origin and meaning of literature (and art) in terms of evolutionary fitness theory. According to a reviewer:

> "It has been primarily through the emerging field of [EP] that the arts have come to the respectful attention of the sciences. Itself undergoing an evolution from what had sometimes seemed to be 'just so' hypotheses to a refined theory with explanatory muscle, [EP] offers the means of bringing together the arts and sciences ..."[36]

Don't hold your breath waiting for this old chestnut to put down roots. The big issue here seems to be how or why the arts arose in the ostensibly desperate circumstances of our ancestors: "An issue that divides [the EP community] is whether the arts are adaptations to the biological exigencies of surviving ... or are by-products of adaptations that evolved to perform other functions." The reviewer notes that the single chapter which analyzes a specific piece of literature, Jane Austen's *Pride and Prejudice*, comes across as good but ordinary literary criticism, without any particular EP spin. The review concludes: "It remains to be seen whether [EP] can spawn a true literary theory that helps us understand narrative." It may be, and probably is, good literary analysis, but fitness theory?

There is a lesson to be learned from these fringes of evolutionary theory, and it is this: once a general theory becomes widely accepted, such as evolutionary theory, it becomes like a black hole or proto-planet in the sense that ever more ideas tend to accrete to it, or get sucked into it. The reason is clear: if evolution theory is a general explanation for all life, then everything about life must be explained in terms of it. The failure of the theory is exposed by the absurdity of trying to explain religion, art, music, literature, etc., in those terms. One is reminded of the search by physicists for a "theory of everything": we are made of atoms and physics includes the theory of atoms, *ergo*, physics must be the underlying explanation for our existence and for evolution, and thus, for art, religion, literature, music, and so on.

Never mind that the theory is all hogwash, comparable to any other primitive myth; it serves to nicely illustrate a general principle about systems of ideas, cultures, personalities, and religions (including science): they all work exactly the same. Just as a proto-planet grows, attracting more and more stuff to it, ever faster and with greater adhesion, so does a core of beliefs, e.g., the theory of evolution, attract more and more fields of thought to it. They accrete and grow in the same way, as strictly logical systems, from axioms to innumerable theorems and corollaries. The bigger they get, the harder they are to topple, no matter how preposterous. The bigger they get, the harder it becomes for those inside the logical sphere to see outside of it, to see that other systems might be just as good or better, again like the black hole from which no light can escape. This is true of every philosophy, of every religion (including evolutionary theory), of every culture; and yet, a real and true system must exist – and it does exist, or will exist, as soon as it is realized that any such system must have *mind and consciousness* in its roots.

8. LANGUAGE GENES. Language is the topic of a later book in this series, but a few words about it here are apropos in view of the big stir about the "language gene." Three generations of a family with severe speech and language impairments were all found in 2001, by Monaco's group, to have a mutation in a gene called FOXP2[37]. This turned out to be a transcription factor regulating about 250 genes expressed in the nervous system and were involved in development. It was promptly called "the language gene" in such popular media as *US News and World Report*[38], further popularized by Steven Pinker, a speech therapist, who wrote a series of best-selling books about language, in a style not unlike that of Richard Dawkins (glib and facile), very much in the spirit of EP, which is to say, featuring "modules" of brain function, reminiscent of Sigmund Freud's fancied divisions of the mind into numerous compartments (id, ego, superego, etc). This gene and others involved in language were recently reviewed (2010) by the original discoverer[39]. By the way, it is noted there that a related gene, FOXP1, is required for songbirds to sing. Has anybody checked the genes of operatic superstars?

This happens to be an excellent example for the debate we are engaged in, since the orthodox scenario would view the human form of this gene as an accidental mutation which facilitated the acquisition of language, followed by natural selection to refine it. A big problem with that scenario is the very large number of additional modifications required, including not only modified shape of the mouth (pharynx, voice box), vocal cords, etc., but also many new muscles for controlling the tongue, and the nerves to operate them, each of which is optimally attached to special points of bone, not found in early hominids. These are only a few of the changes needed; see any specialized book on the subject to appreciate the full extent of modifications for human language. In sum, we hold that the orthodox scenario is not believable.

The alternative? As can be shown, and will be shown later in this series, language was actually an overnight human invention, or discovery, built on crude beginnings, and all of the further anatomical refinements arose as later consequences. This happened rapidly, doubtless facilitated by strong sexual selection for speech ability, since true human language cannot be more than about 100 thousand years old (since otherwise civilizations would have emerged much sooner). As before, it's a matter of faith. You either believe in the orthodox scenario or you don't. Your call.

9. CONCLUSION. Is EP exciting stuff, or old-fashioned hokum? Again, your call. I'm just a reporter. Personally, I find it depressing that scientific reason, that paragon of virtue in my childhood, alongside the Lone Ranger, has fallen so low, so cheapened, so corrupted by charlatans and poseurs. Accordingly, what I find most interesting about EP is not EP *per se* but is EP viewed as a social phenomenon, akin to saucer sightings, alien abductions, and visions of Elvis or the Virgin Mary. For one glimpses here another manifestation of the desperate thirst for rational understanding, call it the "religion instinct," always and everywhere cast in whatever logical frame is taught by the High Priests, no matter how wrong-headed when viewed from outside the sphere, the black hole.

Julian Jaynes wrote a beautiful book back in the 1970s, *The Origin of Consciousness in the Breakdown of the Bicameral Mind* (1976), explaining how humans began to learn to think for themselves. It looks like we're still working on that.

16.

More Stupid Animal Tricks

1. MIND CONTROL. It is well known that many infectious organisms (pathogens) induce responses in their hosts which favor their transmission. Thus, a common cold induces coughing and sneezing, releasing a spray of infectious droplets; and cholera, which is spread by fecal matter, induces diarrhea[1]. It seems quite reasonable to explain these examples as the result of natural selection acting on random variants, favoring those which happen to promote their own transmission. However, many cases are so remarkable as to demand other considerations.

For openers, it has been shown that mosquitoes infected by the malaria parasite become more aggressive feeders, "insatiable ... the insect equivalents of blood-thirsty Count Draculas"[2]. It was later shown that the same parasite, *P. falciparum,* causes the infected human to also become more attractive to other mosquitos[3], clearly promoting the spread of infection, a finding that will surely require tweaking the models of malaria epidemiology[4]. This reminds us of the rabies virus, which causes infected animals to become fearlessly aggressive, as required by a virus that is transmitted by biting. Similarly, rats infected with a certain parasite lose their natural fear of the odor of cats, which is plainly adaptive since cats are the final host of *T. gondii*[5]. In this connection, it may be relevant to note that the gene tinkerers have discovered that mice lacking a certain brain protein, *strathmin*, become fearless[6], hinting that the parasite might be tinkering with strathmin in the brain of the host. Again, consider the parasitic flatworm which develops in a snail (which it castrates), then swims to invade a certain fish for its next stage, causing the fish to thrash, lure-like at the water's surface, attracting the predatory birds in which it completes its life cycle[7].

However, most instructive for present purposes are the dozen or so species of parasitic thorny-headed worms thus far studied (of 1200 species of *Acanthocephela*), described by Janice Moore, all of which alter the behavior of their hosts, but each in a specific way to favor its own transmission[8]. Each has a life cycle requiring that it develop first in an "intermediate" host (an arthropod such as an insect or crustacean), then pass to the intestine of the "definitive" host (a vertebrate such as a duck, rat, beaver, songbird, or fish, according to worm species) where it matures and releases eggs to the feces, to be eaten by another intermediate host. One of these species infects a small and normally shy shrimp-like amphipod which inhabits muddy pond bottoms, avoiding light, but this parasite causes the arthropod to swim to the surface, favoring predation by its definitive host, a certain duck that eats from the surface. However, another type causes it to swim to open light beneath the surface, favoring predation by its definitive host, a diving duck. Another develops in a cockroach, which it causes to move into light to be eaten by a rat, in which it matures. Still another develops in a pill bug, causing it to abandon its usual habit of hiding under leaf litter, so as to be eaten by a starling or other songbird, where it matures; and so on, species by species.

Now, what captures our attention here is the parallel between these species and those of venomous snakes or cone snails earlier discussed. Although the "venoms" in these worms are unknown, it is clear that they have mastered the ability to control the behavior of their hosts by biochemical means so as to facilitate their transmission, but each in a different way. This again suggests that the original ancestor of these species hit upon this highly successful biochemical tactic, leading to the many copy-cat species which modified the strategy as per the examples. Accordingly, we may predict with some confidence that if and when the biochemical genetics of this strategy becomes known, it will exhibit certain commonalities among these species and the same accelerated mutation rate during the period of perfecting it as did snake venoms.

It certainly cannot be explained by the orthodox theory, at least not *plausibly*, leading us to conclude that this strategy is another innovation, i.e. that the worms more-or-less "knew what they were doing" in figuring out exactly what brain button to push to bring about the observed result. It's another case of speciation by "variations on a theme," but this time the theme is *behavioral control*. As an additional item of evidence, it turns out that the worms invariably induce these altered behaviors only at the exact moment they are ready for transmission. (It should be added that behavior, and ordinary conscious awareness, is largely biological and brain seated, yet is derived from the primeval consciousness of which living things are but one expression or reflection. Therefore, biochemical "mind control" is better stated as *behavior* control, not necessarily control of the underlying will.)

There are many more examples of this kind, probably thousands. I have continued to collect additional references on this since 2006 but lack the patience to dig them out, nor have we the space. But from the original list, it may be mentioned that the venoms of certain wasps and spiders do not simply kill or paralyze their prey, but render them docile enough to be led helplessly back to the lair, like a tame cow, to be slaughtered and eaten at leisure – or used to lay eggs in. The larva of *A. nigripes* causes its aphid host to hide, but only if and when hiding benefits the parasite, not otherwise[9]. Then there is the oft-told tale of the liver fluke that begins its life cycle in ants and then must pass to a sheep but, since sheep do not deliberately eat ants, the fluke causes the ant to climb to the top of a blade of grass and to lock onto it with its mandibles, so as to be eaten by the grazing sheep. More such examples are found in Moore's book[10] and related biochemical strategies in Eisner's book, earlier cited.

2. THE ADAPTIVE MUTATIONS FLAP. We have already given evidence that many mutations are adaptively directed, and propose that essentially all mutations may be viewed as "hunting" for an improved ability. This is supported by the numerous cases where mutation rates correlate with adaptive need (the example of venoms earlier discussed), the highly variable rates between species and within gene regions, and many examples where mutations are clearly functional, e.g., in producing antibodies.

By way of update (2010) on snake venoms, recent work from Japan describes the evolution of a set of venom components, called *phospholipases*, so called because these enzymes break down phospholipids. They are normal and essential in vertebrates, but analysis of those in the venom glands revealed a cluster of them with distinct actions, one attacking muscles, another nerves, another causing rapid edema (swelling), and others [11]. It is shown that this gene cluster arose by gene duplication of an ancestral gene, but the question is, how? The old theory held that clusters of gene duplicates were acted upon by natural selection in a "concerted" fashion, meaning all together, which is reasonable since they all did the same thing. It was expected that most of the duplicates would dwindle and die away as redundant. But later work showed this was not always or usually true, requiring a new theory, called "birth-and-death" evolution[12]. What actually happened in this snake was that the duplicated genes rapidly diverged from the original and from one another at an abnormally accelerated mutational pace, giving rise to this particularly toxic group of phospholipases. They also somehow came to be secreted specifically in the venom glands. This certainly looks like another example of "directed evolution" to me. The same is likely true of the hundreds of proteins in scorpion venoms[13-15].

Nevertheless, it is instructive to review the big controversy about "adaptive mutations," also called *Cairn mutations*, or directed mutations, from a few years back. It began with a 1988 paper by John Cairns et al., in *Nature*, plus a similar report by Barry G. Hall in the same year in *Genetics*, both implying that gene mutations could be more-or-less *directed,* or at least "facilitated", or accelerated, to favor the needs of the organism (*E. coli* bacteria were studied). Now, it must be cautioned that much of the ensuing firestorm hinged on definitions, terminology, interpretations, and words like "Lamarckian." Some quotes from a non-technical 1993 review, titled *A challenge to evolutionary biology*, helps to set the stage:

> "[Work by Cairns and by Hall] suggest that organisms can respond to environmental stress by reorganizing their genomes in a purposeful way. ... 'Organisms can sometimes control their own evolution,' says Cairns. ... 'Every other aspect of biology is expected to be responsive to environmental signals,' adds Hall. 'We are just saying that mutations are like everything else ...' 'Many people have made such observations, but they have problems getting them published,' [says Benson]. 'It is important that someone with Cairns'

stature brought the problem to the forefront. ...' As the controversy brews, more supporters of directed mutation are coming forward. [J. Shapiro] points out that research in the last decade has shown that the genome has a 'fantastic system for proof-reading, editing, and repairing itself.'... Adds Cairns, 'We now know that in the processing of biological information, almost anything is possible. Sequences [of DNA] are spliced, rearranged, cast aside, resurrected, and to a limited extent may even be invented when the need arises ...' He might have added that in the course of the history of science, dogma has been spliced, rearranged, or cast aside as well." [16]

The reviewer mentions that Cairns predicted that "the issue will be fought out in a year or two," while Hall predicted that "within five years we can establish ..." what's really going on. Well, it's now 2011 and the issue has gone to a back burner. Actually, as we have shown, it is settled. But nobody seems willing to admit it. Why not? Because the implications are too revolutionary. Among other things, it would imply "teleology": that the evolution of an organism can be goal-directed, which is a huge no-no. Shapiro is quoted as saying that the significance of Cairns' work "is not in the presentation of new data but in the framing of the questions and in changing the psychology of the situation."

The debate tended to be shrouded in forbiddingly complex technical issues, such as, for example, in some 1993 letters[17] on a review of the controversy by Lenski and Mittler[18], followed by letters on letters[19], and still another later letter[20] on another paper[21], and two adjacent papers addressing the issue[22-24], all very dense, to the point of obscurantism, obfuscation.

As the dust began to settle, it emerged that this controversy was productive in stimulating some genuine advances. By 1997, B.A. Bridges, commenting on two then-recent papers, and prior observations by Foster on acquired resistance to an antibiotic (tetracycline), updated the situation[25] and stressed evidence that "only a small proportion of the population [of the bacteria] is mutating at a high rate." But this is exactly what we expect: only the innovators of a strategy will be motivated to develop it.

The status of the debate in 1995 was well summarized by Shapiro, in light of two papers he was reviewing:

"A major conclusion of all these results is that adaptive lacI-Z33 reversion represents more than malfunction of the basic replication machinery under stress. Additional specific molecules participate in selection-induced mutations that restore growth on lactose, namely, homologous recombination and plasmid transfer ... A second conclusion ... is that genetic change in bacteria is often multicellular. DNA rearrangements can occur in one cell and be transferred to another ... Such 'altruistic' mutational events had been predicted by Redfield and Higgins, and Radicella et al., now prove their existence. ... Thus, adaptive mutation has returned to the mainstream of molecular genetics. Does anything novel then remain for evolutionary theory? Most emphatically, yes. The discovery that cells use biochemical systems to change their DNA in response to physiological inputs moves mutation beyond the realm of 'blind' stochastic events ... [In sum] there is no unicorn in the genomic garden. But we have found a genetic engineer there, and she has an impressive toolbox ..."[26]

As far as I can tell, the whole issue then faded away, probably because these notions were just too revolutionary to assimilate. To be more specific, the implication is exactly what we are proposing: a *mental* component is pulling the strings.

3. ALCHEMY. Let us pause to illustrate by a human cultural example. For thousands of years, humans have been experimenting with all sorts of potions and brews and mixtures of things. By the middle ages, this had evolved into alchemy, the search for the Philosopher's Stone, or a means of transmuting base metals into gold. No less than Isaac Newton himself devoted his later years to such pursuits, which were mystical in essence. As it turned out, they never found exactly what they were looking for, but they did discover many interesting and useful things. For example, by boiling down huge pots of piss, Brandt discovered at the bottom a tiny pellet of a mysterious glowing substance: the element, phosphorous.

What is not widely appreciated is that those efforts foreshadowed and directly resulted in modern chemistry, by which we now make useful chemicals as easily as working with tinker-toys: medicines, explosives, poisons, dyes, plastics, etc. So, in a very real sense, those millennia of fumbling efforts paid off: the secret of chemical transformations is secret no more. We suggest that all living things are likewise constantly feeling around in the dark for that light switch on the wall, illuminating greater knowledge and control. Their immediate aims are usually simple: getting more food, more sex, or simply surviving in a harsh environment – such as by better venoms, novel sex-appeal, etc. But beyond that, like a rainbow forever over the horizon, is a deeper thirst (Ch. 2). Of course, one cannot pursue knowledge which one does not yet possess; and yet, if and when one stumbles upon it, an *irreversible* evolutionary advance has been accomplished.

4. BIRD MIGRATIONS. All animals must find their way about. Ants find their way to and from home by the geometry of their forking paths[27] and by other means[28], aside from scent marking their trails. Crabs, too, are good at geometry[29]. A great many creatures make use of magnetic compasses wired into their brains, including honeybees[30], lobsters [31], sea turtles[32, 33], mole rats[34], newts[35], and birds (see below). Hsu and Li reference also salamanders, hornets, salmon, whales, dolphins, alga, and mollusks among those having this gift[30]. The list might now include humans since the same type of magnetite crystals occur in the human brain[36-38]. Brains are not necessary, however, since magnetic navigation was first discovered in bacteria[39]. Insofar as the chains of crystals and their capsular sheaths are the same in all organisms studied, one suspects horizontal gene transfer of the requisite enzymatic equipment in vectoring this technology around the biosphere. In higher vertebrates, however, the geomagnetic sense is more complex, depending also on light[35] and a certain chemical state[41], explaining many of the confusing results of earlier experiments. Many other modes of animal navigation are known, including the widespread use of the sun as a compass, and polarized skylight[42, 43], such as used by migratory butterflies[43], all of which depend, of course, on the well-known clocks that are built into all living things, termed "circadian rhythms."

However, what strikes us as particularly challenging to the standard theory is celestial navigation by migratory birds. That birds do, in fact, navigate by the stars has been documented in dozens of experiments, such as in planetariums, or by moving the birds to a distant and unfamiliar locations, etc.[44]. The problem posed by this feat is that it means the birds must have a map of the stars in their heads. The question is, how did the map get there? It is left as an exercise for the reader to calculate how many random mutations would be needed to produce an acceptably accurate map of, say, just six points of light in the bird's brain corresponding to actual stars. But as usual, that is only the beginning of the questions, for the map by itself would be useless without (a) knowing how to use it, and (b) desiring to use it for long-distance travel.

Hundreds of experiments are described in that 1968 book, using captured birds, hand-raised birds, old birds, young birds, birds in planetariums, birds time shifted by artificial light, birds held behind the rest of the flock or released early, at all hours of day and night, in all kinds of weather, and so on, to a myriad of fascinating details. For present purposes, however, the bottom line is that birds do recognize patterns of stars and navigate by them. Matthews' book was written before decisive proof of the magnetic compass of birds, but the subject has since been updated, e.g., by Alerstam[45]. Vignettes from more recent work:

> "Migrating birds have an embarrassing number of compasses. ... One is magnetic, the other celestial. This redundancy is important because though celestial cues can be very precise, they are unavailable when the sky is overcast. The young sparrow is presented with two problems. The first is to calibrate its celestial vector against the dusk and night skies. The key factor here is to find true north ..." [46]

Here is same commentator writing the next year on another piece of the puzzle:

> "Just when it seemed that we understood the main principles by which migrating birds navigate, along comes a new set of experiments to make us think again. ... They compared the effect on hand-reared [garden warblers] of early exposure to the normal combination of stars and magnetic direction, with exposure to the stars alone [by blocking the earth's magnetic field] ..."[47]

What they found by this ingenious experiment was that blocking the earth's magnetism from the young birds caused them to lose this sense, whence they migrated directly from England toward central Africa rather than taking their usual safer route, which is a dog's leg to the east, avoiding the Alps and the Sahara desert.

The jaw-dropping feats of migrating birds remain under well-deserved investigation, e.g., navigation in high latitudes presents special difficulties[48].

New technologies have aided in these studies, such as mass spectrometry of isotopes in the feathers, which can tell fairly accurately where a bird migrated from. It was shown by this method, combined with gene analysis, that mated pairs of a certain species winter apart, yet return to meet within a couple of days of each other[49]. Getting back to our main question, how the heritable star maps get into the bird's brains, the standard reply would be that given enough time – millions of years is usual – natural selection of random mutations could do the job. As noted above, that is flatly incredible. A related simple question is this: why do the birds migrate at all? The standard reply is that it is a fitness-oriented adaptation, done for safety, or more food, and so forth, and came about by natural selection. Get real! They do it because they can, because they love to fly, love to travel, and love to show off. Besides, not all birds of the same species bother to migrate, some stay behind and seem to suffer no ill effects. Indeed, the nutritional costs of migrations are huge, even life threatening.

Yet, it is definitely instinctive, i.e. inborn. We have argued at several points that mechanisms must exist for learned behavior to become heritable, as Darwin believed, and which seems practically self-evident; and if more proof is needed, it turns out that birds can alter their migratory routes, and such alterations become instinctive (inherited) within a few years, not millions. For a specific example, it was reported in 1992 that a population of blackcaps had significantly shifted their migratory route within 30 years, probably in response to growing numbers of feeding stations provided by British bird lovers; and this shift was *already heritable*[50]. The story was picked up by the *New York Times*[51]. In trying to explain this in conventional terms, it was stated that "this shift must be due to an increased frequency of distinct genotypes," citing their reference 15, to the effect that migratory preferences can be altered by selective breeding. Similarly, Sutherland remarked that "for such a rapid change to take place, the selection pressure must have been strong"[50]. In other words, they struggle to explain the shift in terms of the classical paradigm: natural selection of random variants, fitness theory, etc. Can we be blamed for jeering?

Perhaps aware that this case presents a serious theoretical challenge, some of the same authors revisited it in 2005 with a modified explanation[52], to the effect that birds with genes for greater "behavioral plasticity" are those which made the adjustment to a new migratory route and, indeed, raised larger and more successful broods ("more fit"), whence it is said that "heritable plasticity" is now spreading through the population under natural selection. But "plasticity" by itself – a jargonesque term, perhaps purposely vague – fails to explain how the new migratory route was selected in the first place (against "instinct"), much less how the new map then got into the birds' heads, and genes. Furthermore, if "plasticity" is such a valuable trait, one must wonder why it is being selected for only now, after millions of years without it; and finally, these birds are by no means unusual in having recently altered their migratory route, as many other such cases are known. Are we to imagine that in each such case genes for "plasticity" constitute the "explanation"? The logic here is plainly tortured to comply with the paradigmatic dogma. There is only one explanation that makes real sense: the birds altered their route by intelligent choice, and this alteration became rapidly heritable.

The rigidity of resistance of orthodox biologists to the common-sense explanation, in the face of overwhelming evidence against them, is a great testimony to the tremendous power of cultural tradition, to a logic that has outlined the regime of its validity. The theory of evolution as we have it today will go down in history as the biggest scientific fiasco of all time.

5. THE STUPID BIRD THEORY. Another theory of bird migrations is quite the opposite, arguing that only the stupid birds migrate. This was based on the finding that those birds which stayed behind had slightly larger brains than those which migrated, implying higher intelligence, and therefore were able to find food at home in the cold weather, avoiding the taxing journey[53]. This is sure to win an Ignobel Prize and brings up not only the question of human intelligence, but the whole question of brain size. By way of background, it was long believed that birds were incapable of intelligence because they lacked a cerebral cortex; but it was shown in the 1960s that other parts of bird brains could serve their self-evident intelligence[54]. Not only intelligence, but phenomenal feats of memory. For example, Clark's nutcracker hides about 25,000 seeds per season in up to 2,500 locations and recovers some 2/3 of them at various times, up to 13 months later[55]. Similar abilities are seen in western scrub jays and chickadees. It has been known for a long time that many of the corvids (blackbirds, ravens, jays, magpies) have remarkably high intelligence[56, 57] and that some of them fashion a variety of clever tools. Only more recently has it been shown that this technology is *genetically inherited*[58], while at the same time being subject to regional styles and cultural diffusion[59]. (Likewise, corvid dialects.) Again, consider that certain birds in England made the discovery that they could steal milk out of milk bottles left in doorways:

> "The bottles are usually attacked within a few minutes of being left at the door. There are some reports of parties of tits following the milkman's cart down the street while removing tops from bottles in the cart. ... The method of opening the tops varies greatly... sometimes the whole cap is removed and sometimes only a small hole is made. .. Bottles containing milk of different grades are distinguished by having caps of different colors. [Many observers] reported that the tits attacked only bottles of one preferred type." [60]

One suspects that this feeding habit would have become heritable in a few generations, except that the bottle caps were changed to make them bird proof. Indeed, there is evidence that homing behavior can become genetic after just *a single generation,* since honeybees often return to their original nest several years later, up to a 200 km distance, after all the original bees died out[61]. Monarch butterflies, with brains the size of a pinhead, manage to migrate some thousands of miles, from as far away as Canada, to a place in central Mexico[62], because they are too stupid to survive back home? One is reminded of modern computers which fit in a shirt pocket, yet perform much better than old ones the size of a refrigerator.

6. DRUG RESISTANCE. A few paragraphs on drug resistance are appropriate here. Everyone knows that antibiotics are becoming ineffective due to the rise of resistant strains of pathogens. Not so widely appreciated is how stubborn and complex this problem is, or that it is also a main reason for our inability to cure cancers. The problem extends to the development of insect resistance to pesticides and plant resistance to herbicides, and beyond.

An early hint at understanding was the discovery in 1986 of a gene for multidrug resistance (*MDR1*), coding for an efflux pump in the cell membrane that literally pumps noxious substances out. This is now known to be just one of a family of some dozens of such transporters in mammalian cells, "ATP binding cassette transporters," prominently including the "breast cancer resistance protein" (BCRP). All efforts to outwit these pumps, by thousands of research man-years, have thus far failed.

However, these pumps are only part of the story of resistance, possibly only a small part. For example, enzymes often arise which specifically degrade the drugs. The important class of antibiotics known as beta-lactams (e.g., penicillin) is destroyed by some 200 types of beta-lactamases. In fact, microbes have devised means of eliminating or inactivating every single one of the hundreds of antibiotics thus far developed[63]. This is nothing short of astounding in view of how these drugs work, targeting as they do all kinds of components of microbial biochemistry thought to be vital to them.

Resistance to pesticides and herbicides is a similar story[64, 65]. So now, the explanation? Well, the conventional wisdom is the same old refrain: natural selection from random mutants. But that litany is sufficiently vague as to cover all bases. What does "random" mean? It means simple *unpredictable*. But the disease, hemophilia, seemed to be random, until it was noticed that it runs in families. Indeed, all disease seemed to strike randomly, until their causes became known. Nor are we alone in our skepticism:

> "Conventional wisdom says that any population of a hundred million bacteria would have some resistant individuals just by chance. But Romesberg says the process can't be that passive. 'Resistance is able to evolve efficiently against anything,' he says, 'so the diversity would have to be infinite.'" [66]

So it would. What Romesberg discovered was that genes controlling the so-called "SOS response" in effect promote rapid mutation, and that by suppressing this response, adaptive resistance to drugs is abolished (but details are hazy). In a related development, Cowen and Lindquist demonstrated that a heat-shock protein Hsp90 "can act in diverse ways to couple environmental contingency to the emergence and fixation of new traits," to wit, drug resistance in fungal pathogens[67]. That is to say, many researchers are rethinking the problem of resistance, suspecting that "facilitated variation"[68] must be closer to the truth.

Incidentally, we humans acquire resistance to all sorts of poisons and toxins within mere weeks or months. Heroin addicts enjoy a dose that would promptly kill a neophyte. This ability may be related to points in this article, such as the discovery of a soil-dwelling bacterium with the ability to degrade a poison used against agricultural pests, which the bacterium accomplished by modifying an enzyme it already possessed[69].

7. PARTING SHOTS. This chapter reiterates, with additional examples, many of the points made in previous chapters; notably, (i) that originally intelligent innovations can become heritable in a short time, (ii) that "intelligent innovations" must be extended all the way down, at least to bacteria, (iii) that the fundamental motivational dynamics of all living things is the same as we observe in the human sphere, and (iv) all of the above is quintessentially mental in etiology.

What about "fitness theory" in all of this? Well, a journey of thousands of kilometers to a Pacific atoll is certainly an "honest indicator" of fitness, if ever there was one. But is this any sort of explanation for why they do it? Other birds show off in other ways. Lyrebirds in Australia mimic with startling perfection the ringing of cell phones, vehicle back-up beeps, chainsaws, alarm clocks, motor bikes, humming generators, the click-whir of cameras, etc.[70]. Other such imitators include starlings, magpies, bowerbirds, butcher birds, spangled drongos and, of course, parrots. Then there are the many species of bowerbirds, all having in common the construction of artfully constructed and decorated bowers aimed at pleasing the females who come to inspect and judge them[71-75]. Their bowers are carefully decorated with hundreds of items, and if a human hand rearranges them, the birds restore their design.

All of this figures into the mating game, of course, but is that an explanation? They don't do it for fitness. It is almost the opposite of "natural selection"; call it *unnatural* selection, as if the creatures are deliberately pushing the envelope, *defying* the prudent demands of practical frugality. What we see in the natural work is almost the exact opposite of what the Protestant work ethic – and the classical theory of evolution – would lead us to expect. Until these facts are recognized, the classical theory will remain a laughable hoax; laughable, except that it is poisoning so many young minds with fake science.

17.

The Ball is in Your Court

1. THE THEME. Consider the following comment on the state of modern physics, suggesting that something big and important is missing, similar to the situation back in 1911:

> "He compared the state of physics today to that during the first Solvay conference in 1911. Then, physicists were mystified by the discovery of radioactivity. ... 'They were missing something absolutely fundamental,' he [Gross] said. 'We are [today] missing perhaps something as profound as they were back then." [1]

We propose in this book that much the same is true in modern biology: it is missing something absolutely fundamental. But, unlike the situation in physics, we don't need any huge breakthrough in mathematical or laboratory science, for that missing something is with us always, being nothing more or less than *mentality, consciousness.*

Without it, nothing is explained. With it, everything is explained. More accurately, there are two sides to every explanation. One is the objective or external side: food energy, brood size, calories, etc., which is currently viewed as the *entire* explanation, because that has been the tradition of science all along. It worked okay, or seemed to work okay, in the physical sciences because nobody saw any need for a *mental* explanation of why a stone falls, or why a chemical reaction proceeds as it does. But, as we have seen, that tradition fails in biology. There is no way to explain, in the terms of physics and chemistry, why birds migrate, say – except in human terms. Yet, we are a natural part of the physical cosmos.

That brings us to the other side of the explanation: the inner, personal, motivational, subjective, *mental* side of why things happen. You can study how a horse runs, but not why it runs.

2. THE ANTHROPOMORPHIC SOLUTION. Anthropomorphism is the habit of attributing human qualities to non-human things. It is taboo in science. However, as pointed out in book I, all of science, from bottom to top, is absolutely pure anthropomorphism. There is no way to be rid of it, no way to avoid it short of being dead. We see through human eyes, think with human mind, period. Our every thought and utterance is a *human* thought and utterance.

On the other hand, the injunction against anthropomorphism originated for the good reason that in pre-scientific times, everything was too easily "explained" in human terms, or else in terms of God (also inevitably conceived in human terms). For example, the properties of metals and other substances were formerly explained in terms of their spirit qualities, but that approach did not lead to productive advances.

Now that objective science is reasonably mature; however, we may lay aside this shibboleth. Few will deny that our fellow mammals have minds very much like ours, and probably birds, too; or that snakes and fish eat because they feel hungry, and can suffer pain; or that the sex urge is experienced in roughly the same way by all living things. Perhaps instinctive proclivities modify the mental landscape somewhat; but then again, imagine transplanting the brain of a mouse into the body of a lion: would it continue to act like a mouse?

The point of these remarks is to suggest that with carefully disciplined reason, we can learn much about the wider biosphere from the fact that we are members of it, not essentially different from all the others, differing chiefly in the bodily equipment with which we are endowed.

3. CULTURAL PARALLELS. Among the more fruitful avenues of this approach is to recognize that we ourselves are being constantly driven in the direction of speciation. This is seen in the continuous fragmentation of human populations into ethnicities, nations, political persuasions, religions, and languages. But at the same time, other forces are at work uniting such factions into ever larger and more embracing wholes.

This will be illustrated in a sequel to this book concerning the origin and nature of human language, and the reasons for its subsequent fragmentation into a thousand tongues. Another sequel, on the origin of matrimony, devotes a portion to the bewildering variety of our pre-modern hominid ancestors, all of which were candidate incipient species.

These dynamics within humanity supply us with an "insider's view" of the forces motivating speciation, i.e. every ethnicity is, or might have been, an incipient new species of genus *Homo*. As explained in book I, these dynamics reflect what we have called the "big drives," as distinct from the "little drives" of purely biological origin and function. The big drives are of cosmic origin, operate in all living things, and are otherwise reflected in quantum mechanics and thermodynamics.

4. PERSISTING MYSTERY. This book claims no more than to crack open the door to understanding biology "from the inside." Among the endless mysteries is trying to understand hierarchical systems, meaning exactly how organisms are structured to act as a unity. Even a bacterium is constructed of thousands, millions of parts, which somehow function as a unity of cooperating parts. Higher animals such as ourselves appear to be a hierarchy of systems, imagined to be organized "top-down," as in a kingship. But there are many forms of government, all aiming to be the "most natural" and efficient, and each of which may be viewed as a candidate model of biological organization. The reality might be more like a democracy whose members are all symbionts, serving the nation and being served by it, yet oblivious to the overall will of the nation.

By "will of the nation" we are referring to the Superorganic hypothesis of anthropologist A. Kroeber, championed also by Leslie White. Kroeber argued persuasively that a given culture (or nation, society) has a will and a direction, "*sui generis*," unto itself, above and beyond the understanding of its members. He gave many illustrations, but language shifts will serve here. It so happens that features of a language, such as English, are known to evolve in a constant direction over time scales of centuries, greatly exceeding the lifespan of its individual speakers, who are entirely unaware of any such direction. One of Kroeber's favored examples was the stately, cyclical movement of clothing fashions such as dress hemlines, again exceeding the human lifespan, as if controlled by some unseen celestial force (his metaphor).

Kroeber was eventually forced to recant the "excesses" of his Superorganic hypothesis, having exceeded the proper bounds of science. We do not necessarily advocate his idea as a means of gaining insight into biological organization, but mention it only as one example of the kind of novel thinking that may be needed. There are many possible perspectives, and the best or "true" one remains to be seen. But for example, since DNA evidently contains the entire plan for constructing a human being, from toe to brain, and since every cell contains the identical DNA, then perhaps there is no need to imagine that the parts of the body are "controlled by" the brain; for all of the parts share the same will, and sink or swim together, like the crew of a schooner. Under that conception, there is no need for a harsh captain barking commands.

For that matter, all humanity shares very similar DNA, so it is more than possible that we are all working together in ways much deeper than we know. Indeed, our study of language suggests the existence of covert or cryptic channels of communication embedded within ordinary language, suggesting an explanation for Kroeber's Superorganic phenomena. Here we are reminded of the remarkable parallels between language and the genetic code, including the existence of non-coding or "dark genes."

These parallels, such as letters making words to making phrases to making sentences to making *ideas*, complete with punctuation, are so striking as to suggest direct recapitulation, as opposed to mere analogy. If so, then deep knowledge of our inner workings, normally hidden from conscious awareness, has somehow made its way into our faculty of speech. There may be other evidences of deep knowledge rising to the fore. The famed physicist, Max Planck, is said to have devoted his life to the subject after having a certain epiphany, namely, the far from obvious realization that we humans have been able to gain remarkable insights into the workings of atoms, the movement of planets, the energy of the sun. How did we acquire this ability? It has nothing to do with survival. It must derive from a deep inner understanding, going back to the detailed workings of the cells in our bodies, atom by atom.

It seems paradoxical that at the same time that we higher organisms are becoming farther removed from control over the detailed workings of our cellular machinery; we are also being enabled to gain such knowledge indirectly *via* scientific rationality. But make no mistake, all of that science is rooted in anthropomorphism, meaning based on concepts like *attraction, repulsion, forces, positive vs. negative, energy, gravity*, and so on; all of which are based on everyday personal and subjective experiences. So the seemingly arcane realm of physical science is not really far removed from everyday intuition (as Einstein often pointed out). In fact, intuition was the root of it all, corrected by experiment.

5. FAREWELL. Enough rambling. It is hoped that the reader found this book stimulating. Somehow, we must think our way out of this mess. The ball is in your court.

APPENDIX

Historical Notes on Darwin's Era

This appendix was part of Chapter 1 in the original manuscript but was moved here to avoid digression. It is a thumbnail sketch supplying some historical and social background on the theory. In-depth accounts are cited below.

A.1. A NOT-SO-NEW IDEA. Charles Darwin was perfectly aware that the idea that all life descended from a common ancestor was not original with him. As he himself put it:

> "It is rather a singular instance of the manner in which similar views arise at about the same time, that Goethe in Germany, Dr. Darwin [grandfather of Charles] in England and Geoffroy Saint-Hilaire in France, came to the same conclusion on the origin of species in the years 1794-5." Darwin[1], quoted by Harris[2], pg 36.

Harris goes on to note that the French encyclopedist, Denis Diderot, expressed the same idea in 1774: "The vegetable kingdom might well be and have been the first source of the animal kingdom, and have had its own source in the mineral kingdom." Likewise, Immanuel Kant, in the late 18th century: "The agreement of so many genera of animals in a common scheme ... which with all their differences seem to have been produced according to a common original type, strengthen our suspicions of an actual relationship between them in their production from a common parent," quoted by Harris[2]. Herbert Spencer, too, so greatly admired by Darwin, is shown by Harris to have written similar ideas well before 1859, inspiring Darwin as much as the other way around – as well as coining the term, "survival of the fittest," adopted by Darwin in later editions, and popularizing the word *evolution*.

Not only that, but John Baptiste Lamarck, back in 1802, had advanced a fairly cogent and explicit account of the descent of species from common ancestors, opening the subject to discussion. Lamarck made other key contributions in his *Hydrogeology,* such as his bold defense of the vast age of the earth, putting it correctly in the billions of years (but how he arrived at this I do not know). Mention might also be made of Robert Chambers' immensely popular book of 1844, *Vestiges of the Natural History of Creation,* which continued to out-sell Darwin's *Origin* until the 1880s.

> "Among Chambers' evidences was the 'parallelism' between embryonic development, the stratigraphic record [of geology], and the principle features of animal classification. ... Chambers opted for anonymity because his cosmic evolutionism – with spontaneous generation and a gradual 'unfolding' of 'higher' life forms in a 'gestatory' process – was incompatible with the deemed truths of revealed religion and potentially subversive social order ..." From review of Secord's study[3] .

This "gradual unfolding" seems reminiscent of Hegel's evolutionary philosophy[4], so influential to so many. Although Secord's study of *Vestiges* is doubtless definitive, Eiseley's whets our appetite by observing that Chambers "had hoped for a scientific hearing but was promptly shouted down," ridiculed and scorned in particular by Huxley – it did contain many errors – but all of this fuss had the unexpected effect of making *Vestiges* all the more appealing to the general public.

> "Thus, as Draper commented many years ago, 'happily the whole subject was brought into such prominence that it could be withdrawn into obscurity no more.' The increasing growth of literacy among the working classes was contributing to a widespread interest in the new ideas of science." Eiseley, Ch. V[5]

On top of all this is the well-known fact that Alfred R. Wallace arrived at exactly the same theory as Darwin did, at exactly the same time, and by the same inspirations, scientific expeditions, and reading Malthus and Lyell[6, 7], and other parallels. This circumstance might have led to an ugly quarrel over priority were it not settled amicably[5, 8].

According to Shapin, in the course of reviewing Desmond's book[9], the puzzle about how to classify the newly discovered duck-billed platypus (which lays eggs, suckles it young, has venom, etc.), along with arguments about fossils, were actually debates about evolution. "In short," says Shapin, "the debate about platypuses and fossils was a debate about *evolution*, twenty years before" Darwin's *Origin*. Incidentally, the duck-billed platypus appears to have played a rather important role in setting the stage for *The Origin*[10].

A.2. ORIGIN OF THE ORIGIN. Accordingly, our remark in Ch. 1 that the theory was "in the air" is perhaps an understatement. It was already in books. Indeed, it has been said that the central theme of the entire Enlightenment was evolution. That is to say, this period was greatly interested in the progress of humanity – the rise of civilization from savage origins, the evolution of the human races, notions of a grandly progressive movement from barbarism to the high ideals of the French and American revolutions[11]. Naive it certainly often was, yet it inevitably raised the question of *biological* evolution.

In other words, the cultural climate behind the many theories of evolution may be dissected into two parallel, or helically interwoven threads, one being *social theories* of human progress, the other being advances in the *natural sciences,* especially geology.

A.3. THE SCIENCE THREAD. The core dispute began as that between Biblical creationism and the then-new science of geology. This preoccupation spanned the whole of the 19th century. The nub of the debate was the explanation for geological features, including fossils, and the related matter of the age of the earth, questions which challenged Christian faith in much the way that the Copernican theory of the earth's motion had some centuries earlier – and as evolution does even today. (The difference between then and now is that back then, Christian fundamentalism was the scientific mainstream.) More specifically, the Biblical flood was then the accepted explanation for geological features – including fossils – and the age of the earth was calculated from scriptural authority at 6,280 years, by Bishop Usher[12]. All of the earth's features, like all of the living things on it, were attributed to acts of divine creation. But the "scientists" – a term coined by William Whewell only in 1834, thanks to Mary Somerville[13] – kept extending this age, initially by simple measurements such as the annual rate of silt sedimentation in the Nile Delta. Eiseley speaks of "the discovery of time." Doubts about the scriptural scenario began creeping in as the geologists increasingly found naturalistic explanations, notably in James Hutton's 1788 *Theory of the Earth,* and then, so famously crucial to Darwin (and to Wallace), Lyell's 1830 *Principles of Geology.* Darwin:

> "I always feel as if my book came half out of Lyell's brain, and that I never acknowledged this sufficiently ... I have always thought that the great merit of the *Principles* was that it altered the whole tone of one's mind." Quoted by Harris[2], pg 112.

Lyell did not limit his book entirely to geology, taking the occasion to also criticize Lamarck's theory of evolution, postulating instead a series of divine creations of species. We have no space for the details, but Harris sums it up: "Thus, Lyell's rejection of Lamarck made confirmed evolutionists out of both Darwin and Spencer!" [2] (pg 113) – and, we may add, of Wallace.

Other areas of scientific advance further fed the climate favorable to evolutionism. For example, the systematic classification of plants and animals by Carl Linnaeus in 1737-1753 practically shouted the family tree of living things – or at least, so it seems from our perspective today. Back then, the species were accepted to be as fixed and immutable as the stars above. (However, Linnaeus himself, in his later years, increasingly doubted the permanence of species.) Likewise, studies of developing embryos – which figured into Chambers' *Vestiges* – further hinted at the same conclusion, since embryos of mammals seem to pass through a fish-like stage. In short, then, the Newtonian tradition of science was insistently pressing forward to replace *revealed* truths with *reasoned* truths. Eiseley:

> "The uniformitarian school [after Hutton], in other words, is essentially a revolt against the Christian conception of time as limited ... with supernatural intervention constantly immanent. Rather, this philosophy involves the idea of the Newtonian

machine, self-sustaining and forever operating according to the same principles." [5], pg 114.

The world came to be seen as a puzzle to be figured out by us. Thus, according to Eiseley, "a whole philosophical school in Germany came to regard the world 'as a gigantic system of hieroglyphics, as the language of God or the book of nature'" – to be read by man *directly from nature*, not from a book allegedly written by God. After all, is it not written that we shall know Him by His Works?

A.4. THE SOCIAL THREAD. As earlier noted, the entire Enlightenment may be viewed as having had as its central concern the progress of man and the improvement of society. Marvin Harris:

> "The fact that needs to be established here is that Darwin's principles were an application of social-science concepts to biology. The discussion of socio-cultural progress and evolution among such social theorists as Monboddo, Turgot, Condorcet, Millar, Ferguson, Helvetius, and D'Holbach provided the material for the discussion of biological evolution by Geoffroy St.-Hilaire, Erasmus Darwin [Charles' grandfather], and Lamarck." [2](pg123).

Much of the impetus behind all this social theorizing was the intense curiosity of Europeans towards the then-recently discovered "primitive savages" of the world – in the Americas, Australia, the Arctic, other far-flung places – leading to numerous theories of *racial and cultural* evolution. Most of this theorizing was optimistic and idealistic in the extreme, although discordant elements could not forever be ignored – for example, the high idealism of Condorcet and his faith in the principles of the French revolution, only to fall victim himself to the bloody vengeance of Robespierre.

> "To see the human mind, in one of the most enlightened nations in the world [France], debased by such a fermentation of disgusting passions, of fear, cruelty, malice, revenge, ambition, madness and folly, as would have disgraced the most savage nations in the most barbarous age, must have been such a tremendous shock to his [Condorcet's] ideas of the necessary and inevitable progress of the human mind, that nothing but the firmest conviction of the truth of his principles, in spite of all appearances, could have withstood." [2], Marvin Harris, pg. 115

Thomas Malthus' 1798 *Essay on Population,* which directly addresses Condorcet in its full title[14], provided a much-needed counter-weight to the boundless optimism of the day, pointing out so forcefully that human populations tend to increase exponentially while food resources are limited, and hence, that every society lives at the expense of some other – that life is a bitterly competitive struggle for existence, that nature is "red in tooth and claw." According to Harris (but denied by Eiseley), Darwin's reading of Malthus was as pivotal to him as was his reading of Lyell:

> "On Darwin's own authority we may accept Malthus' discussion of the struggle for survival as the inspiration for natural selection. As Darwin put it, 'This is the doctrine of Malthus, applied to the whole animal and vegetable kingdoms.'" [2]

Wallace is quoted to similar effect, having received his epiphany "in 1855, while recuperating from an illness on the island of Ternate, near New Guinea," in the light of Malthus:

> "Then it suddenly flashed upon me that this self-acting process would necessarily *improve the race,* because in every generation the inferior would be killed off and the superior would remain – that is, *the fittest would survive.* ... The more I thought over it the more I became convinced that I had at length found the long-sought-for law of nature that solved the problem of the origin of species." A.R. Wallace, quoted in Harris[2].

A great discovery, to be sure, no matter obvious it may now appear. Yet, like all great discoveries in science – think of astronomy, that great chain from Aristarchus, Hipparchus and Ptolemy to Galileo, Copernicus, Brahe, Kepler, Newton, Einstein, and still the problem is not solved – it was not complete in itself, but merely led to the next stage of inquiry: how and why do new species arise?

A.5. VIEW FROM THE ACADEMIC LEFT. In book I, we discussed "The Science Wars," being the hot dispute between the orthodox scientific community and the "academic left." Hints of this "academic left", which views science as an instrument of social control as much as an innocent search for truth, were already evident in the 1960s, when communist leanings were fashionable in the academic community. For example, Marvin Harris, whose survey of anthropology[2] would have been peerless were it not so heavily tainted with his opinions (such as that Karl Marx was the greatest anthropologist), opens his chapter 5 with the following, where we have added italics to highlight his attribution of ideological overtones in the theory of evolution:

> "To propose that Herbert Spencer's and Charles Darwin's theories were an inevitable product of a certain phase of Western history is not to deny the contribution of cumulative scientific advances ... The technological pay-off resulting from this inquiry was *essential to the maintenance of capitalism.* Although theological dogmas continued to be *useful in the control and discipline of the masses,* a veritable cornucopia of technological advances forced the theological establishment into an essentially rear-guard action. Finally, in 1859, Darwin's materialist explanation of the origin of species *broke the authority of the theologians* in the domain of the life sciences." [2]

This train of thought subsequently blossomed into the view that scientific theories function to aid, and abet the ideological order, a kind of hidden agenda (but exactly who is running the show is always left obscure). Throughout Harris' book, the view is expressed that early anthropologists, the British ones especially, propounded theories designed to defend the Imperialist policy of exploiting non-western peoples, i.e., stressing their presumed racial inferiority, stupidity, etc., all presented as scientific facts. Growing awareness of these hidden agendas in so-called scientific writing led to the extreme views of the recent "social studies" groups – a loose-knit clique of certain historians, literary critics, etc. – termed the "academic left," discussed in the previous volume, who hold essentially that all of science, even physics and mathematics, is a social construction of this sort, functioning to channel our minds and hobble our freedom.

We disputed part but not all of that view and in this present book have drawn attention at several points to the social and moral implications of the New Synthesis (which are not the views of Darwin himself). For example, the views and policies of Adolf Hitler, and practice of eugenics in the USA and in other western nations[15-17] stemmed directly from neo-Darwinism. Leaving further comments for another venue, one thing is clear from this present book: the theory of evolution as we have it today is a very poor imitation of the exact sciences that it apes, and certainly raises the question of how such hare-brained theories manage to rise to the level of authority and prestige that they have.

A.6. OPINIONS ON DARWIN'S STATURE. Although now lionized in the popular press, with endless portraits of The Great Man, opinions on his stature have actually been rather polarized from the beginning, ranging from placing him on a pedestal as the "Newton of biology" to trivializing his contribution as little more that an obvious application of the ideas of Malthus, Spencer and others. Thus, on the iconoclastic side, see for example Løvtrup[18]. Sulloway's insight on birth order[19, 20] was inspired by his epiphany that Darwin was "just an ordinary person ... wasn't capable of anything I couldn't understand myself"[20], a remark that reveals whiz-kid Sulloway as slightly sophomoric. (All major scientific advances seem easy and obvious when explained to us, as do solutions to otherwise challenging math problems.) Landeau would reduce Darwin's theory, and perhaps all scientific theories, to the status of mythology[21]. Also fitting in here is another of Richards' books[22], and this is only a small sampler of Darwiniana.

A.6. OTHER RESOURCES. The present writer, who still has a "day job", has relied heavily on book reviews in getting his arms around this vast subject matter, since years of full-time reading would be required to read even half the books cited. The following are sources collected in that way, along with a few from the writer's shelf. Some brief notes are supplied at the references.

Many biographies of Darwin have appeared[23-25] but differ widely in their emphasis or approach. For example, Bowler traces out the history of evolutionary thought[26]. Kohn profiles six major figures, from Darwin himself to more recent shapers of the field[27]. Kardiner profiles Darwin and Spencer along with seven leaders in anthropology, including Sigmund Freud[28]. A descendant of Charles Darwin has focused on Darwin's daughter, Annie[29]. It is widely stated that the definitive Darwin biography is that by Desmond and Moore, "setting a new and higher standard"[30].

A number have sought to frame psychology (mind, behavior) in terms of evolution[22, 31, 32], and at least one looks at psychosis in those terms[33]. Application of Darwinian principles to anthropology (human cultures) has also been undertaken by several[34-36], sometimes in the terms of "memes," which is Richard Dawkins' catch-word for concepts transmitted culturally, rather like genes. Another has attempted to phrase the theory of evolution in mythic terms[21].

Several books have been written about important peripheral figures, notably about T.H. Huxley[37-39]. Books about Alfred Wallace were cited above and are tempting. We conclude with a nice quotation, which conveys some of Darwin's sensibilities:

> "If we choose to let conjecture run wild, then animals, our fellow brethren in pain, disease, suffering, and famine – our slaves in the most laborious works, our companions in our amusements – they may partake of our origin in one common ancestor – we may be all melted together." [5], from Darwin's *Notebook* of 1837.

References:

Ch 1-17.

Chapter 1.

1 Fortey, R. **Life: An Unauthorized Biography**. NY: Harper Collins, (1997). <> (Reviewed in *Nature*, 388:731, by A.H. Kroll, who calls this "the best account of life's history that I know of." Now also by Knopf under title, "Life: A Natural History of the First Four Billion Years of Life on Earth," reviewed in *Science*, 280:1542, 1998.)

2 Fortey, R. **Fossils: The Key to the Past**. Cambridge, MA: Harvard Univ. Press, (1991) <> (Reviewed in *Amer Scientist* 81:89, Jan/Feb 1993, where it is recommended as a good introduction to the geological basis of evolution theory.)

3 Eiseley, L. **Darwin's Century (1st ed'n.)**. Garden City, N.Y.: Doubleday Inc. / Anchor Books; 1st ed'n., (1958)

4 Greene, M.: **Hierarchies in biology**. *Amer Scientist* 75 (5, Sep/Oct), 504 (1987).

5 Darwin, C. **On The Origin of Species, or the Preservation of Favored Races in the Struggle for Life**. London: John Murray, Albemarle Street, (1859) <> (The library copy I am now using [Random House / Modern Library, 1993] does not specify which edition it is but is probably the 6th of 1872. The later editions became increasingly complicated and confusing as Darwin sought to reconcile his theory to various critical attacks.)

6 Watson, L.: **Jungle fever**. *The Sciences* (May/Jun), 43 (1989). <> (Short essay describing Wallace's discovery of the principle of natural selection.)

7 Commins, S. and Linscott, R. N., eds. **Man and the Universe** (vol 4): The Philosophers of Science. N.Y.: Random House/Washington Square Press, 1947. (Paperback, 1964.)

8 Ghiselin, M. T.: **Darwin and evolutionary psychology**. *Science* 179 (Mar 9), 964 (1973). <> (possible error in this reference?)

9 Darwin, C. **The Descent of Man and Selection in Relation to Sex**. London: John Murray (2nd ed'n), (1874)

10 Darwin, C. **The Expression of the Emotions in Men and Animals**. London: John Murray (original), (1872) <> (1998 ed'n, Oxford Univ. Press / Harper Collins N.Y. Reviewed in *Nature,* 392:459. By the way, it is noted that a new and annotated edition of Charles Lyells' *Principles of Geology,* famed for its influence on Darwin and Wallace, is now available from Penguin Books.)

11 Richards, R. J. **Darwin and the Emergence of Evolutionary Theories of Mind and Behavior**. Chicago, IL: Univ. Chicago Press, (1988) <> (Lead review in *Science,* 239:198, by historian J.C. Greene, under caption, 'Darwinism and Moral Purpose.' We object only to the reviewer's perception that this is the first book to recognize the importance of Spencer, long appreciated in anthropological circles.)

12 Orel, V. **Gregor Mendel: The First Geneticist**. N.Y.: Oxford Univ. Press, (1996) <> (Reviewed *Science,* 275:1438, Mar 7 1997.)

13 Stebbins, G. L. and Ayala, F. J.: **The evolution of Darwinism**. *Sci American* July, 72-83 (1985).

14 Adams, M. D., ed. **The Evolution of Theodosius Dobshansky**. Princeton, N.J.: Princeton Univ. Press, 1990. <> (Reviewed in *Science,* 266:1589, by David L. Hull. A collection of essays giving "a remarkably well-rounded portrait of one of the most important evolutionary biologists of the century,")

15 Mayr, E. **What Evolution Is**. N.Y.: Basic Books, (2001) <> (Reviewed in *Nature,* 417:223, by Mark Ridley, and in *Science,* 295:50, by Menno Schilthuizen. Intended for a popular audience. Written at age 97. Mayr was the last surviving architect of the New Synthesis, best known for his 1942 book, *Systematics and the Origin of Species.* Book said to be well written, remarkably up-to-date, and bristling with good-natured jousts against rivals such as Steven J Gould. The *Science* review, however, finds it too polemical, and too thin on developments since the 1970s. Although Mayr now admits to reality of sympatric speciation, he barely mentions sexual selection. Ridley mentions that Mayr recently coauthored with Jared Diamond, a superb work of ornithology, his specialty, *The Birds of Northern Melanesia,* reviewed *Nature* 415:959.)

16 Mayr, E.: **80 years of watching the evolutionary scenery**. *Science* 305 (2 Jul), 46-47 (2004). <> (Short but pithy historical overview by the recently deceased architect of the New Synthesis, at age 100, in which he now admits the reality of sympatric speciation in a few cases.)

17 Stearns, S. C. **The Evolution of Life Histories**. N.Y.: Oxford Univ. Press, (1992) <> (Lead review in *Science,* 258:1820, Dec 11, 1992, by S.M. Scheiner, under the wry caption, "Grand Synthesis in the Making." Reviewer treats us to a brief run-down of major post-Neo-Darwinian theoretical positions, notably including the 'Modern Synthesis,' said to have been "mostly completed by the mid-1950's," followed by, in the wake of opposition and the rise of the Watson-Crick model of DNA, what he calls

the 'Molecular Synthesis,' of which he says "I feel safe in predicting that its broad outlines will be in place within the next 15 years." Wrong. Beyond that, he says, "we are moreover on the threshold of the Final Synthesis in biology ..." Book focuses on "life-history traits [which] include number and size of offspring, the timing of maturity, the timing of reproductive episodes, and the process of aging and death.")

18 Konner, M.: **Weaving life's pattern**. *Nature* 418 (18 Jul), 279 (2002). < > (Short but incisive essay on status of developmental biology, especially the topic of psychological maturation. Mentions Changeux's paradox, how a small number of human genes (he says 30,000) can specify the architecture of a vast number of brain cells, etc. We quoted last paragraph. Another pearl: "A main key to development is this on-off pattern [of genes], a pulsing, embryo-wide light-show that turns genetic instructions into animals." Duh.)

19 Bell, G. **Selection: The Mechanism of Evolution**. N.Y.: Chapman and Hall, (1997) < > (Reviewed in *Science,* 277:1255, Aug 29, 1997, by J.A. Endler, along with a simplified edition, *The Basics of Selection.* Also in *Nature,* 384:526, Dec 12, 1996, by J.F. Crow, along with book by G.C. Williams. From Endler's review, this appears to be a thorough statement of the orthodox theory. Author draws heavily on laboratory experiments with microbes. Discusses distinction between selection of existing alleles ("sorting") and new mutations, the latter being a much slower source of variation. Discusses divergence of isolated populations, because each subgroup by chance "represents only a sample of all possible allele combinations" (a.k.a. founder's effect). Reviewer spends long paragraph trying to correct a "misunderstanding" by creationists and other critics about the probability of producing the observed structures by random changes, but not very convincingly. "Bell's literature survey suggests that populations of all kinds of organisms at all scales are kept in flux by continual change in the intensity and direction of selection" – leading us to wonder why so many organisms have been so stable for so many tens of millions of years.)

20 Williams, G. C. **Plan and Purpose in Nature**. N.Y.: Basic Books, (1997) < > (Reviewed in *Nature,* 384:526, 12 Dec 1996, by J.F. Crow, along with two by Graham Bell. Reviewer quotes 10 basic principles of Bell. Reviewer reminds us that Williams is noted for his 1966 book, *Adaptation* ..., and that this book is an update; those by Bell, though also focusing on natural selection, barely touch on the topic of adaptations.)

Chapter2.

1 Simpson, G. G. **The Meaning of Evolution**. New Haven, CT: Yale Univ. Press, (1949).

2 Gee, H.: **Aspirational thinking: progressive evolution**. *Nature* 420 (Dec 12), 611 (2002).

3 Ruse, M. **Monad to Man: The Concept of Progress in Evolutionary Biology.** Cambridge MA: Harvard University Press, (1997) <> (Reviewed by David L. Hull in *Nature,* 385:497, Feb 6 1997; also by F. Ayala in *Science,* 275:495, Jan 24 1997; and widely elsewhere.)

4 Adams, M. D., ed. **The Evolution of Theodosius Dobshansky.** Princeton, N.J.: Princeton Univ. Press, 1990. <> (Reviewed in *Science,* 266:1589, by David L. Hull. A collection of 16 essays giving "a remarkably well-rounded portrait of one of the most important evolutionary biologists of the century," beginning with his training under T.H. Morgan, the pioneer of fruit fly genetics. Includes essays on his personal inspirations, such as the philosophy of Father Teilhard de Chardin and his support of limited programs in eugenics.)

5 Parker, J.: **Richard Dawkin's Evolution.** *New Yorker* (Sep), 41-45 (1996). <> (Wry account of the influential author of *The Selfish Gene* (1989), *The Blind Watchmaker* (1996), *River out of Eden* (1995), *Climbing Mount Improbable* (1996), etc. Alludes to rivalries, such as with S.J. Gould.)

6 Stebbins, G. L. and Ayala, F. J.: **The evolution of Darwinism.** *Sci American* July, 72-83 (1985). <> (Stebbins was a founder of the New Synthesis. They frankly admit to many new uncertainties, e.g.: "Until the relation of genes to development [and behavior, etc.] is better understood ... the full impact of molecular biology on evolutionary theory cannot be assessed." However, in the closing section they express confidence that future modifications will consist only of minor details.)

7 Ayala, F. J.: **The mechanisms of evolution.** *Sci Amer* 239 (#3, Sept), 56-68 (1978). <> (Special issue of *Scientific American* on evolution. Opening article is by Ernst Mayr.)

8 Ryan, F. J.: **Evolution observed.** *Sci Amer* 189 (Oct 4), 78-82 (1953). <> (He notes that a human generation is about 20 years, compared to 20 minutes for the bacteria he uses (*E. coli*). Thus, "in two years, bacteria can grow through more generations than man has in a million or so years." He experimented with up to 7,000 generations.)

9 Apenzeller, T.: **Test-tube evolution catches time in a bottle.** *Science* 284 (Jun 25), 2108-2110 (1999). <> (In special issue on evolution. Features work by Lenski *et al* who subjected *E. coli* cultures to various nutrient and environmental conditions (selection pressures) over a period of 11 years, giving 24,000 generations. Related work also cited, e.g. by Rainey and Travisano.)

10 Dawkins, R. **The Blind Watchmaker.** New York: W.W. Norton, (1987).

11 Zimmer, C. **Evolution: The Triumph of an Idea.** N.Y.: Harper Collins pub., (2001). <> (Reviewed in *Science,* 293:2209, Sep 21 2001, by T.H. Goldsmith. Book is companion to Public TV 'documentary' in 7 episodes. Said to give many "fascinating examples of how rapidly our understanding of this process is growing," e.g. on

evolution of the eye: "Recent computer models validate how easy this is." Appears aimed at creating the illusion that it's all explained.)

12 Gardner, A. and West, S. A.: **Spite among siblings**. *Science* 305 (Sep 3), 1413-1414 (2004).

13 Queller, D. C. and Strassman, J. E.: **The many selves of social insects**. *Science* 296 (Apr 12), 311-313 (2002).

14 Olsson, M., Shine, R., Madsen, T., Gullberg, A. and Tagelstrom, H.: **Sperm selection by females**. *Nature* 383 (Oct 17), 585 (1996). <> (An early paper on what is now an accepted phenomenon, sperm selection, and competition. Purports to show that sand lizard females, which seem to indiscriminately accept mating from males, store the sperm and exert control over which sperm cell is used to fertilize her eggs.)

15 Simmons, L. W. **Sperm Competition and its Evolutionary Consequences in the Insects**. Princeton, N.J.: Princeton Univ. Press, (2001). <> (Reviewed *Nature*, 417:122, May 2, by M.T. Siva-Jothy, who notes that three other books surveying this field already exist, but that Simmons offers a novel perspective and is really good.)

16 Short, R. V.: **Do the locomotion**. *Nature* 418 (Jul 11), 137 (2002). <> (Note on report by H. Moore et al, p174-7 this issue, showing that hundreds of sperm can cooperate by linking together to swim at speeds twice as fast as single sperm in this species of woods mouse, noted for being polygamous and the large testes of males. Writer refers to another paper, *Nature* 418:174, concerning competition for sperm swimming power.)

17 Dawkins, R. **The Selfish Gene**. New York: Oxford Univ. Press (2nd ed'n; 1st 1976), (1989) <> (This best-seller may be taken as representative of the "orthodox position," though perhaps slightly right-wing, despite his sharp disagreements with his equally famed and also orthodox counterpart in the USA, the late Steven J. Gould. It should be added that recent advances in genetic science undermine many of his arguments, which now seem naive.)

18 Peterson, H.: **Beyond the double helix**. *Nature* 421 (Jan 23), 310-312 (2003). <> (This news item mentions how DNA can adopt additional unexpected shapes or conformations, such as the Z-structure, having added twists opposite to the usual, and a propeller-like form near the telomeres, etc.)

19 Wright, R.: **The accidental creationist**. *New Yorker* (Dec 13), 56-65 (1999). <> (According to Wright, Gould's stance was radically anti-progressive in his 1989 book *Wonderful Life* (unfavorably reviewed by Wright) but softened slightly in his 1996 book, *Full House*. My own readings confirm Wright's appraisal: I subscribed to *Natural History* for five years, every issue of which contained a Gould essay; and most

touched on the issue of progress. This article drew many letters of reply, pro and con, printed in next issue.)

20 Holden, C.: **Defying Darwin**. *Science* 305 (17 Sep), 1711 (2004). <> (News item on outrage over publication in a respectable journal of a critique of evolutionary theory by Stephen Meyer, director of Seattle's Discovery Institute, a creationist think-tank. Holden's reporting of this incident is, like most others, acidly biased, using words like "gloating" and "crowing" and casting the skeptics as pseudo-scientists by placing their "science" in quotes. See also *The Scientist*, Oct 11, 2004, pg 12, on this flap.)

21 Lotka, A. J.: **Contribution to the energetics of evolution**. *Proc Natl Acad Sci* 8, 147-151 (1922). <> (Includes references to other of his publications on this topic back to 1907. Article is followed by another by A.J. Lotka, *Natural selection as a physical principle*, p151.)

22 Lotka, A. J.: **The law of evolution as a maximal principle**. *Human Biology* 17 (3), 167-194 (1945). <> (Footnotes include references to his other relevant papers, notably, "Evolution and irreversibility," *Science Progress*, 1920, v14:406-12; the PNAS article of 1922; his main book, *Elements of Physical Biology*, and some mathematical work. It is clear that he was much inspired by Wilhelm Ostwald.)

23 Silver, B. **The Ascent of Science**. Oxford: Oxford Univ. Press, (1998). <> (Reviewed in *Nature*, 392:241, Mar 19, 1998. Said to skillfully introduce for non-scientists many basics, including thermodynamic irreversibility. Written by an experienced teacher of physical chemistry. The reviewer's only caveat concerns Silver's philosophical remarks, said to be annoyingly naive.)

24 Schulman, L. S. **Time's Arrow and Quantum Mechanics**. NY: Cambridge Univ. Press, (1997). <> (Many physicists accept thermodynamic irreversibility as defining the direction of time. Reviewed in *Nature*, 391:453.)

25 White, L. A. **The Evolution of Culture**. New York: McGraw Hill Inc. (paperback), (1959) <>

26 Sahlins, M. D., and Service, E. R., eds. **Evolution and Culture**. Detroit, MI: Univ. of Michigan Press (1966).

27 Laporte, L. F. **George Gaylord Simpson: Paleontologist and Evolutionist**. N.Y.: Columbia Univ. Press, (2000) <> (Reviewed in *Nature*, 409:768, Feb 15, by M.J. Novacek, who complains about lack of depth and details on several matters, which seems unfair, so my guess is that most readers – as opposed to professional historians – will enjoy learning about this man and his role in the New Synthesis.)

28 Richards, R. J. **The Meaning of Evolution: The Morphological Construction and Ideological Reconstruction of Darwin's Theory**. Chicago, IL: Univ. Chicago Press, (1992)<> (Lead review in *Science*, 256:1223, 22 May 1992, by P.R. Sloan, who

notes that the title alludes to G.G. Simpson's classic, which "many of us cut our evo-lutionary teeth on," by way of saying that "in spite of the wide agreement I find with [Richards], the more polemical point ... requires comment. ... Richards' book is more broadly an attack upon the historiography adopted by neo-selectionist evolutionary theory – especially Stephen J. Gould and Peter Bowler – and behind them Simpson and Ernst Mayr ..." The reviewer attempts to mediate, but appears to be won over. These remarks again highlight divisions within the field.)

29 Bower, B.: **One-celled socialites.** *Science News* 166 (Nov 20), 330-332 (2004). <> (Reviews recent advances in the field of microbial cooperation, congregation, nepo-tism, division of labor, boil formation, quorum sensing, etc. Also touches on group selection theory as an explanation for altruism, alternative to kin selection theory. Cites physicist Ben Jacob's radical theory that such phenomena reflect emergence of social intelligence.)

30 Pray, L.: **Microbial multicellularity.** *The Scientist* (Dec 1), 20-23 (2003). <> (Notes several eminent biologists who now regard bacteria as "social creatures" but, as the title implies, they may also be viewed as "multicellular organisms." As usual, quorem sensing – meaning the ability of bacteria to detect and respond cooperatively to addi-tive signals secreted by a certain number of others nearby – was originally thought to be rare and specific but is now seen to be widespread.)

31 Russo, E.: **Microbial co-op in evolution.** *The Scientist* (Oct 26), 25 (2003). <> (Short review of efforts by Paul Rainey, Gregory Velicer, and others to understand complex "social" interactions of one-celled organisms including bacteria, certain fungi, and protozoans, to gain insight on key events in evolution, notably the rise of multicellular forms. Of special interest is that after deleting the genes for pro-ducing the pili they use for social motility, they found another way of doing the same within about 36 weeks. Also touches on theory of cooperation and group selection.)

32 Spake, A.: **Taking aim at the citadels of slime.** *U.S. News & World Report* (Aug 14), 42 (2000). <> (Short review of the newly recognized complexities of bacterial com-munities, with emphasis on those causing disease.)

33 Douglas, K.: **Superorganisms.** *New Scientist* (Aug 9), 34 (2005). <> (Exhibits Portuguese man-of-war jellyfish as example of primitive multicellularity, viewing it as a colony of one-celled organisms which somehow assemble and specialize. Cites other classic examples, e.g. the amebic slime molds, and higher on the scale, colonies of bees, ants, wasps, some other eusocial species, as superorganisms.)

34 Margulis, L. and Sagan, D. **Acquiring Genomes: A Theory of the Origin of Species.** N.Y.: Basic Books, (2002) <> (Reviewed *Nature*, 418:275, July 18 2002, by Axel Myer, under headline, "Viewing life as cooperation: can symbiosis and genome acqui-sition account for all speciation?" Reviewer reports "they argue that all speciation

events are caused by symbiosis, cooperation and the reticulation of genomes, questioning some of Charles Darwin's central." Meyer is intrigued but doesn't buy it.)

35 Sagan, L. M.: **On the origin of mitosing cells.** *Journal of NIH Research* 5 (Mar), 65-72 (1993). <> (Condensed reprint of her classic 1967 paper in *J Theoret Biol*, 14:225-274, featured in the historical "Landmarks" column of this issue. Her three major books are also referenced. Now Margulis, her name was then Sagan.)

36 Provine, W. B., ed. **Selected Papers by Sewell Wright.** Chicago, IL: Univ. of Chicago Press, 1986.<> (Reviewed in *Amer Scientist,* 75:292, May/Jun 1987, by D. Charlesworth, along with a biography of Wright by the same author and publisher. Wright wrote the main article on *Evolution* in the 1968 Encyclopedia Britannica (vol 8, p917), a good summary of the subject then, and including some of his central ideas such as of fitness landscapes and group selection theory.)

37 Dawkins, R. **Climbing Mount Improbable.** N.Y.: Viking, (1996) <> (Lead review in *Nature,* 382:309, 25 Jul 1996, by S. Kauffman: "the central image of the book is ... a multi-peaked fitness landscape ..." For insights on reviewer Kauffman, see his *The Origin of Order*.)

38 Kauffman, S. A. **The Origins of Order: Self-Organization and Selection in Evolution.** N.Y.: Oxford Univ. Press, (1993) <> (Lead review in *Science,* 260:1531, 4 Jun 1993, by G.P. Wagner, under the caption, "Final Theory in Biology" – facetiously? The first chapter is said to offer good historical background. After that is a lot of complex theory, but boiling down to S. Wright's fitness landscape, with its peaks of adaptive perfection, etc.)

Chapter 3.

1 Staff: **Color genes help mice and lizards.** *Science* 309 (Jul 15), 374-375 (2005). <> (News item showing that several populations of deer mice and lizards in Florida and New Mexico living on sand dunes have acquired markedly lighter colorations compared to neighboring groups in more concealed habitats. Authors estimate that about 100 genes are involved. The changes occurred over periods of 600 to 5000 years.)

2 Wright, S.: **Evolution, organic.** *Encyclopedia Britannica* 8, 917-929 (1968).

3 Grant, P. R. **Ecology and Evolution of Darwin's Finches.** Princeton, N.J.: Princeton Univ. Press, (1988) <> (Reviewed in *Amer Scientist,* 76:287, May/Jun 1988, where it is said to be "the first major synthesis on Darwin's finches since Lack's influential book of 1947.")

4 Grant, P. R.: **Natural selection and Darwin's finches.** *Sci American* (Oct), 82-87 (1991).

5 Weiner, J. **The Beak of the Finch**. NY / London: Knopf / Jonathan Cape, (1994) < > (Reviewed in *Nature* 371:27, Sep 1, 1994, by J.J.D. Greenwood, under caption "Evolution in Real Time," and in *The Sciences,* Sep/Oct 1994, by L.A. Marscall. The latter review drew a letter in *The Sciences,* Jan/Feb 1995, p5, by G.L. Schroeder, who says that Marscall incorrectly implies that the observed changes are related to mutation-driven evolution, as admitted in Marscall's reply.)

6 Pennisi, E.: **Finches adapt rapidly to new homes**. *Science* 295 (Jan 11), 249-250 (2002). < > (Note on full article, p316 this issue, to the effect that certain finches have taken up residence in human houses in the 20th century and already show distinctive features. One authority, D. Reznick, is quoted as saying that "the time scale of decades [not centuries] is really enough for animals to evolve." That statement evoked controversy.)

7 Bell, G. **Selection: The Mechanism of Evolution**. N.Y.: Chapman and Hall, (1997) < > (Reviewed in *Science,* 277:1255, Aug 29, 1997, by J.A. Endler, along with a simplified edition, *The Basics of Selection.* Also in *Nature,* 384:526, Dec 12, 1996, by J.F. Crow, along with book by G.C. Williams.)

8 Williams, G. C. **Plan and Purpose in Nature**. N.Y.: Basic Books, (1997) < > (Reviewed in *Nature,* 384:526, 12 Dec 1996, by J.F. Crow, along with two by Graham Bell. Reviewer quotes 10 basic principles of Bell.)

9 Williams, G. C. **Adaptation and Natural Selection**. Princeton, NJ: Princeton Univ. Press, (1966)

10 Kerr, R. A.: **Did Darwin get it all right?** *Science* 267 (Mar 10), 1421-1422 (1995) < > (Reports on additional fossil evidence supporting 'punctuated evolution,' which holds that new lineages tend to appear abruptly in the fossil record, geologically speaking, and thereafter often persist unchanged for long periods.)

11 Nelson, G.: **Older than that**. *Nature* 367 (Jan 13), 108 (1994) < > (In a letter, Nelson complains about a prior article, "Punctuated Equilibrium Comes of Age," 366:223-227, pointing out that the age of this theory "is not 21 but at least 128" years, citing a reference of 1865, also cited by him in the book edited by Otte and Erdler, *Speciation and its Consequences.* "The wonder is," Nelson writes, "that this continental recipe of the mid-nineteenth century should today be relished as innovation ..." It may be added that heated debate between gradualists and saltationists took place also in founding the New Synthesis.)

12 Koehn, R. J. and Hilbish, T. J.: **The adaptive importance of genetic variation**. *Amer Scientist* 75 (Mar/Apr), 134-140 (1987).

13 Kettleworth, H. B. D.: **Darwin's missing evidence**. *Sci American* (Mar), 48 (1959).

14 Hooper, J. **Of Moths and Men: Intrigue, Tragedy, and the Peppered Moth**: Fourth Estate Press, (2002) < > (Review *Nature,* 418:19-20, July 4 2002, by Jerry A. Coyne,

who readily admits serious problems with the experimental side of support for industrial melanism, yet argues that the naturalistic evidence strongly supports it, and accuses the author of exploiting and dramatizing the whole affair, and smearing the well-deserved high repute of Kettleworth. He also notes how the creationists jumped on this story as ammunition for their anti-evolution campaign.)

15 Hopkin, M.: **Dark prospects.** *Nature* 430 (29 Jul), 522 (2004) <> (Briefly noted, referring to work by L.M. Cook et al (*Biol J Linn Soc* 82:359) on moth *Bistron* and vari-colored relatives. Among other things, refutes notion that the coloration change was due to diet.)

16 Allison, A. C.: **Sickle cells and evolution.** *Sci American* (Aug), 87 (1956).

17 Miller, L. H.: **Protective selective pressure.** *Nature* 383, 480-481 (1996) <> (Perspective on report p522 this issue by Williams *et al,* which appears to extend the classic case of malaria *vs.* the sickle cell trait in Africa to malaria in the South Pacific by the putative malaria-protective genetic defect of thalassemia, an inherited blood disorder, by showing that the gene frequencies of that disorder parallel the intensity of malarial endemicity.)

18 Ryan, F. J.: **Evolution observed.** *Sci Amer* 189 (Oct 4), 78-82 (1953) <> (He experimented with up to 7,000 generations.)

19 Apenzeller, T.: **Test-tube evolution catches time in a bottle.** *Science* 284 (Jun 25), 2108-2110 (1999). (In special issue on evolution. Features work by Lenski *et al* who subjected *E. coli* cultures to various nutrient and environmental conditions over a period of 11 years, giving 24,000 generations. Related work by others also cited, e.g. by Rainey and Travisano.)

20 Weiner, J.: **Evolution made visible.** *Science* 267 (Jan 6), 30-33 (1995).

21 Mlot, C.: **Microbes hint at a mechanism behind punctuated evolution.** *Science* 272 (Jun 21), 1741 (1996) <> (Reports on a 20 yr experiment on bacterial evolution in the laboratory by Lenski *et al,* p1802 this issue, purporting to 'watch evolution happen' in bacteria. Results appear to indicate bursts of adaptive change interspersed with periods of little change.)

22 Coyne, J. A. and Charlesworth, B.: **Mechanisms of punctuated evolution.** *Science* 274 (Dec 6), 1748-1750 (1996) <> (Comments critical of report by Lenski *et al.,* and perspective by Mlot claiming to exhibit certain evolutionary process by bacterial cultures, with response; and followed a few issues later by a letter critical of it, by Eldredge and Gould – with sharp response by Coyne and Charlesworth. Debate hinges on why evolution of lineages proceeds in spurts, and why the spurts are followed by long periods of stasis.)

23 Hori, M.: **Frequency-dependent natural selection in the handedness of scale-eating cichlid fish.** *Science* 260 (Apr 9), 216-219 (1993).

24 Case, T. J.: **Natural selection out on a limb.** *Nature* 387 (May 1), 15-16 (1997) <> (Addresses the fact that there are "few well-documented cases in which evolution by natural selection has actually been witnessed outside the laboratory," and comments on the new case reported by Losos *et al.,* p70 this issue, of the 10-14 year experiment of introducing *Anolis* lizards to 14 small Bahamian islands. The experiment is widely hailed as a great success in demonstrating similar adaptations to similar habitats, e.g., see also *Science,* 276:682, "Catching Lizards in the Act of Adapting," and *Science,* 279:2043, "For Island Lizards, History Repeats Itself". Personally, I see no great significance in this work.)

25 Morell, V.: **Catching lizards in the act of adapting.** *Science* 276 (May 2), 682-683 (1997).

26 Harvey, P. H. and Partridge, L.: **Different routes to similar ends.** *Nature* 392, 552-553 (1998). <> (It is argued that the adaptations of *Anolis* lizard species to similar niches, e.g., tree tops, etc., on different islands have been closely convergent though independent. Okay. But is that a breakthrough?)

27 Morell, V.: **Predator-free guppies take an evolutionary leap forward.** *Science* 275 (Mar 28), 1880 (1997) <> (Perspective on report p1934 by Reznick *et al.* In this experiment in a Trinidad river, guppies from downstream pools were transplanted to upstream pools where predators were rarer. After 4 years, and again at 7 years, many changes were noted in the transplanted population, consistent with being predator-free.)

28 Reporter, T. M. B.: **Snail tale.** *Sci American* (May), 32 (1990) <> (Reports work by Crowl and Covich in a then-recent issue of *Science,* showing that certain pond snails in waters where crayfish are actively preying on them grow too large to be eaten, but not in waters where the predators are absent. The same was observed in a tropical species in response to predation by a shrimp. The assumed chemical cues are not yet identified. The postulated trade-off here is that the cost of growing larger is delayed reproductive maturity.)

29 Holmes, B.: **In the blink of an eye.** *New Scientist* (Jul 9), 28 (2005). (Caption: "If you thought evolution was slow and gradual, think again." Focus is on "rapid evolution" forced by human effects such as trophy-hunting for the finest bighorn sheep rams, elephants with the largest tusks, and the largest fish, in each case resulting in smaller average sizes and reduced fitness and fecundity. Also mentioned, somewhat incongruously, is the interesting case of the side-blotched lizard, which practices three different mating strategies, too complex to summarize here, but resulting in 3-4 year cycles of predominant type – like the scale-eating cichlid fishes – likened to a game of rock-paper-scissors.)

30 Levinton, J. S.: **The big bang of animal evolution.** *Sci Amer* (Nov), 84-91 (1992).

31 Rokas, A., Kruger, D. and Carroll, S. B.: **Animal evolution and the molecular signature of radiations compressed in time.** *Science* 310 (Dec 23), 1933-1938 (2005) <> (See also perspective, p1910, "Is the 'big-bang' in animal evolution real?" by L.S. Jermlin, L. Poladian, and M.A. Charleston, who are impressed, yet cautious because of important technical questions about mutational clocks and other reservations.)

32 Richardson, J. E. et al: **Rapid diversification of a species-rich genus of neotropical rain forest trees.** *Science* 293 (Sep 21), 2242-2245 (2001). (Discusses explosive speciation of trees in South America, all within last 10 million years, some within 2 million, mostly sympatrically, and attributes this to events related to Andes Mountains uplift, and by implication, climate. See also perspective, p2214.)

33 Mendelson, T. C. and Shaw, K. L.: **Rapid speciation in an arthropod.** *Science* 433 (Jan 27), 375 (2005) <> (The highest rate so far recorded for an arthropod: mean value, 4.2 species per million years for crickets on Hawaiian Islands, said to be "explosive and ongoing." Many seem to differ only in their mating songs.)

34 Schubert, C. D., Diesel, R. and Hedges, S. B.: **Rapid evolution to terrestrial life in Jamaican crabs.** *Nature* 393 (May 28), 363-365 (1998) <> (Shows by DNA that the 9 species of land crabs on Jamaica appeared about 2 million years ago and rapidly diverged, and do not derive from similar land crabs in Oceania – some of which, as I recall, climb palms to cut down coconuts. "For example, the bromeliad crab ... raises its young in water-filled bromeliad leaf axils. The mother crab manipulates water quality by removing detritus, circulating the water to oxygenate it, and carrying empty snail shells into leaf axis as both a calcium source and a pH buffer. She also protects the leaf axil against ... predators. The snail-shell crab [another species] breeds in empty shells of the snail [which she turns] upside down to collect rainwater or carries water to the shell. Both species feed their offspring, which remain in the nursery habitat for several months ...")

35 Berthold, P., Helbig, A. J., Mohr, G. and Querner, U.: **Rapid microevolution of migratory behavior in a wild bird species.** *Nature* 360 (Dec 17), 668-670 (1992) <> (Demonstrates genetic basis of recently altered migration habit of a portion of the European blackcap population. For commentary by W.J. Sutherland, see p625, "Genes mark the migratory route.")

36 Sniegowski, P. D., Gerrish, P. J. and Lenski, R. E.: **Evolution of high mutation rates in experimental populations of *E. coli*.** *Nature* 387 (Jun 12), 703-705 (1997).

37 Taddel, F., Radman, M., Maynard-Smioth, J., Toupance, B., Gouyon, P. H. and Godelle, B.: **Role of mutator alleles in adaptive evolution.** *Nature* 387 (12 Jun), 700- (1997).

38 Attenborough, D. **The Living Planet**. Boston, Toronto: Little, Brown, (1984) <> (See Ch. 6 of this companion book to the PBS TV series of that name. Although broad and wide-ranging and aimed at laymen, it is mainly fact-based, a refreshing departure from many other such series which constantly supply odiously panglossian explanations for everything, e.g. those narrated by Suzuki, as if there were no more questions or doubts.)

39 Gould, S. J. and Lewontin, R.: **The spandrels of San Marco and the Panglossian paradigm: a critique of the adaptationist programme**. *Proc Roy Soc Lund* B205, 581-598 (1979).

40 Voltaire: **Candied**. (Pangloss was Candide's sidekick in this classic, who constantly proclaimed that "this is the best of all possible worlds" – a satire on Leibnitz' philosophy – and who by that logic, supplied maddeningly simplistic answers for all of Candide's agonized questions.)

41 Jack Selzer, E. **Understanding Scientific Prose**. Madison, WI: Univ. Wisconsin Press, (1993) <> (Reviewed in *Science,* 263:697, Feb 4, 1994, by Keith Stewart Thomson, being essays from diverse viewpoints on "The spandrels of San Marco ..." by Gould and Lewontin.)

42 Palevitz, B. A.: **Love or hate him, Stephen J. Gould made a difference**. *The Scientist* (Jun 10), 12-13 (2002) <> (On the occasion of his passing, by lung cancer. See also Commentary by Ricki Lewis, p10 this issue.)

43 Briggs, D. E. G.: **Stephen J Gould (1941-2002)**. *Nature* 417 (Jun 13), 706 (2002) <> (Briefly recounts his main achievements, notably evidence for "punctuated equilibrium" ca. 1972 with Niles Eldredge. His passing also noted in the May 23 issue, p371. His last major work was *The Structure of Evolutionary Theory*, a huge book written over 20 years, reviewed in *Nature* 416:787, 2002.)

44 Rose, M. R. and Lauder, G. V., eds. **Adaptation**. San Diego, CA: Academic Press (1996) <> (Reviewed in *Science,* 277:189, Jul 11, 1997, by M.E. Feder.)

45 Young, K. V., Brodie Jr., E. D. and Brodie Sr., E. D.: **How the horned lizard got its horns**. *Science* 304 (Apr 2), 65 (2004) <> (Shows that longer horns protect against predation by loggerhead shrikes. Article drew letters in issue of Sep 24 (305:1909) generally praising but raising questions of interpretation; with response.)

46 Edwards, A. W. E.: **Hamilton built on work by Haldane and Fisher (Letter)**. *Nature* 416 (Apr 11), 581 (2002) <> (Writer complains that O. Judson attributed "inclusive fitness" to Hamilton in her review of vol. II of his *Collected Papers*, citing Fisher's *The Genetical Theory of Natural Selection* (Oxford, 1930) and Haldane's *The Causes of Evolution* (Longman, London, 1932) as both clearly stating the principle. This fact is

acknowledged in Triver's obituary of Hamilton, but indicates that Hamilton much improved on this principle.)

47 Trivers, R.: **William Donald Hamilton (1936-2000).** *Nature* 404 (20 Apr), 828 (2000) < > (Obituary. Briefly reviews his main achievements, notably *Inclusive Fitness* (1964); social insects; game-theory on "reciprocal altruism" (*Prisoner's Dilemma,* 1981); role of parasites in evolution, with M. Zuk (1981); etc. Died of malaria contracted during fieldwork.)

48 Real, L. A., ed. **Ecological Genetics.** Princeton, NJ: Princeton Univ. Press (1994) < > (Reviewed in *Science,* 266:468-470, 21 Oct 1994, by R.E. Michod, who portrays it as a fine and fascinating book, but feels that it "lacks a central focus or problem," which he sees is that of *fitness,* "perhaps the only concept that is uniquely biological." Though we doubt that, he raises a crucial question: "What is fitness? How do we measure fitness? These questions are the specific concerns of ecological genetics [as distinct from] population genetics or evolutionary genetics ..." He continues in his dismal appraisal of 'fitness' for several paragraphs, but concludes on a more upbeat note: "Does this mean there can be no general theory or framework? I think not. Therein lies the task...")

49 Keller, E. F. and Lloyd, E. A. L. **Keywords in Evolutionary Biology.** Cambridge, MA: Harvard Univ. Press, (1992) < > (Reviewed in *Science,* 260:1153, 21 May 1993, by D.F. Futuyma: "This book consists of 51 essays, about half by biologists and half by philosophers and historians of science, on 37 keywords or topics in evolutionary biology and ecology, alphabetically arranged from 'adaptation' to 'unit of selection.'")

50 Luling, K. H.: **Archerfish.** *Sci American* (Jul), 100 ff (1963).

51 Eisner, T. **For Love of Insects.** Cambridge, MA: Harvard Univ. Press, (2004) < > (Reviewed *Nature,* 428:368, Mar 25, by J.N. McNeil. "We learn how many insects defend themselves by spraying aggressors with noxious secretions ... that insects may produce their own compounds or get them from their food. In some instances, females may receive them from males during mating, and the quantity of a male's defense compounds may determine whether he gets a mate. ... We also learn of the ingenious ways in which defenses may be overcome, such as the herbivorous insects that can feed on milkweed leaves by severing the veins that carry latex to the leaf, and how some predators eat only part of their prey to avoid toxins.")

52 Kaiser, J.: **Mad about mites.** *Science* 286 (Nov 5), 1047 (1999) < > (Note on website devoted to mites set up by Australian mite expert ['acarologist'] David Walter, stating that a handful of forest humus can contain 100 different mite species, and mentions among examples the basket mite "which camouflages itself by toting around soil in an exoskeletal structure on its back.")

53 Griffen, D. R. **The Question of Animal Awareness**. N.Y.: Rockefeller Univ. Press, (1976)

54 Griffen, D. R. **Animal Thinking**: Harvard Univ. Press, (1984)

55 Dawkins, M. S. **Through Our Eyes Only? The Search for Animal Consciousness**: W.H. Freeman Co., (1993) < > (Reviewed in *Nature,* 364:398, 29 Jul 1993, by Sara J. Shettleworth, who notes that Dawkins' earlier book, *Animal Suffering* (Chapman and Hall, 1980) is "now unfortunately out of print," and cites the latest Griffen book.)

56 Hansell, M. **Bird Nests and Construction Behavior**. Boston: Cambridge Univ. Press, (2000) < > (Reviewed *Nature,* 409:18, Jan 4, 2001, by T. Slagsvold, who notes that this book includes much new material since two other classics on the subject from the 1980's, which he names, one by the same author.)

57 Holden, C.: **Telling an ovenbird by its nest**. *Science* 286 (Dec 3), 1843 (1999) < > (Short note on paper in October issue of *The Auk,* being a survey of nest differences among some 230 species of ovenbird, with the aim of placing them in sequence of origin. Since they range from the dunes of Chile to the forests of Mexico, it is not surprising that they use different materials, including adobe-like mud houses resembling a small oven.)

58 Staff: **Quick-drying foam**. *Science* 307 (Jan 7), 19 (2005) < > (Note on paper in *J Exp Biol* 207:4727 concerning the sandcastle worm, which constructs a tube-like house by gluing together sand grains; this paper examines the sophisticated biochemistry of the glue produced by the worm for this purpose.)

59 Milius, S.: **Ambush ants**. *Science News* 167 (Apr 23), 260 (2005) < > (Note on paper by J Orivel's team in Apr 21 issue of *Nature* discovering that these ants of French Guiana, in addition to sheltering in conveniently rolled leaves of a certain tree whose secretions they feed upon, also trap insects, up to grasshopper size, by fashioning well-concealed loose-knit platforms of fungi threads which when stepped on by the prey allow the ants to seize its legs and antennae from underneath, immobilizing it for an easy kill. The fungi is apparently imported by the ants and carefully cultivated.)

60 Chin, G.: **Taking in the welcome mat**. *Science* 311 (Jan 27), 437 (2006) < > (Short note on paper in *Biotropica* 37:670, by Longino et al, on novel nesting habits of two species of the ant genus *Stenamma,* said to be remarkable partly because this genus has been intensively studied before.)

61 Dawkins, R. **The Blind Watchmaker**. New York: W.W. Norton, (1987)

62 Dawkins, R. **The Selfish Gene**. New York: Oxford Univ. Press (2nd ed'n; 1st 1976), (1989)

Chapter 4.

1 Eiseley, L. **Darwin's Century (1st ed'n.).** Garden City, N.Y.: Doubleday Inc. / Anchor Books; 1st ed'n., (1958)

2 Thomson, K. **Before Darwin: Reconciling God and Nature.** New Haven, CT: Yale Univ. Press, (2005) < > (Reviewed in *Science,* 309:1493, Sep 2, 2005, by Alan Cutler, under caption, '200 Years of Accommodation' between religion and evolution. Also in London by Harper-Collins under title, *The Watch on the Heath: Science and Religion Before Darwin,* in reference to Paley's watch-maker metaphor.)

3 Hoyle, S. F. and Wickramasinge, N. C. **Evolution from Space: A Theory of Cosmic Creationism.** New York: Simon & Schuster, (1981) < > (These eminent astronomers are frankly contemptuous of evolutionary theory. and so argue that life on earth must have come from outer space. But that merely removes the problem to another planet.)

4 Toumey, C. P.: **God's own scientists.** *Natural History* 103 (Jul), 4-9 (1994).(Toumey, an anthropologist, reports on attending regular meetings of a group of educated Christians, many of them scientists, teachers or physicians, who are attempting to rationally defend creationism as part of their Bible studies. The next article deals with "The Struggle for the Schools," by E.C. Scott, anthropologist and prominent opponent of teaching creationism. Page 79 provides sketches of the authors, mentioning Toumey's *God's Own Scientists* [1994] and others he recommends: Tom McIver's *Anti-Evolution* [1988] and George E. Webb's *The Evolutionary Controversy in America* [1994]. Scott recommends Ronald Numbers' *The Creationists* [1992], Arthur Strahler's *Science and Earth History: The Creation-Evolution Controversy* [1988], and *The Creationist Movement in Modern America* [1990] by R.A. Eve and F.B. Harrold.)

5 Scott, E. C.: **Monkey business.** *The Sciences* (Jan/Feb), 20-25 (1996). < > (Subtitle "Rebuffed in the courts, anti-evolutionists are seeking a niche in the schools." This updates her article in *Natural History,* July 1994.)

6 Schmidt, K.: **Creationists evolve new strategy.** *Science* 273 (Jul 26), 420-423 (1996). < > (Subtitle "Creationists are taking their fight to the legislatures from Georgia to Ohio." Includes struggle over textbooks and list of key court cases since 1925. Article touched off series of letters, see 273:1321, 274:904, 274:1993, 275:142, several objecting to the smug and pompous tone of the article, others debating such matters as the meaning of a theory, whether it is worth debating the creationists, etc.)

7 Zimmerman, M. **Science, Nonscience and Nonsense: Approaching Environmental Literacy.** Baltimore, MD: Johns Hopkins Univ. Press, (1996) < > (Reviewed in *Nature,* 381:125, 1996, where it is said that this account of the battle between creationists and evolutionists "is fascinating and at times frightening." Paperback, 1998.)

8 Pennock, R. T. **Tower of Babel: The Evidence Against the New Creationsim.** Cambridge, MA: MIT Press, (1999)

9 Pennock, R. T., ed. **Intelligent Design, Creationism and its Critics.** Cambridge, MA: MIT Press, (2002) <> (Subtitle, "Philosophical, Theological, and Scientific Perspectives." Reviewed in *Science,* 295:2373, 29 Mar 2002, by paleontologist Kevin Padian, who mentions Pennock's previous book on this, *Tower of Babel: The Evidence Against the New Creationism* (MIT, 1999), this one differing in presenting chapters by representatives of both sides, likened to a "smack-down" according to Padian.)

10 Holden, C.: **Anti-evolution TV show prompts furor.** *Science* 271 (Mar 8), 1357 (1996). <> (Many far-fetched notions were apparently presented, including the flatly false one that humans were living among the dinosaurs.)

11 Pockley, P.: **Geologist set to challenge 'creationism' verdict.** *Nature* 387 (Jun 26), 837 (1997).(One of several reports on a creationism stir in Australia; see also *Nature* 386:638, 17 Apr 1997, "'Ark evidence' challenged in Sydney court," and *Nature* 387:540, 5 Jun 1997, "Geologist loses 'creationism' challenge.")

12 Greenspan, N. S.: **Not-so-intelligent design.** *The Scientist* (Mar 4), 12-13 (2002) <> (Editorial opinion on the debate in Ohio about teaching I.D., with 10 references, including to Michael Behe, by the usual rationality that I.D. is not science.)

13 Rennie, J.: **15 answers to creationist nonsense.** *Sci Amer* (July), 78-85 (2002). (Intended to help school teachers respond to questions from skeptical kids, i.e. those taught to believe creationism, intelligent design, etc. It also cites other resources such as *How to Debate a Creationist, Defending Evolution in the Classroom,* Niles Eldredge's book, *The Triumph of Evolution,* and web sites.)

14 Shermer, M.: **Vox populi: ... why evolution remains controversial.** *Sci Amer* (Jul), 37 (2002).<> (Shermer, publisher of *Skeptic* magazine and a biographer of Alfred Wallace, has a regular column and is a perfect heir to myth-buster Martin Gardner, long-time math columnist and author of the genre-setting *Fads & Fallacies,* a best-seller of the 1960's, bashing so-called pseudo-science but back-firing due to Gardner's excesses. Here, Shermer reflects on why he received hundreds of letters responding to his February column on evolution, compared to the usual dozen or so.)

15 Olson, S.: **Shapes of a wedge.** *Science* 304 (May 7), 825-827 (2004). <> (Review of three books listed below, the one edited by Campbell and Meyes defending I.D., the other two seeking to expose it as nonsense. Title refers to the 'Wedge Strategy' of the creationists, leaked from their headquarters in 1999. Reviews ooze contempt for doubters of orthodoxy.)

16 Forrest, B. and Gross, P. R. **Creationism's Trojan Horse: The Wedge of Intelligent Design**. N.Y.: Oxford Univ. Press, (2004) <> (Reviewed *Science*, 304:825, 7 May 2004, by Steve Olson, with two others.)

17 Shanks, N. **God, the Devil, and Darwin**. N.Y.: Oxford Univ. Press, (2004) <> (Reviewed *Science*, 304:825, 7 May 2004, by Steve Olson, with two others.)

18 Campbell, J. A. and Meyer, S. C., eds. **Darwinism, Design, and Public Education**. East Lansing, MI: Michigan State Univ. Press, (2004) <> (Reviewed *Science*, 304:825, 7 May 2004, by Steve Olson, with two other related titles.)

19 Staff: **Bush and Kerry offer their views on science**. *Science* 306 (Oct 1), 46-53 (2004). <> (See pg 51 for stance on creationism.)

20 Editor: **Keeping religion out of the classroom**. *Nature* 436 (Aug 11), 753 (2005). <> (Editorial on Kansas fracas, as earlier reported in *Nature* 406:552, 2000, and introducing news item p761, "Scientists attack Bush over intelligent design.")

21 McKenzie, D.: **End of the enlightenment**. *New Scientist* (Oct 8), 39-51 (2005). <> (First of five articles in this "Special Report on Fundamentalism: Reality Wars." Includes "Enemy at the Gates," by Mike Holderness, p47-49 – as well as Brian Appleyard's "People in Glass Houses," p50-51, for a touch of much needed balance.)

22 Stokes, T.: **Pro-intelligent design thesis stalls**. *The Scientist* (Jul 4), 12-13 (2005). <> (Note on dispute about a PhD thesis in education defending I.D. at Ohio State University, the question being whether faculty sympathetic to I.D. should be allowed to judge it. The next note on the page tells of a flap at the Smithsonian Institute, concerning a film produced in part by a big-shot in the I.D. community. This piece drew letters printed in the Aug 29 issue, one from a faculty person at Ohio State denying the accuracy of the reports.)

23 Editor: **Vatican astronomer rebuts Cardinal's attack on Darwinism**. *Science* 309 (Aug 12), 996-997 (2005).

24 Mervis, J.: **Dover teachers want no part of intelligent design statement**. *Science* 307 (Jan 28), 505 (2005).

25 Anderson, G.: **Keep censorship out of schools (letter)**. *Science* 308 (Apr 22), 495 (2005). <> (One of four letters concerning creationism / I.D., mainly focusing on Pennsylvania; in response to article by J. Mervis, *Science* 307:505, on the Dover fracas.)

26 Biever, C.: **God goes to court in all but name**. *New Scientist* (Oct 29), 6-9 (2005).

27 Talbot, M.: **Darwin in the dock**. *New Yorker* (Dec 5), 66-77 (2005). <> (The usual excellent journalism, with good historical background such as legal precedents, covering the courtroom proceedings over I.D. in the schools of Dover, PA.)

28 Editor: **Judge Jones defines science.** *Science* 311 (Jan 6), 34 (2006). < > (Title continues, "and Why Intelligent Design Isn't." Excerpts three main points from his 139-page ruling: (1) "I.D. violates centuries-old ground rules of science by invoking and permitting supernatural causes," (2) "The argument of irreducible complexity ..." was previously judged as nonsense, (3) "I.D.'s negative attacks on evolution have been refuted by the scientific community.")

29 Kennedy, D.: **Breakthrough of the year: Evolution.** *Science* 310 (Dec 29), 1869 (2005). < > (Kennedy is chief editor. For text, p1878.)

30 Kennedy, D. and Norman, C.: **What don't we know?** *Science* 309 (Jul 1), 75 (2005). < > (To mark the 125th anniversary of AAAS, the editors list the top 125 unsolved problems in science, 25 in detail. The second, after dark matter, is the problem of consciousness. Thirteen of the remaining 23 are in biology, most of which pertain to evolution, e.g. #13 p90 "What determines species diversity?" #14 p91 "What genetic changes made us uniquely human?")

31 Dawkins, R. **The Blind Watchmaker.** New York: W.W. Norton, (1987)

32 Keynes, R. **Annie's Box: Charles Darwin, His Daughter and Human** Evolution: Fourth Estate Press, (2001) < > (Reviewed in *Nature,* 411:739, Jun 14, 2001, by Bruce Weber, who notes that Keynes, although not an historian, descended from Charles Darwin and so has access to rich sources. Reviewer feels that this book nicely complements the two recent major biographies, one by Desmond and Moore, *Darwin, the Life of a Tormented Evolutionist* [1994], the other by Janet Browne, *Charles Darwin: Voyaging* [1996]. The Annie of the title is Darwin's daughter, whose death agonized him. Mentions Darwin's shift in temperament as revealed from his notebooks of 1839 vs. 1859, where he speaks of "Nature's face" in whose ecosystems wedges were constantly being hammered in, causing others to pop out.)

33 Dembski, W. A. **No Free Lunch: Why Specified Complexity Cannot be Purchased without Intelligence:** Rowman & Littlefield, (2002) < > (Reviewed *Nature,* 418:129, Jul 11, 2002, by Brian Charlesworth, in predictably disparaging terms.)

34 Biever, C.: **Creationists seek redesign.** *New Scientist* (Jan 7), 8 (2006). < > (Comments on aftermath of the verdict ruling that ID cannot be taught in Pennsylvania, with remarks by leading losers, William Dembski of the Discovery Institute and Michael Behe, biochemist. Biologist Kenneth Miller, who testified against ID, states that "if there was genuine scientific evidence for intelligent design, it would simply become a part of normal science.")

35 Ruse, M. **The Evolution-Creation Struggle.** Cambridge, MA: Harvard Univ. Press, (2005) < > (Reviewed at length in *Nature,* 437:815 PAGE?, Oct. 6, 2005, by John Hedley Brooks, who remarks that he received a similar insight from an older book,

E.L. Tusevon's *Millenium and Utopia* [Univ. of California Press, 1947]. Also reviewed in *Science,* 309:560, Jul 22, 2005, by S. Sarkar.)

36 Ruse, M. **Monad to Man: The Concept of Progress in Evolutionary Biology.** Cambridge MA: Harvard University Press, (1997) <> (Reviewed by David L. Hull in *Nature,* 385:497, Feb 6 1997; also by F. Ayala in *Science,* 275:495, Jan 24 1997; and widely elsewhere.)

37 Ruse, M. **Can a Darwinian be a Christian? The Relationship Between Science and Religion.** New York: Columbia Univ. Press, (2000) <> (Reviewed in *Nature,* 411:239, May 17, 2001, by Charles L. Harper, along with related book by Arthur Peacock, *Paths from Science Towards God: The End of all our Exploring.* Ruse is described as "the distinguished philosopher of evolutionary biology," while Peacock, described as a biochemist who also trained as a theologian, won the annual Templeton Prize for Progress in Religion. Nearly all of the issues mentioned were fully discussed in the 19th century. The reviewer is worried that as more and more people become more literate and educated in science, they will dispense with religion entirely; but others opine that the opposite is more likely and that they will dispense with science entirely. Review concludes with a quotation from 1889 cited in both books: "Darwinism appeared, and under the guise of a foe, did the work of a friend.")

38 Kosik, K. S.: **Neuroscience gears up for duel on the issue of brain vs. deity** [Letter]. *Nature* 439 (Jan 12), 138 (2006).

39 Brumfiel, G.: **Darwin skeptic says views cost tenure** [news item]. *Nature* 447 (May 24), 364 (2007). <> (Astronomer G. Gonzalez, whose work is highly respected, is being denied advancement (tenure), allegedly because of his Christian faith, which includes a version of ID, and co-authoring the book, *The Privileged Planet.* Physicist Robert Park agrees with the decision, saying "he has established that he does not understand the scientific process.")

40 Monmaney, T.: **Marshall's hunch.** *New Yorker* (Sep 20), 64-72 (1993). (Tells of remarkably rigid resistance to hypothesis that ulcers are caused by infection by *H. pylori,* now established.)

41 Dawkins, R. **The God Delusion.** New York: Bantam / Houghton Mifflin, (2006) <> (Reviewed in *Nature,* 443:914-5, by L.M. Krauss, clearly a Dawkins fan, who describes the book as a "sermon." Also in *New Scientist,* Oct. 7, 2006, by Mary Midgley, who finds it small-minded.)

42 Dennett, D. C. **Breaking the Spell: Religion as a Natural Phenomenon.** New York: Viking / Allen Lane, (2006) <> (Lead book review in *Nature* 439:535, Feb. 2, 2005, by Michael Ruse, under the leader, "A Darwinian philosopher turns his attention to the strength of religion" in the USA. Says "a major plank in Dennett's discussion is that religion is all smoke and mirrors," which strikes me, in view of Dennett's other

books, a case of the pot calling the kettle black. Ruse concludes that this is a worthy topic "but we need better books than this to address the issues." Appears to echo the century-old atheist refrain, that religion is a delusional and irrational state of mind, as opposed to the supposedly clear and truthful vision afforded by the scientific perspective of DD. By the way, Sigmund Freud fancied himself a "true scientist.")

43 Gottlieb, A.: **Atheists with attitude**. *New Yorker* (May 21), 77-80 (2007). <> (Discusses *God is Not Great: How Religion Poison's Everything*, by Christopher Hitchens; *The God Delusion*, by Richard Dawkins; *The End of Faith*, by Sam Harris; *Breaking the Spell; Religion as a Natural Phenomenon*, by Daniel Dennett, and some others.)

Chapter 5.

1 Rickles, R. E. and Schluter, D., eds. **Species Diversity in Ecological Communities**. Chicago, IL: Univ. Chicago Press, (1994).

2 Otte, D. and Endler, J. A., eds. **Speciation and its Consequences**. Sunderland, MA: Sinauer Press, (1989). <> (Reviewed in *Science*, 245:872, 25 Aug 1989, by H.L. Carson. Critically summarizes major issues and controversies, and remarks: "Advances do not come from confusing 'brainstorming' that goes on in big symposia like the one reported here. Needed are crafted books in the tradition of Darwin, Simpson, Mayr, and Dobzhansky." The reviewer finds the book pretentious, and comments that "most 'populationist' authors still consider the origin of 'isolating mechanisms' to be crucial to speciation process. But who said science must advance by majority vote?")

3 Zimmer, C. **Evolution: The Triumph of an Idea**. N.Y.: Harper Collins pub., (2001) <> (Reviewed in *Science*, 293:2209, Sep 21 2001, by T.H. Goldsmith. Book is companion to Public TV 'documentary.)

4 Endler, J. A. **Monographs in Population Biology** 21, (1986) <> (Reviewed *Amer Scientist*, 75:524, Sept/Oct 1987 by D.W. Winkler, under heading, Natural Selection in the Wild. Deals with controversy about natural selection as sole agency of speciation. The reviewer reminds us of related issues, e.g. problem of definition of species, phenotypic *vs.* genotypic selection, etc.)

5 Butlin, R. K. and Tregenza, T.: **Is speciation no accident?** *Nature* 387 (Jun 5), 551-553 (1997). <> (Perspective on article on p589 defending natural selection as the agent of speciation. By the way, notice the title, implying that sexual selection is not an "accident" but is deliberate, as distinct from other causes.)

6 Hendry, A. P.: **The power of natural selection**. *Nature* 433 (Feb 17), 694 (2005). <> ("The primary mechanism of adaptive evolution is natural selection ... Yet perceptions of the power of selection have recently swung at the end of a pendulum." The "power" seems to refer to whether it is strong (fast) or weak (slow, gradual). Refers to

245

classic case of industrial melanism in the moth and Endler's classic, *Natural Selection in the Wild.* Conclusion: "Meanwhile, we are only deluding ourselves that we have a good handle on the typical power of selection in nature.")

7 Mayr, E. **Systematics and the Origin of Species**. N.Y.: Columbia Univ. Press, (1942)

8 Mayr, E.: **Evolution**. *Sci Amer* 239 (#3, Sept), 46-55 (1978).

9 Tautz, D.: **Splitting in space**. *Nature* 421 (16 Jan), 225-226 (2003). <> (Perspective on report p259 by Doebeli and Dieckmann.)

10 Pilkey, O. and Pilkey, L. **Useless Arithmetic: What Environmental Scientists Can't Predict the Future**: Columbia Univ. Press, (2007) <> (Reviewed in *Nature* v447:35 and in *Science* v316:202. Book exposes the follies and fallacies of mathematical models. Focus is on climate change but the same principles apply to modeling biological ecosystems.)

11 Mayr, E. **What Evolution Is**. N.Y.: Basic Books, (2001) <> (Reviewed in *Nature,* 417:223, by Mark Ridley, and in *Science,* 295:50, by Menno Schilthuizen. Intended for a popular audience. Written at age 97.)

12 Coyne, J. A. and Orr, H. A. **Speciation**. Sunderland, MA: Sinauer, (2004) <> (Reviewed *Science,* 305:612, Jul 30, 2004, by B.K. Blackman and L.H. Rieseberg, with accolades. It adopts "a relaxed version of the BSC [biological species concept] that allows for some gene flow among species ..." and "concentrates almost exclusively on reproductive isolation," which to them is "almost synonymous with speciation," following Mayr in holding that speciation is almost exclusively allopatric. However, they accept "without enthusiasm" a few cases of sympatric speciation. It "convincingly presents evidence for several once-unpopular theories that have returned to dominate current thinking. Most important among these is the primacy of natural and sexual selection over drift ..." Includes "the genetics of hybrid incompatibilities" leading the authors to reaffirm the Dobzhansky-Muller model. "Chromosomal rearrangements are now cast as facilitators, rather than causal agents, of reproductive isolation. ... The authors take cautious views on controversial questions like ... sympatric speciation." Authors argue that "shift from outbreeding to selfing is not a kind of reproductive isolation because gene flow is reduced as much within as among taxa." Etc. This book of 557 pgs is said to be unusually up-to-date and "sure to become a classic."

Also reviewed in *Nature,* 431:399, Sep 21, 2004, by Axel Meyer, who calls it a "joyous duet by two of the field's leading soloists." Appears to be the latest and best summary of the field. Reviewer regards this as a major book. However, no clear conclusions – no decisive solution to the problem – is intimated.)

13 Pray, L.: **Mechanisms of speciation**. *The Scientist* 17 (Sep 17), 14-17 (2003). <>

14 Shoemaker, D. D. and Ross, K. G.: **Effects of social organization on gene flow in the fire ant** *Solenopsis invicta*. *Nature* 383, 613-616 (1996 17 Oct). <> (This article addresses "the controversial model of speciation [which] proposes that the development of alternative social organizations within populations of group-living animals may drive the inception of reproductive isolation," i.e. sympatric speciation. The authors examine certain DNA portions of variant social forms in the named ant, "demonstrating the potential for social selection to generate significant barriers to gene flow. ..." Not addressed is how or why "alternative social organizations" might arise.)

15 Mayr, E.: **80 years of watching the evolutionary scenery.** *Science* 305 (2 Jul), 46-47 (2004).

16 Wu, C. I. and Johnson, N. A.: **Endless forms, several powers.** *Nature* 382 (Jul 25), 298-299 (1996). <> (Notes on conference on "Species and Speciation." The dominant topic appears to be sympatric speciation. Raises issue of "how many genes does it take to make a new species?" Speaks of the example of cichlid fishes, also the 26 species of sculpins in Lake Baikal. Inquires "what of evolutionary forces driving speciation – sexual selection in particular?" Another barrier leading to reproductive isolation is differential timing of foods such as fruit maturation. Etc.)

17 Whitfield, J.: **Species, more or less.** *Nature* 422 (6 Mar), 33 (2003). <> (Short note on M. Cardillo *et al* writing in *J Evol Biol* 16:282, 2003, documenting that "the number of species in a genus correlates most closely with the geographical size of the groups' range and with its litter size." No correlation with body size but distribution is non-random. Oldest groups do not contain the most species, contrary to expectation of some.)

18 Hanski, I. and Gyllenberg, M.: **Uniting two general patterns in the distribution of species.** *Science* 275 (Jan 17), 397 (1997). <> (Opens saying "The species-area curve is one of the few universally accepted generalizations in community ecology" but goes on to speak of different theories to explain it.)

19 Rosenzweig, M. L. **Species Diversity in Space and Time.** N.Y.: Cambridge Univ. Press, (1995) <> (Reviewed in *Science* 270:503, Oct 20, 1995, said to be a major work.)

20 Miya, M. and Nishida, M.: **Speciation in the open ocean.** *Nature* 389 (23 Oct), 803 (1997). <> (Evidence is presented of unexpected speciation in this 'sympatric' environment: "We have now found large, localized genetic differences within a circumglobal, monotypic species of dee-sea fish." They conclude: "Numerous questions remain unanswered ...")

21 Schliewen, U. K., Tautz, D. and Paabo, S.: **Sympatric speciation suggested by monophyly of Crater Lake cichlids.** *Nature* 368 (Apr 14), 629-632 (1994). <> (Shows by

DNA analysis that the fish in each lake arose from single ancestral type, demonstrating sympatric speciation, i.e. divergence in absence of isolating boundaries.)

22 Atkinson, N.: **Darwin meets Chomsky**. *The Scientist* (Dec 20), 16-18 (2004). < > (Quotes from Darwin's *The Descent of Man:* "The formation of different languages and of distinct species and the proofs that both have developed through a gradual process are clearly the same."

23 Bradbury, J. W. and Andersson, M. B., eds. **Sexual Selection: Testing the Alternatives.** NY: Wiley, (1987). < > (Reviewed *Amer Scientist* 77:181, Mar/Apr 1989. Proceedings of the Dahlem Workshop. Berlin, 1986.)

24 Tautz, D.: **Selector genes, polymorphisms, and evolution.** *Science* 271 (12 Jan), 160- (1996).

25 Løvtrup, S. **Darwinism: Refutation of a Myth**: Croom Helm, (1987) < > (Reviewed in *Amer Scientist,* 76:394, by Keith Stewart Thomson. This book by a distinguished developmental biologist is apparently highly iconoclastic – probably with some justice. But the reviewer is probably right in remarking that "the reader will find a lot of this rhetoric irritating," even while admitting that a lot of it is on the money. Author is an "avowed epigenetic macromutationist.")

26 Elena, S. F. and Lenski, R. E.: **Test of synergistic interactions among deleterious mutations in bacteria.** *Nature* 390 (Nov 27), 395-398 (1997). (With perspective by S.P. Otto, p343.)

27 Meyer, A.: **Hox gene variation and evolution.** *Nature* 391 (Jan 15), 225-229 (1998).

28 Brooke, N. M., Garcia-Fernandez, J. and Holland, P. W. H.: **The ParaHox gene cluster is an evolutionary sister of the Hox gene cluster.** *Nature* 392 (Apr 30), 920-922 (1998).

29 Sun, S., Ting, C. T. and Wu, C. I.: **The normal function of a speciation gene, Odysseus, and its hybrid sterility effects.** *Science* 305 (Jul 2), 81 (2004). < > (By gene knockout in fruit fly they deduce "the implication of a weak effect of OdsH on the normal phenotype but a strong influence on hybrid male sterility" and this is discussed "in light of Haldane's rule of postmating isolation.")

30 Ueshima, R. and Assami, T.: **Single gene speciation by left-right reversal.** *Nature* 425 (Oct 16), 679 (2003).

31 Rennie, J.: **Kissing cousins: a DNA repair system stops species from interbreeding.** *Sci American* (Feb), 22-23 (1990). < > (Survey of studies with bacteria reported in *Nature,* Nov 1989; in this theory, defects in the DNA mismatch repair system "may help create new species by erecting a reproductive barrier between groups ... with slightly different DNA even if there are no geographical obstacles.")

32 Butlin, R. and Ritchie, M. G.: **Searching for speciation genes**. *Nature* 412 (5 Jul), 31 (2001). <> (Perspective on two recent articles in PNAS, by Ting *et al* and by Doi *et al.* They search for gene loci affecting – "governing" – male and female behavior. However, "the conclusions of the two papers are strikingly different.")

33 Wheeler, D. A. and al, E.: **Molecular transfer of a species-specific behavior from** *Drosophilia simulans* **to** *Drosophilia melanogaster*. *Science* 251 (Mar 1), 1082-1085 (1991). <> (The gene *per* was introduced artificially to transfer the courtship song from one fruit fly species to another.)

34 Wilson, E. O. **Sociobiology: The New Synthesis**. N.Y.: Cambridge Univ. Press, (1975)

35 King, M. **Species Evolution: The Role of Chromosome Change**. N.Y.: Cambridge Univ. Press, (1993) <> (Reviewed under caption "Barriers to Gene Flow" in *Nature*, 366:27, Nov 4, 1993, where it is said that this book was inspired by that of Michael White 15 years before, *Modes of Speciation,* arguing the importance of chromosome change. The reviewer notes that King laments "'a failure to integrate specific findings into a common evolutionary perspective,'" then accuses King of doing the same. Although admitting he makes a good case, "he does not provide arguments for its importance relative to other modes [theories] of speciation.")

36 Garagna, S., Zuccotti, M. and Redi, C. A.: **Trapping speciation**. *Nature* 390 (Nov 20), 241 (1997). <> (Opens by asserting that "chromosome rearrangements are often an important factor in the origin of new species," citing book by M. King, *The Role of Chromosome Change,* Cambridge, 1993. Their evidence is study of local Italian races of mice. By the way, I recall similar more recent study of races of mice living in Chicago.)

37 Berger, S. L.: **Local or global?** *Nature* 408 (Nov 23), 412-414 (2000). <> (Perspective on article p495 this issue in where appears the startling discovery that modifications to the histone proteins, which act like spools around which DNA is compactly wound in the chromosomes, appear to have very widespread effects, genome-wide or "global," not limited to the local region of DNA where the modification occurs. This is a further challenge to the old assumption that each gene acts independently, By the way, it was not long ago that these "spools" were thought to be nothing more than that; it is now known that they play active roles in gene expression.)

38 Pennisi, E.: **Mutterings from the silenced X chromosome**. *Science* 307 (Mar 18), 1708 (2005). <> (Summary of a report in *Nature* that "genes once thought to be silenced in women are sometimes expressed – and that their degree of expression varies from woman to woman.")

39 Kioussis, D.: **Kissing chromosomes**. *Nature* 435 (Jun 2), 579 (2005).

40 Jackson, J. B. C. and Cheetham, A. H.: **Evolutionary significance of morphospecies: A test with Cheilostome Bryozoa.** *Science* 248 (May 4), 579-582 (1990). <> (Addresses question concerning the "punctuated equilibrium" hypothesis, to wit, "are fossil species really biologic species?" and "to what extent can we recognize biologic species using only morphology [appearance]?" Careful microscope study comparing fossil forms with living forms confirms validity of assumptions, hence evidence for the hypothesis "should be taken at face value.")

41 Mitchell, R. and al, E.: **Species concepts [Letter].** *Nature* 364 (Jul 1), 20 (1993). <> (Eleven scientists sign this letter objecting to the "biological species concept," BSC [due to Mayr], as used in a previous article. They admit they "offer no alternative" but seem to suggest "the need for different species concepts for different taxa.")

42 Baumberg, S., Young, J. P. W., Wellington, E. M. H. and Sanders, J. R. **Population Genetics of Bacteria.** N.Y.: Cambridge Univ. Press, (1995) <> (Reviewed *Science,* 270:1233, Nov 17, 1995, where it is noted that this is "only the third [book on this subject] to appear in the last two decades", then asking, "So what's new? A lot! ... Perhaps the most significant conclusion that emerges ... is that bacterial populations are clearly not the homogeneous entities that classical models of evolution in asexual populations would have us believe.")

43 Lucentini, J.: **Sex in the media.** *The Scientist* (Jan 19), 31 (2004). <> (Short note on paper by J.P. Gratia and M. Thiery (*Microbiology,* 149:2571) tending to confirm that bacteria "really do have sex." They "liken the phenomenon to egg fertilization and distinguish it from bacterial conjugation ..." It has been proposed that recombination, the hallmark of sexual reproduction "dates back to life's origins.".)

44 Papke, R. T. et al: **Frequent recombination in a saltern population of** *Halorubrum*. *Science* 306 (Dec 10), 1928-1929 (2004). <> (The first such demonstration in the primitive group known as *Archaea*. From the abstract: "We used multi-locus sequence typing to demonstrate that haloarchaea exchange genetic information promiscuously, exhibiting a degree of linkage equilibrium approaching that of a sexual population.")

45 Brownlee, C.: *Giardia* **bares all.** *Science News* 167 (Jan 29), 67 (2005). <> (Note on paper by J. Logsdon et al writing in Jan 26 issue of *Current Biology*. Shows that this primitive unicellular parasite *Giardia intestinalis* possesses genes for meiosis, used in higher forms for producing germ cells, suggesting they can reproduce in that way, though previously thought to reproduce only asexually.)

46 Wright, S.: **Evolution, organic.** *Encyclopedia Britannica* 8, 917-929 (1968).

47 Rieseberg, L. H., Fossen, C. V. and Desrochers, A. M.: **Hybrid speciation accompanied by genomic reorganization in wild sunflowers.** *Nature* 375 (May 25), 313-316 (1995). <> (For later findings by the same authors, see *Science,* 272:741, May 3, 1996, with comment p700.)

48 Ives, A. R. and Whitlock, M. C.: **Inbreeding and metapopulations.** *Science* 295 (Jan 18), 454 (2002). <> (Experiment on water flea, *Daphnia,* living in tide pools indicated that hybrids were 30% more fit. Full report p485.)

49 Roush, W.: **Hybrids consummate species invasion.** *Science* 277 (Jul 18), 316 (1997). <> (Report on effect of newly introduced crayfish species on established ones in midwestern lakes, resulting in fertile hybrids. The hybrids are proving successful.)

50 Frankham, R. and Ralls, K.: **Inbreeding leads to extinction.** *Nature* 392 (Apr 2), 441 (1998). <> (Perspective on report by Saccheri *et al,* p491, purporting to show a correlation between rate of extinction of locally isolated breeding communities of butterflies with extent of inbreeding.)

51 Davies, K.: **Costs of consanguinity.** *Nature* 371 (Oct 13), 630 (1994). <> (A traditional argument for the harmful effects of consanguineous couplings in humans, buttressed with new data.)

52 McKusick, V. A.: **The royal hemophilia.** *Sci American* (Aug), 88 (1965). <> (Article traces hemophilia through European royalty.)

53 Lawn, R. M. and Vehar, G. A.: **The molecular genetics of hemophilia.** *Sci American* (May), 48 (1986).

54 Cavalli-Sforza, L. L., Moroni, A. and Zei, G. **Consanguinity, Inbreeding and Genetic Drift in Italy.** Princeton, N.J.: Princeton Univ. Press, (2004) <> (Sum of 50 years of research by eminent author, reviewed in *Nature,* 432:554, Dec. 2, 2004, by Walter Bodmer.)

55 Press, A.: **Birth-defect risk lower than expected for kids of cousins.** *Miami Herald* (Apr 4), 23A (2002). <> (Report of study in April issue of *Journal of Genetic Counseling* showing that risk of defects is normally 3% to 4% and that cousin marriage adds 1.7% to 2.8% to this, but may be highly variable depending on genetics of the families.)

56 Thornhill, N. W., ed. **The Natural History of Inbreeding and Outbreeding: Theoretical and Empirical Perspectives.** Chicago, IL: Univ. Chicago Press, (1993). <> (Reviewed in *Science,* 263:107-108, Jan 7, 1994, with praise.)

57 Caro, T. M. and Laurenson, M. K.: **Ecological and genetic factors in conservation: a cautionary tale.** *Science* 263, 485-586 (1994). <> (Refutes the supposition that inbreeding – or low genetic polymorphism – is harmful, for the example of cheetahs. They find the population is endangered mainly because of lion and hyena predation, not because of inbreeding or lack of disease resistance.)

58 Levy, S.: **Die hard.** *Nature* 434 (Mar 17), 268 (2005). <> (Study of mona monkeys on the island of Grenada, West Indies, after decimation by Hurricane Ivan. It is noted

that the whole population is descended from just a few founders, possibly just one pregnant female brought by a sailor on the slave trade in the 1700's, but rose to become a thriving population despite absence of genetic diversity.)

59 Reporter: **In Sweden, the beavers are all cousins.** *Sci News* Oct 9, 235 (1993). <> (News item referring to Ellegran *et al* writing in Sept 1, 1993 issue of PNAS. "Although biologists have long thought that a species needs genetic variation in order to thrive, beavers are proving this assumption wrong." Reason is that beavers in Sweden went extinct but conservationists reintroduced a small few in 1920 and they are now thriving despite very limited genetic diversity. Effort is made to explain how this could possibly be but sounds lame.)

60 Visscher, P. M. et al: **A viable herd of genetically uniform cattle.** *Nature* 409 (Jan 18), 303 (2001).

61 Hurst, L. D. and McVean, G. T.: **...and scandalous symbionts.** *Nature* 381 (Jun 20), 650-651 (1996). <> (Reviews theories of the supposed need for sexual exchange to prevent extinction, in light of a bacterium adapted to the gut of aphids, without benefit of sexual exchange, which has nevertheless survived for about 100 million years, in conflict with theoretical expectation.)

62 Gaidos, S.: **Keeping it in the family.** *New Scientist* (Aug 13), 41-43 (2005). <> (Reviews recent findings that "inbreeding may not be such a bad thing. ... Such a view would once have been heresy.")

63 Wake, D. B.: **Speciation in the round.** *Nature* 409 (Jan 18), 299 (2001). <> (Perspective on article by Irwin *et al*, pg 333. One view holds that speciation can occur in the ring around the range of a population – ring species – where hybridization is most likely. Article tends to confirm that a classic example – a warbler of the Tibetan plateau – does in fact "meet most of the requirements of the classic model.")

64 Harrison, R. G., ed. **Hybrid Zones and the Evolutionary Process.** NY: Oxford Univ. Press, (1993). <> (Reviewed in *Science,* 264:727, 29 Apr 1994.)

65 Hasselquist, D.: **Hybrid costs avoided.** *Nature* 411 (May 3), 34 (2001). <> (Perspective on article by Veen et al, pg 45, on hybrid theory of speciation, examining two closely related bird species, pied and collared fly-catchers, based on large data set. Too complicated to briefly summarize, but conclusion is "the longer two species come into contact, mixing and exchanging genes, the more likely it is that major and rapid evolutionary changes will occur.")

66 Butlin, R. M., and Neems, R. M.: **Hybrid zones and sexual selection.** *Science* 265 (1 Jul), 122 (1994). <> (Technical comment critical of a report in a prior issue, with response. Deals with distance gradation of secondary sexual plumage of manakins in the "hybrid region.")

67 Enserink, M.: **Life on the edge: Rainforest margins may spawn species.** *Science* 276 (Jun 20), 1791-1792 (1997). < > (Introduces report on p1855 defending the theory that speciation is promoted at the interface between two geographical types.)

68 Stearns, S. C. **The Evolution of Life Histories.** N.Y.: Oxford Univ. Press, (1992) < > (This was cited in a previous chapter; speaks of a "final synthesis" in the making.)

69 Carroll, S. B. **Endless Forms Most Beautiful: the New Science of Evo Devo and the Making of the Animal Kingdom.** N.Y.: W.W. Norton & Co., (2005) < > (Reviewed in *Science,* 308:955, May 13, 2005, by Denis Deboule; and in *Nature,* 435:1029, June 23, by Jerry A. Coyne. Both reviews are respectful but luke-warm, yet agree that this is a good introductory survey of this approach.)

70 Jablonka, E. and Lamb, M. L. **Evolution in Four Dimensions: Genetic, Epigenetic, Behavioral, and Symbolic Variation in the History of Life:** Bradford Books, (2005) < > (Lead review in *Nature,* 435:565, June 2, 2005, by Massimo Pigliucci, who cites four prior books, including his own, leading up to this one, representing a new New Synthesis. One must distinguish between a synthesis of ideas and a hodge-podge.)

71 Ren, D.: **Flower-associated Brachycera flies as fossil evidence for Jurassic angiosperm origins.** *Science* 280 (Apr 3), 85-89 (1998). < > (Fossil evidence given to indicate that flowering plants arose hand-in-hand with their pollinators. For background, see p57, "How Old is the Flower and the Fly?")

72 Moore, P. D.: **Insect pollinators see the light.** *Nature* 387 (19 Jun), 759-760 (1997).

73 Heinrich, B.: **Of Bedouins, beetles and blooms.** *Natural History* (May), 52-59 (1994). < > (Account of studies by Avishai Shmida on plants in the Judean desert, the "poppy guild." Includes remarkable examples of plant mimics, such as tulips and buttercups that mimic a poppy and an orchid that looks and smells like a female bee to attract its pollinator. It appears that the many plant species which have become red here have done so by "imitating one another" to sexually arouse and attract certain beetle pollinators.)

74 Margulis, L. and McMenamin, M., eds. **Concepts of Symbiogenesis. A Historical and Critical Study of the Research of Russian Botanists [L N Khakhina].** New Haven, CT: Yale Univ. Press, (1993). < > (Reviewed in *Science,* 261:628, Jul 30, 1993, with perspective on western advocates Wallin, Margulus.)

75 Boucher, D. H., ed. **The Biology of Mutualism.** Cambridge, MA: Oxford Univ. Press, (1985). < > (Reviewed *Amer Scientist,* 75:537, Sep/Oct 1987.)

76 Brooks, D. R. and McLennan, D. A. **Parascript: Parasites and The Language of Evolution.** Washington, DC: Smithsonian Institution Press, (1993) < > (Lead review in *Science,* 261:927, Aug 13, 1993.)

77 Douglas, A. E. **Symbiotic Interactions**. N.Y. / London: Oxford Univ. Press, (1994) <> (Reviewed in *Nature,* 371:570, Oct 13, 1994, by D.L. Hawksworth. Reviewer refers to a 1987 book by this author with Sir David Smith, *The Biology of Symbiosis.* Said to be packed with fascinating examples, but reviewer not satisfied with efforts at definitions of terms.)

78 Smetacek, V.: **A watery arms race.** *Nature* 411 (Jun 14), 745 (2001). <> (Essay. Focus is on defenses of plankton, and theme is that they are shaped by defense rather than competition. However, he also broaches larger issues, concluding by wondering why the ocean is not entirely covered by a thick scum of green algae, which is certainly a plausible outcome, consistent with present theory, i.e., existing theory gives us no idea of why that did not happen as a stable and final outcome.)

79 Jr, E. G. L. and Ziegler, C. **A Magic Web**. New York: Oxford Univ. Press, (2003) <> (Reviewed in *Nature,* 424:132, July 10, 2003, by Lawrence Sack and Dina Dechmann. The authors, tropical biologists, spent 15 months on a protected island in Panama making the splendid photographs in this book, intended to acquaint general readers with the intricate web of ecological relationships in the forest. Reviewer describes it as far more than a coffee-table book. "The book will enrich any scientist's view of biology. The tropical biologist will crave returning to the field." Though I am not a tropical biologist, this book makes me wish I was.)

80 Sudd, J. H. and Franks, N. R. **The Behavioral Ecology of Ants**. N.Y.: Chapman and Hall, (1987) <> (Reviewed in *Amer Scientist,* 76:406, Jul/Aug 1988, by N.F. Carlin: "From the outset, ant sociality is presented as an evolutionary enigma to be resolved in terms of the workers' proxy reproductive success." Reviewer notes that this book has been deeply influenced by "Hamilton's landmark proposal that hymenopteran workers are more closely related to their sisters than to their own offspring, and that they increase the propagation of their genes by aiding the reproduction of their mother queen. ... is therefore quite different in its interpretation as compared with Sudd's book on ants of 1967.")

81 Holldobler, B. and Wilson, E. O. **The Ants**. Cambridge, MA: Balknap / Harvard Univ. Press, (1990) <> (Widely reviewed including in *New York Times* Book Reviews, July 29, 1990, by T.E. Lovejoy. All agree it is the definitive book on the subject.)

82 Paulson, G. and Akre, R. D.: **A fly in ant's clothing.** *Natural History* 1/94 (Jan), 57-58 (1994).

83 Becerra, J. X.: **Insects on plants: macroevolutionary chemical trends in host use.** *Science* 276 (Apr 11), 253-256 (1997). <> (Argues that the "historical patterns of host shifts strongly correspond to the patterns of host chemical [defense] similarity, indicating that [the evolution of] plant chemistry has played a significant role in the evolution of host shifts by phytophagous insects.")

84 DeVries, P. J.: **Singing caterpillars, ants, and symbiosis.** *Sci American* (Oct), 76-81 (1992). <>

85 Moore, P. D.: **Green policies for defense spending.** *Nature* 391 (Feb 26), 838-839 (1998). <> (Survey of recent studies on the endless 'weapons race' between plants and herbivores.)

86 Handel, S. H. and Beattie, A. J.: **Seed dispersal by ants.** *Sci American* (Aug), 76-83 (1990).

87 Lucentini, J.: **Secondary endosymbiosis exposed.** *The Scientist* (June 6), 22-23 (2005). <> (Excerpt: "Beneath their sometimes striking enclosures, diatoms and many other algae reveal a sordid past of endosymbiotic events. They are, in essence, three organisms in one – or more precisely, one within another within another.")

88 Bushman, F. D.: **Evolutionary teamwork.** *The Scientist* (May 10), 33-34 (2004). <> (Begins: "The endosymbiotic theory, which posits that organelles such as mitochondria and chloroplasts descended from formerly free-living cells, has received wide acceptance ... But recent findings suggest that endosymbiotic process may have contributed still more cellular components." Like the cell nucleus.)

89 Martin, W. et al: **Gene transfer to the nucleus and the evolution of chloroplasts.** *Nature* 393 (May 14), 162-165 (2002).

90 Sagan, L. M.: **On the origin of mitosing cells.** *Journal of NIH Research* 5 (Mar), 65-72 (1993). <> (This was cited with notes in a previous chapter.)

91 Doolittle, W. F.: **A paradigm gets shifty.** *Nature* 392, 15-16 (1998 5 Mar). <> (It has long been accepted that mitochondria, which in part characterize eucaryotic cells, originated as endo-symbiotic bacteria [put forth by Margulis some 30 years ago], recently supported by molecular data. However, many questions remain, such as how and when actin filaments, microtrubes, etc., were also acquired by eucaryotes; and what were these cells before they were eucaryotes?" Piece supplies background for the report on p37 this issue by Martin and Muller in which a novel theory is tendered.)

92 Dyer, B. D. and Obar, R. A. **Tracing the History of Eukaryotic Cells: The Enigmatic Smile.** NY: Columbia Univ. Press, (1994) <> (Reviewed in *Nature,* 369:451, 9 Jun 1994, by L.D. Hurst, under caption, "A penchant for protists." By the way, the reviewer concurs with Dyer and Obar in his admiration for the 1991 book edited by Margulis *et al,* "Handbook of Protoctista," which he describes as "possibly her second great contribution to science." We note this by way of defending her status as a major biologist, opposing the ridicule to which she has often been subjected because of her 'radical' views – meaning her rejection of many orthodox evolutionary dogmas.)

93 Editor: **Lynn Margulis [Interview].** *The Scientist* (Jun 30), 11 (2003). <> (Interview with this "maverick" tries to find out what irks her, and several things stand out, first,

rule of science by fashion, second, erosion of classical learning, third, major scientists overlooked, and fourth, "I am so frustrated by the lack of attention given to slime molds, all the algae, all the traditional protozoa ... they are all protocxtists. ... Nobody sees the importance" of these creatures. She prefers students who are not straight-A, saying "real scientists ... had to rebel against their teachers.")

94 Margulis, L. and Sagan, D. **What Is Life?** N.Y.: Simon & Schuster, (1995) < > (Reviewed in *Nature*, 379:409, Feb 1, 1996, by Christian de Duve, who (predictably) finds it loaded with corny sentiments and vague metaphors, and objects to the "philosophical message," that purpose, choice and consciousness are innate to all life and are factors in evolution; and that some elements of learning may become heritable, favoring the arguments of Lamarck and Samuel Butler. My reading of the book is favorable, except that she overlooks Lotka.)

95 Margulis, L. and Sagan, D. **Acquiring Genomes: A Theory of the Origin of Species.** N.Y.: Basic Books, (2002) < > (Reviewed *Nature*, 418:275, July 18 2002, by Axel Myer, under headline, "Viewing life as cooperation: can symbiosis and genome acquisition account for all speciation?" Reviewer reports "they argue that all speciation events are caused by symbiosis, cooperation and the reticulation of genomes, questioning some of Charles Darwin's central idea[s]." Meyer is intrigued but doesn't buy it.)

96 Margulis, L.: **Essay (on kefir).** *Sci American* 271 (Aug), 96 (1994). < > (Margulis here explores a 'pre-sexual society of microbes' in the curds of kefir, a nutritious drink popular in the Caucasus of southern Russian and Georgia, made by fermenting milk with a traditional culture of yeast and other microbes, about 2,000 years old, analogous to cultures for sourdough bread. "After the curds die, kefir individuals become an arbitrary mix of fermenting microbes rather than the specific combination of bacteria and yeasts that form each curd." She goes on to cogently speculate on the rise of sexuality, and concludes by saying that "Kefir is a sparkling demonstration that the integration processes by which our cells evolved still occur.")

97 Ryan, F. **Darwin's Blind Spot: Evolution Beyond Natural Selection.** N.Y.: Houghton & Mifflin / Texere, (2003) < > (Reviewed *Nature*, 421:579, Feb 6, 2003, by Steven A Frank, under wry headline, "Mergers and Acquisitions," because the book proposes that symbiotic relationships, along with horizontal gene transfer, are *the major* explanation for species and evolution. "Ryan has built an exciting story of heroic outsiders and fierce conflicts over the nature of evolutionary innovation" but review goes on to suggest that Ryan has exaggerated the importance of symbiosis and sensationalized the schisms among evolutionists. Frank sees Ryan as a "radical symbiologist.")

98 McFall-Ngai, M. J. and Ruby, E. G.: **Symbiont recognition and subsequent morphogenesis as early events in an animal-bacterial mutualism.** *Science* 254 (Dec 6), 1491-1494 (1991). < > (This article, featured on the cover, details how the light

organ of a certain squid – and presumably all squid and other creatures whose luminous organs depend on bacterial symbionts – arises in the juvenile squid host following infection, causing major morphological and other changes, "suggesting that the initiation of symbiosis influences, and is perhaps a prerequisite for, the normal developmental program of the ... host.")

99 Boekhout, T.: **Gut feeling for yeasts**. *Nature* 434 (Mar 24), 449 (2005). < > (Summary of report in *Mycological Reseach* revealing the presence of some 200 new species of fungi in the guts of certain beetles.)

100 Bry, L., Falk, P. G., Midvedt, T. and Gordon, J. I.: **A model of host-microbial interactions in an open mammalian ecosystem**. *Science* 273 (Sep 6), 1380-1383 (1996). < > (Approaches the difficult "ecology" of intestinal microflora, which in humans "involves more than 400 bacterial species," by raising mice under germ-free conditions and introducing specific species to them.]

101 Dawkins, R. **The Selfish Gene**. New York: Oxford Univ. Press (2nd ed'n; 1st 1976), (1989) < > (This was cited with notes in an earlier chapter.)

102 Morell, V.: **Genes vs. teams: weighing group tactics in evolution**. *Science* 273 (Aug 9), 739-740 (1996). < > (Highlights from upcoming symposium. "Although the idea of group selection was discredited more than 30 years ago, a growing number of researchers say that it deserves a fresh hearing." She recounts the attack on group selection theory as set forth in V.C. Wynne-Edwards' book of 1962, by George C. Williams in his 1966 book, *Adaptation and Natural Selection,* after which it was ignored, until the current revival – or rather, revision, since the weaknesses in Wynne-Edwards are now avoided. She goes on to explain the ideas of current advocates like David Sloan Wilson, Leo W. Buss, Thomas D. Seeley, and Richard E. Michod, but mentions that others remain skeptical, such as Jerry Coyne, pointing to semantic ambiguities among other problems.)

103 Harris, M. **The Rise of Anthropological Theory**. N.Y.: Thomas Y. Crowell & Co. (1st edn., 1968)

104 Behrensmeyer, A. K.: **Climate change and human evolution**. *Science* 311 (Jan 27), 476-478 (2006). < > (She discusses the "challenge" of proving this hypothesis but seems to believe in it. On pg 472 of the same issue, by the way, is a review of a book claiming the exact opposite, that prehistoric humans had "early and profound" impact on climate, even averting an ice age!)

105 Steinberg, D.: **How did natural selection shape human genes?** *The Scientist* (May 10), 28-30 (2004). < > (Review of work purporting to show that climate was the main agent of hominid evolution, in particular, mitochondria. Source papers include *Science* 303:323.)

106 Klicka, J. and Zink, R. M.: **The importance of recent ice ages in speciation: A failed paradigm.** *Science* 277 (Sep 12), 1666-1669 (1997). <> (Shows by DNA analysis of 35 North American songbird species almost continuous change, with no evidence of ice-age associated bursts or declines.)

107 Rosenzweig, M. L.: **Tempo and mode of speciation.** *Science* 277 (Sep 12), 1622-1623 (1997). <> (Commentary on article p1666 this issue by Klicka and Zink on timing of songbird speciation but with broader perspective, pointing out that this speciation event occurred during the climate shift *before* an ice ace, not during or after. He argues persuasively that speciation in time is fractal – "like the jagged edges of a coastline" – in time.)

108 Schwartz, J. H. **Sudden Origins: Fossils, Genes, and the Emergence of Species.** N.Y.: John Wiley & Sons, (1999) <> (Reviewed in *Nature*, 399:745, Jun 24, 1999, by Eors Szathmary, who remarks that the author "wants to convince the reader that his 'new evolution' is the Holy Grail of the field." Author sets forth series of basic principles but the essence of his theory is macromutations as the cause of new species. Reviewer is not impressed, except by the fascinating wealth of history also given in the book. Was also reviewed briefly in *Sci Amer*, Sept 1999 p103. "Mutations affecting homeobox genes, Schwartz argues, can have dramatic effects, giving rise to new species." Author is anthropologist. "The mutation would take generations to spread, yet the novel feature would appear suddenly in multiple individuals, which sounds like the classical conservative view. The reason that species remain reproductively separate, he suggests, may be that individuals recognize potential mates on the basis of similar new traits" – i.e. sexual selection. Does not sound highly original, but probably makes good reading. Now in paperback from Wiley.)

109 Brown, A. **The Darwin Wars: How Stupid Genes Became Selfish Gods.** N.Y.: Simon and Schuster, (1999) <> (Reviewed in *Nature*, 398:385, Apr 1, 1999, by David L. Hull, under caption, "Evolutionists red in tooth and claw." Said to be a journalist's account of the many rifts among leading evolutionary theorists in the last 25 years, many of them quite acerbic and personal. Hull judges it to be well written and well researched.)

Chapter 6

1 Morell, V. and Gibbons, A.: **On the many origins of species.** *Science* 273 (Sep 13), 1496-1502 (1996). <> (Pithy 7-page report of highlights of a conference, "Endless Forms: Species and Speciation" – "endless forms" being a quotation from the last page of Darwin's *Origin*. Surveys many different approaches, theories, doubts, and controversies, particularly those surrounding sympatric speciation.)

2 Pray, L.: **Mechanisms of speciation.** *The Scientist* 17 (Sep 17), 14-17 (2003).

3 Schwarz, D. et al: **Host shift to an invasive plant triggers rapid animal speciation.** *Nature* 436 (Jul 28), 546-549 (2005). <> (Apparently new *Rhagoletis* fruit fly species in North America whose maggots are adapted to a brushy honeysuckle plant, *Lonicera,* brought from Asia about 250 years ago. Story is similar to the classical apple maggot except the authors find this new species to be a hybrid of the blueberry maggot and the snowberry maggot. The hypothesis given is that the original hybrid freed it from competition with congeners, but it is easy to think of alternative scenarios, such as that the hybridizing copulation took place on the new host. That seems more convincing than the explanation given by Schwartz' interview in *New Scientist* (Jul 30, 2005, p12): "There must be something to hybridisation that makes them colonize honeysuckle.")

4 Morell, V.: **Sex frees virus from genetic 'ratchet'.** *Science* 278 (Nov 28), 1562-1563 (1997). <> (The article indicates that Chao's experiments help settle another dispute in evolutionary theory, that between the stepwise-advance model of R.A. Fisher, and the topographical landscape model of Sewell Wright. As might be expected, his experiments indicate that both have merit, for after all, each is just a slightly different conceptual metaphor. This is another of many examples that could be cited showing that evolutionists often magnify minor issues into high-profile arguments.)

Ch 7 Refs & Notes

1 Darwin, C. **The Descent of Man and Selection in Relation to Sex.** London: John Murray (2nd ed'n), (1874)

2 Bradbury, J. W. and Andersson, M. B., eds. **Sexual Selection: Testing the Alternatives.** NY: Wiley, (1987). <> (Reviewed *Amer Scientist* 77:181, Mar/Apr 1989. Proceedings of the Dahlem Workshop. Berlin, 1986.)

3 Andersson, M. **Sexual Selection.** Princeton, NJ: Princeton Univ. Press, (1994) <> (Reviewed in *Science* 267:712, Feb 3, 1995, by M.J. Ryan; and in *Nature* 371:28, Sep 1, 1994, by J.L. Gould. Supplies 2000+ references.)

4 Kodric-Brown, A. and Brown, J. H.: **Why the fittest are the prettiest.** *The Science* (Sep/Oct), 26-33 (1985). <> (The Browns open with background on the "Chicago School" of followers of Sir Ronald Fisher's theory, holding that exaggerated male display attributes result from "runaway sexual selection." They then critique Zahavi's handicap principle, and finally explain why their own theory is right. Their conclusions were inspired by observing desert pupfish, whose males acquire the brilliant blue color favored by females only if they have triumphed in securing a breeding territory.)

5 Atkinson, N.: **Speciation's defining moment.** *The Scientist* (Apr 11), 30-33 (2005). <> (Subtitle: "Genetics and genomics enliven an old controversy on reinforcement."

Quote: "The whole theory of evolution rests on the notion that species can split and diverge, but the precise mechanisms for the process have been difficult to discern. Reinforcement has long been suspected to be key, but empirical evidence has lagged behind the algebraic march." Piece is an effort to briefly survey some current controversies – actually the same old ones – about sexual selection, with sound-bite interviews with some of the leaders. No clear conclusions.)

6 Hoskin, C. J. and al, E.: **Reinforcement drives rapid allopatric speciation.** *Nature* 437 (Oct 27), 1353-1356 (2005). <> (Nice study of a "mosaic contact zone of two lineages of a rainforest frog." The word 'allopatric' in the title may be misleading since, if I understand correctly, the authors also observe sympatric splitting.)

7 Whitfield, J.: **Flowers to fit the bill.** *Nature* 423 (MAY 1), 30 (2003). <> (Note on report in *Science* 300:630, that males and females of the purple throated carib hummingbird have very different beaks, each for sipping nectar from a different species of *Heliconia* flower, for which they are the sole pollinator.)

8 Henry, R. C.: **The mental universe {Essay}.** *Nature* 436 (Jul 7), 29 (2005). <> (Points out how Galileo and others transformed public awareness but says "the more recent physics revolution of the past 80 years has yet to transform public understanding.")

9 Huxley, J. **Man in The Modern World.** N.Y.: Mentor Paperback / New Amer Library, (1948)

10 Monod, J. **Chance and Necessity.** N.Y.: Vintage Books / Random House, (1972) <>

11 Orians, G. H. **Blackbirds of the Americas.** Seattle, WA: University of Washington Press, (1987) <> (Reviewed *Amer Scientist* 75:206, Mar/Apr 1987, by W.A. Searcy, who speaks of author as "our foremost expert on the blackbirds. ... The approach [of the book] is to assume that all traits are the products of natural selection working at the individual level; given this assumption, the crucial step in analyzing the evolution of a trait is to determine its costs and benefits to the individuals possessing it. This method has been criticized as Panglossian, but it has also proved remarkably heuristic." Reviewer remarks that this book is also visually stunning for its superb illustrations.)

12 Funk, D. H.: **The mating of tree crickets.** *Sci American* (Aug), 50-59 (1989). <>

13 Mendelson, T. C. and Shaw, K. L.: **Rapid speciation in an arthropod.** *Science* 433 (Jan 27), 375 (2005). <> (The highest rate so far recorded for an arthropod: mean value, 4.2 species per million years, for crickets on Hawaiian islands, said to be "explosive and ongoing." Many seem to differ only in their mating songs.)

14 Palumbi, S. R.: **A star is born.** *Nature* 390 (Dec 11), 556 (1997). <> (Details the phlyogenetic relationships among 11 species of a starfish genus, and notes the diversity in reproductive modes: "Some produce tiny eggs that develop first into free-swimming

larvae before transforming into tiny crawling stars. Others produce ... eggs ... that develop directly into starfish. Some adults brood their embryos in the goad where cannibalistic juveniles fight their siblings for the right to be born; others cast their eggs anonymously into the sea. Yet all these species are closely related ...")

15 Bauer, R. T. and Martin, J. W., eds. **Crustacean Sexual Biology.** N.Y.: Columbia Univ. Press, (1991). <> (Reviewed in *Amer Scientist,* 81:290, May/Jun 1993. The best known crustaceans "are the decapods [shrimps, crabs, lobsters] ..." but many other taxa are included among these papers from a symposium. Their sexual habits are known to be highly diverse, but much less is known about them than their cousins, the insects, for the simple reason that they live underwater and are harder to study. But this volume implies that they are equally diverse.)

16 Conniff, R.: **When it comes to moths, nature pulls out all the stops.** *Smithsonian* 26 (Feb), 68-81 (1996). <> (Some 12,000 species in North America alone. Huge range of diets including one that sucks blood. Their caterpillars are similarly diverse and cryptic. The emerald green looper moth has a spring brood of caterpillars mimicking flowers and a later brood posing as oak twigs, adapted for eating tougher vegetation, so distinctive it was previously thought to be a different species. Their pheromones and toxins often or usually derive from host plants. Male moths produce scents too, as they approach females, often resembling jasmine or cinnamon. Some spiders copy moth pheromones to lure moths. In some, female judges male suitor by toxin package presented, to her, to defend her young. Moths can hear bats from some distance, taking evasive action; some send out ultrasound signals to 'jam the radar' of bats - or convey that they are poisonous. See also p129 of issue.)

17 Eberhard, W. G. **Sexual Selection and Animal Genitalia.** Cambridge, MA: Harvard Univ. Press, (1985) <> (Reviewed *Amer Scientist* 75:310, May/Jun 1987, by T. Ghiselin.)

18 Eberhard, W. G. **Female Control: Sexual Selection by Cryptic Female Choice.** Princeton, NJ: Princeton Univ. Press, (1996) <> (Reviewed in *Science,* 275:1075, Feb 21, 1997, by R.H. Wiley, who calls this a "mesmerizing review of the intricacies of mating by diverse animals." A major theme of the book is said to be the many ways in which females affect fertilization after copulation.)

19 Caldwell, J. P.: **Pair bonding in spotted poison frogs.** *Nature* 385 (Jan 16), 211 (1997). <> (Opens by saying, "once thought to interact very little with conspecifics, amphibians are now known to have elaborate mating systems." Goes on to detail how these frogs carefully care for their tadpoles in tiny pools which they dig for the purpose.)

20 Murphy, J. C. **Amphibians and Reptiles of Trinidad and Tobago:** Krieger, (1997) <> (Details nine distinct reproductive methods used by frogs on these islands, and

"the diversity of reptilian and amphibian life in general on these two islands ..." Cited with photo in *Nature,* 388:842.)

21 Pooley, A. C. and Gans, C.: **The nile crocodile.** *Sci American* Apr (Apr), 114- (1976). < > (Demonstrates tender care of the hatchlings, and unexpectedly complex social order, in sharp contrast to other related species.)

22 Ryan, M. J. and Rand, A. S.: **Female response to ancestral advertisement calls in Tungara frogs.** *Science* 269 (Jul 21), 390-392 (1995). < > (Approach was to genetically estimate ancestral distances among 8 related frog species in Venezuela, then see how females responded to calls of ancestrals compared to conspecifics. The clearest conclusion is not in the rather hazy abstract but is on p392: "Thus, we suggest that in this system, stimuli that elicit recognition are widespread throughout the history of these frogs, but that speciation usually results in addition or fine-tuning of stimuli effecting discrimination in favor of conspecific calls ..." A weakness of the study appears to be that only 2 of the species were sympatric, an important variable because widely separated species have no need to distinguish their calls. This may account for some of the ambiguity.)

23 Milius, S.: **Well-tuned bats.** *Science News* 165 (Jun 12), 373 (2004). < > (Describes work by T. Kingston and S. Rossiter in the June 10 issue of *Nature,* indicating that three groups of the large-eared horseshoe bat on certain Indonesian islands share the same range but employ different complex signaling patterns, and seem to be in the process of splitting into distinct species. In the same issue, B. Simmers and H. Schnitzler report similarly on five subspecies of bats in Germany, and suggest they are diverging into species due to distinctive communication channels and slightly different hunting practices.)

24 Staff. **Behavioral neuroscience uncaged.** *Science* 306 (Oct 15), 432-434 (2004). < > (Highlights from a meeting on neuroethology, as on communication by vibrational signals, magnetic direction finding in birds, and signals from electric fish. It was known that various species use their electric discharges for multiple purposes such as "radar", for stunning or confusing prey, and communication. What is new is the finding that pulse frequency and waveshape may also communicate social status and may be a mating signal, being distinctive for each species, many of which are found in the same habitat. The author investigated the biophysics of the discharges, finding that each species has distinctive sodium channels in the electric organ.)

25 Milius, S.: **Electric fish may jam rival's signals.** *Science News* 168 (Nov 19), 324-325 (2005). < > (Note on article in upcoming issue of *Animal Behavior* on the brown knifefish, which creates a weak electric field used for navigation and communication, distinctive for each individual. Evidence shows that in disputes, one may jam the signals of another.)

26 Xavier, K. B. and Bassler, B. L.: **Interference with AI-2-mediated bacterial cell-cell communication.** *Nature* 437 (Sep 29), 750-753 (2005). <> (AI-2 is a group of chemicals produced by bacteria for communication, some of which are species-specific, others being more like a universal language. Some authors have spoken of similar chemical signals, such as in yeasts, as "pheromones" since they appear to be involved in sex-like conjugation. From the abstract: "Here we show that some species of bacteria can manipulate AI-2 signaling and interfere with other species' ability to assess and respond correctly.")

27 Ranz, J. M. et al: **Sex-dependent gene expression and evolution of the *Drosophilia* transcriptome.** *Science* 300 (Jun 13), 1742-1745 (2003). <> (Compared two "nearly identical" fruit fly species and found that most of the differences were in males and sex-linked. States that molecular changes associated with sex and reproduction changed faster than those restricted to survival. Their data suggest that visual cues are more important for *D. simulans* (which do not mate in darkness) while olefaction genes are more prominent in *D. melanogaster,* in which cuticular hydrocarbons [pheromones] may be more important in courtship and are distinctive between these species. They do not discuss genes involved with fruit fly courtship dance and song, but do conclude that sexual selection was the major factor in the species divergence, and that most of the difference is in males.)

28 Holden, C.: **Secret species nabbed in DNA sweep.** *Science* 306 (Oct 8), 224 (2004). <> (Short note on papers in *PNAS* and *PloS Biology* about status of effort to 'bar-code' species by sequencing certain portions of DNA. A spin-off of this effort has been the accidental discovery of many "cryptic species," i.e., which look alike yet have different (assortative) mating preferences. Nine new species of Costa Rican butterflies were found, as well as 'sister species' of four American birds: marsh wren, warbling vireo, eastern meadowlark, solitary sandpiper.)

29 Hobson, K. A.: **Incredible journeys.** *Science* 295 (Feb 8), 981-983 (2002). <> (Perspective on paper by Rubenstein et al, p1062 this issue. Application of isotope ratios to determine that migration patterns of the black-throated blue warbler in North America actually consists of two distinct populations with different migratory habits, one group being from northern U.S. and wintering in the western Caribbean, the other from southern U.S. wintering in the eastern Caribbean. It is implicit that the two populations preferentially inbreed.)

30 Dupre, J., ed. **The Latest on the Best: Essays on Evolution and Optimality.** Cambridge, MA: MIT Press, (1988). <> (Reviewed in *American Scientist,* Jan/Feb 1988, p105. Essays by well-known leaders on controversies in "optimality," a basic tenet in evolutionary theory, to the effect that organisms *optimize* their energies and resources in trade-offs among and between conflicting interests. No surprise, that.)

31 Coleman, S. W., Patricelli, G. L. and Borgia, G.: **Variable female preferences drive complex male displays.** *Nature* 428 (Apr 15), 742-745 (2004). <> From

the abstract: "We suggest that complex male displays might also arise because of variations in female preferences for particular male display traits." However, it seems equally likely that variations in female preferences occur "because of" the variety of innovative displays presented to them by the males. Article notes that female tastes change with age – mature? The male is attempting to appeal to an audience with diverse tastes. See also perspective, p708, *Fickle females?*)

32 Hurst, L. D. and Pomiankowski, A.: **The eyes have it**. *Nature* 391 (Jan 15), 223-224 (1998). < > (Article featured on cover. Supplies background to a report by Wilkinson *et al*, p276 this issue. Rationality is more complex than we could explain in text, involving also selfish gene theory and meiotic drive.)

33 Roush, W.: **Sizing up dung beetle evolution**. *Science* 277 (Jul 11), 184 (1997). < > (Summary of report by Emlen and Nijhout showing that horn size in this beetle, which is a criterion of sexual selection, is apparently genetically linked to eye size, so that bigger horns entail smaller eyes, yet diet also a plays a role. This builds on earlier studies of that beetle by Emlen et al, see *Nature,* 369:359, Jun 2, 1994, on report in *Proc R Soc,* showing that horn length depends on nutrition, not genes, and is thus an 'honest' indicator of fitness.)

34 Young, K. V., Brodie Jr., E. D. and Brodie Sr., E. D.: **How the horned lizard got its horns**. *Science* 304 (Apr 2), 65 (2004). < > (Shows that longer horns protect against predation by loggerhead shrikes. Article drew letters in issue of Sep 24 (305:1909) generally praising but raising questions of interpretation; with response.)

35 Pennisi, E.: **Colorful males flaunt their health**. *Science* 300 (Apr 4), 29 (2003). < > (Perspective on two reports this issue, one on zebra finches, one on blackbirds, both showing that brighter beak is preferred by females, one study by manipulating dietary carotenoids [the chemical which gives carrots their color], the other study by inducing mild illness causing the color to fade. The general intent of both is to show that this feature is indeed an indicator of good health, if not good genes.)

36 Sutherland, W. J. and Reynolds, J. D.: **Honesty in sexual selection**. *Nature* 375 (May 25), 260 (1995). < > (Perspective on report by Gustaffson *et al,* p311.)

37 Cannon, A.: **Going in style: The role of fashion in burial customs**. *The Sciences* (Nov/Dec), 38-42 (1991). < > (After recounting the rise of fancy cemetery monuments, she then notes how they fell out of fashion among the rich, at the same time as the middle and lower classes became able to afford them. "The attempt by lower classes to emulate the gentry was doomed by the very effort. Elaborate mortuary displays became associated with people having the lowest status in society, who were in turn rebuked for their excesses".)

38 Zahavi, A. and Zahavi, A. **The Handicap Principle: A Missing Piece of Darwin's Puzzle**. N.Y.: Oxford Univ. Press, (1997) < > (Noted in *Nature,* 392:36, Mar 5, 1998,

being a book-length update on "a controversial series of papers by Zahavi in the mid-1970's, which remain widely cited. The reviewer has high praise, but some reservations: "The book is extremely tendentious ... rival theories are often unfairly dismissed with glib and specious arguments.")

39 Holden, C.: **Big fish.** *Science* 310 (Oct 28), 616 (2005). <> (Short note on article finding that when the dominant male of a certain cichlid fish is removed, one or more previously subordinate males promptly develops the brilliant colors and other features of the former dominant male. Also noted in *New Scientist*, Oct. 22, pg 23, and *The Scientist*, Nov. 7, pg 22, where it is noted that the color changes are accompanied by rapid changes in gene expression and hormones.)

40 Buston, P.: **Size and growth modification in clownfish.** *Nature* 424 (Jul 10), 145-146 (2003). <> (In a breeding group, females are largest followed by the alpha male, and those of lesser social rank are progressively smaller. However, if an individual is removed the next lower in rank quickly grows larger. The experiments show clearly that the change in size is purely a response to altered social situation. This is reminiscent of bees that change caste to fill the ranks after some of that caste are removed – or, for that matter, the regulated population of types of blood cells.)

41 Starkey, A.: **The butterfly effect.** *New Scientist* (Sep 24), 46-47 (2005). <> (On efforts by the cosmetic industry to manufacture make-up – "eye-candy" – based on newly understood "photonic" principles of butterfly wing colors, especially iridescent blues. Reviews scientific work behind it, and the astonishing intricacy of the wing scale structures responsible for the visual effects.)

42 Vukusic, P. and Hopper, I.: **Directionally controlled fluorescence emission in butterflies.** *Science* 310 (Nov 18), 1151 (2005). <> (Article examines the optical qualities of the microstructure of the wings of the swallowtail butterfly. Noted also in *Nature,* 438:436, and in *Science News,* 168:324, from which summary we quote: "Each colored scale has three distinct tiers. The bottom tier is made up of three layers – cuticle, then air, then cuticle – each roughly 90 nm thick. The middle tier is an air space about 1.5 um thick, studded with columns of cuticle. In the top tier, about 2 um thick, the cuticle forms an imperfect honeycomb pattern and contains thousands of air-filled cylinders, each roughly 240 nm diameter. The fluorescent pigment resides within the walls of those cylinders." The full article explains how this structure gives rise to the observed colors. See also pg 1148 of issue, *Inspirations from Biological Systems for Advanced Photonic Systems.*)

43 Vukusic, P., Sambles, J. R. and Lawrence, P. R.: **Color mixing in wing scales of a butterfly.** *Nature* 404 (Mar 30), 457 (2000). <> (Authors investigate the ultrastructural basis of "the extraordinary combination of both yellow and blue iridescence" of the male *Papilio palinurus* butterfly of Indonesia, and find "a previously undiscovered retro-reflection process." Remarkable details are beyond this summary but the authors note that similar effects, involving both pigments and optical interference

and reflective effects, are seen also in some bird feathers and skin of some reptiles, and, we may add, squid.)

44 Crookes, W. J., et al.: **Reflectins: the unusual proteins of squid reflective tissue.** *Science* 303 (Jan 9), 235-238 (2004). <> (The symbiotic bacteria *Vibrio* which reside in local parts of the squid body are its source of luminosity. The authors discover the existence of stacks of hundreds of tiny reflective plates made of specific proteins (*reflectins*) which serve as mirrors, reflecting, focusing, and intensifying the light emitted.)

45 Sword, G. A., Lorch, P. and Gwynne, D. T.: **Migratory bands give crickets protection.** *Nature* 433 (Feb 17), 703 (2005). <> (Mormon crickets march in huge numbers. Study aims to test the usual theory, that they are thereby protected against predation. Method was to put a few dozen individuals out by themselves. The finding was that a larger fraction of them were eaten by predators than were the billions of marchers. But did they keep accurate count of how many of the billions were eaten? Calculation of the percentage eaten is misleading. Authors conclude, – lamely – that a benefit of marching *en masse* "is by sating local predators," i.e., it's an all-you-can-eat movable feast.)

46 Guilford, T. and Rowe, C.: **Unpalatable evolutionary principles.** *Nature* 382, 667-668 (1996 22 Aug). <> (Explores theory of why certain bad-tasting insects which are conspicuously marked, such as caterpillars of the cinnabar moth, feed in social groups, i.e. are gregarious. Several theories floated. Clever experiments arranged for birds (great tits) to feed on mock-insects consisting of short lengths of straw packed with food, with or without a bad-tasting additive, then observing how quickly the birds learned to recognize marks on the straws, and to generalize their aversion to bad tasting ones to similar looking marks, presented individually or in clusters to duplicate social groups. An incidental finding was that the birds *ate much more* when the food was presented in grouped clusters rather than scattered; see their Fig. 2, p709.)

47 Vermeij, G. J. **Nature: An Economic History.** Princeton, NJ: Princeton Univ. Press, (2004) <> (Reviewed in *Science* 306:1294, by D.H. Erwin who, like Vermeij, is a noted paleobiologist. Appears to be an effort to represent evolution in terms of economics based on energy efficiency principles and "power" in the sense of dominance, as in predator-prey relationships, probably along lines promulgated by W. Ostwald and A.J. Lotka and later applied to anthropology by Leslie White and his students, Sahlins and Service.)

48 Staff: **Nature vs. culture: a lesson from the guppy.** *Science* 272 (Apr 12), 203 (1996). <> (Brief summary of experiments by L. Dugatkin writing in recent issue of PNAS.)

49 Marston, W.: **The way of all fish: jealousy rekindles desire.** *The Sciences* (May/Jun), 11 (1994). <> (Reports work showing that jealousy is a mechanism of sexual stimulation in certain mollys. Details of their sexual habits are too complex to sketch here.)

50 Schlupp, I., Marler, C. and Ryan, M. J.: **Benefit to male sailfin mollies of mating with heterospecific females.** *Science* 263 (Jan 21), 373- (1994).

51 Hart, C. W. M. and Pilling, A. R. **The Tiwi of North Australia**. NY: Holt, Rinehart and Winston, (1979) <> (Fieldwork edition. First edition 1960. Of series "Case Studies in Cultural Anthropology"; George and Louise Spindler, ed's, Stanford Univ.)

52 Branam, M. A. and Greenfield, M. D.: **Flashing males win mate success** [Scientific Correspondence]. *Nature* 381 (Jun 27), 745 (1996).

53 Buck, J. and Buck, E.: **Synchronous fireflies.** *Sci American* May, 74 (1976). <> (From the blurb: "Fireflies in temperate zones flash in unsynchronized fashion [independently] whereas certain species of Asia and the Pacific flash in unison. What is the evolutionary role of this behavior?"]

54 Eisner, T. and *al*, E.: **Firefly 'femme fatales' acquire defensive steroids (lucibufagins) from the firefly prey.** *Proc Natl Acad Sci USA* 94 (Sep 2), 9723-9728 (1997). <> (For short commentary, see *Nature* 389:130, 11 Sep 1997, and also Science, 277:1611, 12 Sep 1997, 'Firefly Seductresses Steal Chemical Defenses.' It appears that the 'motive' is not only for food but is to acquire a toxin, a lucibufagin, from the prey species, not possessed by the predator, which protects it against predation by birds, bats, spiders, and others.)

55 Toner, M.: **When squid shine.** *International Wildlife* 24 (May/Jun) (1994). <> (Popular description of some of the numerous creatures that use bioluminescence.)

56 Fernald, R. D.: **Cichlids in love: what a fish's social caste tells the fish's brain about sex.** *The Sciences* (Jul/Aug), 28-31 (1993).

57 Schliewen, U. K., Tautz, D. and Paabo, S.: **Sympatric speciation suggested by monophyly of Crater Lake cichlids.** *Nature* 368 (Apr 14), 629-632 (1994). <> (Shows by DNA analysis that the fish in each lake arose from single ancestral type, demonstrating sympatric speciation, i.e. divergence in absence of isolating boundaries.)

58 Barleunga, M. and al, E.: **Sympatric speciation in Nicaraguan Crater Lake cichlid fish.** *Nature* 439 (Feb 9), 719-723 (2006). <> (This article is written as if sympatric speciation was still highly controversial and doubtful. The work eliminates doubt for the two species studied. For perspective by M. Hopkin, see pg 640.)

59 Hori, M.: **Frequency-dependent natural selection in the handedness of scale-eating cichlid fish.** *Science* 260 (Apr 9), 216-219 (1993). <>

60 Danchin, E., Giraldeau, L. A., Valone, T. J. and Wagner, R. H.: **Public information: From nosy neighbors to cultural evolution.** *Science* 305 (Jul 23), 487-491 (2004). <> (The gist of this article is to document that animals, too, exchange information (about food locations, etc.) and observe and imitate one another in such features as

mating displays, and that this implies or leads to culture. See also four letters highly critical of this rather pretentious article, with response by the authors, in *Science,* 308:35-6, Apr 15, 2005.)

61 Wilson, J. Q. and Herrnstein, R. J. **Crime and Human Nature**. NY: Simon and Schuster, (1986) <> (Reviewed in *The Sciences,* Mar/Apr 1986, p48, by David Kelley, who makes a good case, apart from an annoying introduction about how philosophy is the "childhood of science" – as if psychology was a mature science, as if it had outgrown the need for philosophical reflection!)

Ch 8 Refs & Notes

1 Hamilton, W. D. **Narrow Roads of Gene Land: The Collected Papers of W.D. Hamilton (v1)**. N.Y.: Freeman / Spektrum, (1996) <> (Lead review in *Nature,* 381:287, 23 May 1996, by Mark Ridley.)

2 Edwards, A. W. E.: **Hamilton built on work by Haldane and Fisher (Letter)**. *Nature* 416 (Apr 11), 581 (2002). <> (Writer complains that O. Judson attributed "inclusive fitness" to Hamilton in her review of vol. II of his *Collected Papers*, citing Fisher's *The Genetical Theory of Natural Selection* (Oxford, 1930) and Haldane's *The Causes of Evolution* (Longman, London, 1932) as both clearly stating the principle. This fact is acknowledged in Triver's obituary of Hamilton, but indicates that Hamilton much improved on this principle.)

3 Trivers, R.: **William Donald Hamilton (1936-2000)**. *Nature* 404 (20 Apr), 828 (2000). <> (Obituary. Briefly reviews his main achievements, notably *Inclusive Fitness* (1964); social insects; game-theory on "reciprocal altruism" (*Prisoner's Dilemma,* 1981); role of parasites in evolution, with M. Zuk (1981); etc. Died of malaria contracted during fieldwork.)

4 Dawkins, R. **The Selfish Gene**. New York: Oxford Univ. Press (2nd ed'n; 1st 1976), (1989) <> (This best-seller may be taken as representative of the "orthodox position," though perhaps slightly more right-wing than most today, despite his sharp disagreements with his equally famed and also orthodox counterpart in the USA, Steven J. Gould.)

5 Dawkins, R. **The Blind Watchmaker**. New York: W.W. Norton, (1987) <>

6 Dawkins, R. **Climbing Mount Improbable**. N.Y.: Viking, (1996) <> (Lead review in *Nature,* 382:309, 25 Jul 1996, by S. Kauffman: "the central image of the book is ... a multi-peaked fitness landscape ..." For insights on reviewer Kauffman, see his *The Origin of Order*.)

7 Parker, J.: **Richard Dawkin's Evolution**. *New Yorker* (Sep), 41-45 (1996). <> (Wry account of the influential author of *The Selfish Gene* (1989), *The Blind Watchmaker* (1996), *River out of Eden* (1995), *Climbing Mount Improbable* (1996), etc.)

8 Fletcher, D. J. and Michener, C. D., eds. **Kin Recognition in Animals**. N.Y.: Wiley, (1989). <> (Reviewed in *Amer Scientist*, 77:393, Jul/Aug 1989. Summarizes work as of then pertaining to W.D. Hamilton's 1964 thesis. The review is favorable but says it is "remarkably uncritical" and points out several lapses and questions.)

9 Hamilton, W. D.: *J Theor Biol* 7, 1-52 (1964). <>

10 Sudd, J. H., and Franks, N. R. **The Behavioral Ecology of Ants**. N.Y.: Chapman and Hall, (1987) <> (Reviewed in *Amer Scientist*, 76:406, Jul/Aug 1988, by N.F. Carlin: "From the outset, ant sociality is presented as an evolutionary enigma to be resolved in terms of the workers' proxy reproductive success." Reviewer notes that this book has been deeply influenced by "Hamilton's landmark proposal that hymenopteran workers are more closely related to their sisters than to their own offspring, and that they increase the propagation of their genes by aiding the reproduction of their mother queen. ... is therefore quite different in its interpretation as compared with Sudd's book on ants of 1967.")

11 Hepper, P. G., ed. **Kin Recognition**. N.Y.: Cambridge Univ. Press, (1991). <> (Reviewed in *Science*, 255:217, 10 Jan 1992. Among the probing comments: "Clearly, whether or not an organism treats kin differently from nonkin is a separate issue from the evolutionary reason it does so. ... Its primary weakness is its failure to integrate the widely disparate subject matter into a unified framework ...")

12 Sundstrom, L.: **Sex ratio bias, relatedness asymmetry, and queen mating frequency in ants**. *Nature* 367 (Jan 20), 266-268 (1994).

13 Page, R. E., Robinson, G. E., and Fondrk, M. K.: **Genetic specialists, kin recognition, and nepotism in honey-bee colonies**. *Nature* 338 (Apr 13), 576-579 (1989).

14 Keller, L. and Reeve, H. K.: **Familiarity breeds cooperation**. *Nature* 394 (Jul 9), 121 (1998). <> (Perspective on report by Roberts and Sheratt, p175 this issue. Caption: "Many theoretical models have been developed to study the conditions under which unrelated individuals should cooperate or not cooperate. ... new mathematical models allow the optimal level of cooperation to be [calculated].")

15 Smith, J. M.: **The evolution of behavior**. *Sci Amer* 239 (Sep), 176-192 (1978). <> (Emphasis is on altruism problem and the theory of W.D. Hamilton about it – he calls it "the solution" – which is well explained. Good historical perspective.)

16 Emlen, S. T. and Wrege, P. H.: **Gender, status, and family fortunes in the white-fronted bee-eater**. *Nature* 367 (Jan 13), 129-132 (1994). <> (Mathematical effort to formulate a model of reproductive decision making in the monogamous but highly social communities of this species, *vis a vis* economic and reproductive optimality. The notions are compared to human economic theory, and are based on Hamilton's

'inclusive fitness.' The analysis resembles certain anthropological approaches to human societies.)

17 Bourke, A. F. G. and Franks, N. R. **Social Evolution in Ants**. Princeton, NJ: Princeton Univ. Press, (1995) <> (Reviewed in *Science*, 271:1682. "Aim of the book is to bring followers of ant sociobiology up to date following the monumental general treatise of Holldobler and Wilson." Authors are "basing their approach firmly on Dawkins' view of the gene as the unit of selection ... [and] argue convincingly that kin selection is the single crucial factor in the evolution of eusociality." Authors show that "apparently rival theories" amount to the same thing. Book further erodes the old idea of the nest as a peaceful cooperating queendom, being in reality a "sea of competing interests;" reviewer cites Proverbs 6:7. Notes that no ant species is known which is not fully eusocial, in contrast to bees and wasps, many of which can switch back and forth between varying degrees of socialization, suggesting absence of strictly genetic compulsion.)

18 Smuts, B. **Sex and Friendship in Baboons**. ?: Aldine Press, (1985) <> (Reviewed in *Amer Scientist,* Sep/Oct 1987, 21. Although Smuts is "avidly selectionist" in her "approach to the interpretation of behavior," she also emphasizes the importance of friendship in baboon social life, and of "what she refers to as the spirit and vitality of the animal and its society." Evolutionist disparage such 'sentiments,' and we must admit that while real enough, they are too vaguely stated at present to serve as a foundation.)

19 Morin, P. A., Moore, J. J., Chakraborty, R., Jin, L., Goodall, J. and Woodruff, D. S.: **Kin selection, social structure, gene flow, and the evolution of chimpanzees**. *Science* 265 (Aug 26), 1193-1201 (1994). <> (For perspective, see p1172 this issue. Note that authors include the famed Jane Goodall. Abstract says "these data support the kin-selection hypothesis for the evolution of cooperation among males." By the way, they also find genetic evidence supporting the bonobo as a distinct species, not subspecies. For critical commentary with response, see letter, *Science,* 268:185.)

20 Pusey, A., Williams, J. and Goodall, J.: **The influence of dominance rank on the reproductive success of female chimpanzees**. *Science* 277 (Aug 8), 828-831 (1997). <> (See also p774 of this issue for background perspective on this article.)

21 Caporael, L. R. and Dawes, R. M.: **Altruism: docility or group identification?** *Science* 252 (Apr 12), 192 (1991). <> (Some cavils on Simon's "docility" model. Aside from praising Simon's "shift ... from the gene as a unit of selection to a social psychological mechanism as the important unit of analysis," they discuss *identification with the group* as an alternative to Simon's tradeoffs against docile submission. Includes response by Simon.)

22 Eldredge, N. **Why We Do It: Rethinking Sex and the Selfish Gene**. New York: W.W. Norton, (2004) <> (Reviewed in *Nature,* 430:613, Aug 5, 2004, by Robert

Foley who says "it sets out to tear down the whole edifice of reproduction-driven neo-Darwinian behavioral and evolutionary ecology. ... At heart, the book is deeply anti-gene. It argues that genes are wrongly placed at the centre of evolutionary theory." Book opposes the ultra-darwinists, chiefly Richard Dawkins and E.O. Wilson. Book "is highly polemical ... Written in a chatty and homely style, which will appeal to some and grate on others. ... If you are in the mood for some relentless Dawkins-bashing, or want a rush of arguments against biological determinism, than you might enjoy it." But, judging only by this review, it strikes me that the book makes points worth pondering.)

23 Davis, B. D. **Storm over Biology: Essays on Science, Sentiment, and Public Policy**: Prometheus Press, (1986) <> (Reviewed in *Amer Scientist,* 76:80, Jan/Feb 1988. According to the reviewer, who hints at acquaintance with Davis, the book illustrates "what happens to professors who become embroiled in reacting to massive multivalent attacks on their attitudes." Briefly, it appears to be a passionate defense of traditional orthodoxy ("positivism") – sociobiology, determinism, etc. – in the face of what he sees as the erosion of these positions. For example, it appears that Davis' response to the alleged amorality of sociobiology is to assert that it actually endorses the advance of altruism. Said by the reviewer to be an "exhausting book" to read".)

24 Lotka, A. J.: **Families of curves of pursuit and their isochrones.** *Amer Math Monthly* 35, 421 (1923).

25 Lotka, A. J.: **Contribution to the mathematical theory of capture.** *Proc Natl Acad Sci* 18, 172 (1932).

26 Merali, Z.: **How African frog hunts with eyes wide shut.** *New Scientist* (Aug 27), 13 (2005). <> (Investigates how this frog can hunt so unerringly, hovering in the water surface, even in total darkness, or even when blinded. The answer is that it is capable of 'calculating' the exact location of prey from patterns of tiny ripples, even when multiple ripple patterns are emanating from several objects in the vicinity or due to wind. Young frogs, however, must learn these skills from visual cues by hunting at twilight, until their skin receptors become 'trained'.)

27 Rinberg, D. and Davidowitz, H.: **Do cockroaches 'know' about fluid dynamics?** *Nature* 405 (Jun 15), 756 (2000). <> (Investigates the ability of cockroaches to detect and analyze the speed and direction of approaching objects – such as a human hand or predator – for the purpose of evasive action, by sensing rapid tiny variations in air pressure. It is clear that this ability requires sophisticated 'calculation'. Of course, the same kind of question could be posed of numerous sensory feats such as bat and dolphin echo finding, winged flight, etc.)

28 Angrlski, D. E., Green, A. M. and Dickman, J. D.: **Neurons compute internal models of the physical laws of motion.** *Nature* 430 (Jul 29), 560-563 (2004).

29 Davies, N. B. **Dunnock Behavior and Social Evolution**. New York: Oxford Univ. Press, (1992) <> (Reviewed in *Science*, 260:374, Apr 16, 1993, by Robert Gibson, under the wry caption, *Menage a Plusieurs*. As the reviewer points out, this bird raises the question of why so many other birds are more strictly confined to one particular mating practice.)

30 Dehaene, S., Izard, V., Pica, P. and Spelke, E.: **Core knowledge of geometry in an Amazonian indigene group**. *Science* 311 (Jan 20), 381-384 (2006). <> (The question arises because of a theory in linguistics, known as the Whorf hypothesis, to the effect that language determines our thinking. The conclusion is not surprising: "Our results provide evidence for geometrical intuitions in the absence of schooling, experience with graphic symbols or maps, or a language rich in geometrical terms.")

31 Sherman, P. W., Jarvis, J. U. M. and Braude, S. H.: **Naked mole rats**. *Sci American* (Aug), 72-78 (1992). <> (Although this article contains a number of "Just-So Stories," such as its explanation for hairlessness and cold-blood, it reports many interesting facts, such as that these animals are careful to avoid completely eating the tuberous plant roots they feed on, allowing them to regenerate.)

32 Braude, S. and Lacey, E.: **The underground society: The secret communal life of the naked mole rat**. *The Sciences* (May/Jun), 23-28 (1992).

33 Hamilton, W. D.: **Oedipal mating [Letter)**. *The Sciences* (Nov/Dec), 5 (1992).

34 Clutton-Brock, T. H. and al, e.: **Selfish sentinels in cooperative mammals**. *Science* 284 (Jun 4), 1640-1644 (1999). <> (See also perspective by D.T. Blumstin, p1633.)

35 McDonald, D. B. and Potts, W. K.: **Cooperative display and relatedness among males in a lek-mating bird**. *Science* 266 (Nov 11), 1030-1032 (1994). <> (Featured on cover. Paper tries to explain altruism in the beta male which "poses a problem for evolutionary theory" because most cases involve "kin selection or reciprocity" but this case does not: the beta male helps the alpha to mate even though unrelated. The author states in his summary that "here it is shown that alpha and beta partners are not relatives ... Instead, direct though long-delayed benefits to beta [altruistic] males are demonstrated, which include rare copulations, ascension to alpha status, and female lek fidelity.")

36 Coghllan, A.: **Love's a fight between the eyes and the nose**. *New Scientist* (Jul 13), 12 (2005). <> (It is noted that "mice prefer the smell of other mice with different variants of MHC genes to their own and tend to confirm this preference by mating with mice with dissimilar MHC genes." Also notes that "studies in which people have been asked to sniff sweaty tee-shirts generally suggest we also find the smell of those with dissimilar MHC genes more attractive.")

37 Koehn, R. J. and Hilbish, T. J.: **The adaptive importance of genetic variation**. *Amer Scientist* 75 (Mar/Apr), 134-140 (1987).

38 Davies, K.: **Costs of consanguinity**. *Nature* 371 (Oct 13), 630 (1994). <> (A traditional argument for the harmful effects of consanguinous couplings in humans, with new data.)

39 Thornhill, N. W., ed. **The Natural History of Inbreeding and Outbreeding: Theoretical and Empirical Perspectives**. Chicago, IL: Univ. Chicago Press, (1993). <> (Reviewed in *Science,* 263:107-108, Jan 7, 1994, with praise for many of the papers in this volume.)

40 Caro, T. M. and Laurenson, M. K.: **Ecological and genetic factors in conservation: a cautionary tale**. *Science* 263, 485-586 (1994). <> (Refutes the supposition that inbreeding – or low genetic polymorphism – is harmful, for the example of cheetahs. They find the population is endangered mainly because of lion and hyena predation, not because of inbreeding or lack of disease resistance.)

41 West, S. A., Pen, I. and Griffen, A. S.: **Cooperation and competition between relatives (Review)**. *Science* 296 (Apr 5), 72-75 (2002). <> (Aims to make kin selection theory more accurate by bringing in additional parameters. From the abstract: "Nevertheless, competition between relatives can reduce, and even totally negate, the kin selected benefits of altruism towards relatives.")

42 Clutton-Brock, T.: **Breeding together: kin selection and mutualism in cooperative vertebrates (Review)**. *Science* 296 (Apr 5), 69-72 (2002).

43 Chin, G.: **Rapid fin movement sleep**. *Science* 305 (Sep 10), 1533 (2004). <> (Note on paper by Goldschmid et al in *Limnol. Oceanogr.* 49:1832, noting that fish defend reefs, clear them of sediment, and help circulate oxygen-rich water by rapid fin movements as they sleep. Authors suggest that this enables reefs to flourish where they otherwise could not.)

44 Neumann, J. v. and Morgenstern, O. **Theory of Games and Economic Behavior: Sixtieth Anniversary Edition**. Princeton, N.J.: Princeton Univ. Press, (2004) <> (Reviewed in *Nature,* 431:511, by Karl Sigmund. Book is a reissue of this classic of 1945. It is noted that John Nash, at age 20 in 1950, made additional seminal contributions. With introduction by Harold Kuhn.)

45 Axelrod, R. and Hamilton, W. D.: **The evolution of cooperation**. *Science* 211 (Mar 27), 1390-1396 (1981).

46 Smith, J. M. and Price, G. R.: **The logic of animal conflict**. *Nature* 246 (Nov 2), 15-18 (1973).

47 May, R. M.: **More evolution of cooperation.** *Nature* 327 (May 7), 15-17 (1987). <> (This somewhat critical commentary on work by Boyd and Lorberbaum on p58 of this issue is cited more often than the work it comments on.)

48 Burt, A. and Trivers, R. **Genes in Conflict: The Biology of Selfish Genetic Elements:** Belknap Press, (2006) <> (Reviewed in *Nature* 440:609, by James F. Crow.)

49 Taylor, P. D. and Day, T.: **Cooperate with thy neighbor?** *Nature* 428 (Apr 8), 611-612 (2004). <> (Perspective on two papers this issue, p643 and p646, on math models of cooperation vs. selfish behavior. These writers propose also the Blizzard Game. Relatedly in this issue, p650, is article entitled "Pre-social benefits of extended parental care," which opens thus: "The evolution of helping, in which some individuals forfeit their own reproduction and help others to reproduce, is a central problem in evolutionary biology." As is often the case, the factual part of the study – parental care of a certain wasp challenged with parasites – is fascinating but the theoretical rationality is less than convincing; to the extent that it is, resembles our everyday judgments of tradeoffs.)

50 Doebeli, M., Hauert, C. and Killingback, T.: **The evolutionary origin of cooperators and defectors.** *Science* 306 (Oct 29), 859-862 (2004).

51 Nowak, M. and Sigmund, K.: **A strategy of win-stay, lose-shift that outperforms tit-for-tat in the prisoner's dilemma game.** *Nature* 364 (Jul 1), 56 (1993). <> (Features his "Pavlov" type of player, referring to "Pavlov's dogs," i.e. 'conditioned reflexes.' With commentary by M. Milinski, p12-13.)

52 Hauert, C. and Doebeli, M.: **Spatial structure often inhibits the evolution of cooperation in the snowdrift game.** *Nature* 428 (Apr 8), 643-646 (2004). <> (With perspective by P.D. Taylor and T. Day, p611-12, "Cooperate with thy neighbor?")

53 Nowak, M. A. and May, R. M.: **Evolutionary games and spatial chaos.** *Nature* 359 (Oct 29), 826-829 (1992).

54 Nowak, M. A. and al, e.: **Emergence of cooperation and evolutionary stability in finite populations.** *Nature* 428 (Apr 8), 646-650 (2004). <> (With perspective by P.D. Taylor and T. Day, p611-12, "Cooperate with thy neighbor?")

55 Jansen, V. A. A. and Baalen, M. v.: **Altruism through beard chromodynamics.** *Nature* 440 (Mar 30), 663-666 (2006).

56 McNamara, J. W., Barta, Z. and Houston, A. I.: **Variation in behavior promotes cooperation in the Prisoner's Dilemma game.** *Nature* 428 (Apr 15), 745-748 (2004). <> (Strikes this writer as a more sophisticated analysis compared to 5-10 years earlier, mainly because it incorporates more realistic complexity, specifically, it allows the group of players to consist of individuals with differing habitual responses, which under some assumed conditions results in ESS cooperation. In concluding paragraph,

authors point out that findings may support group selection theory. Variant indi-vidual behaviors, by the way, are now believed to underlie the robust stability of many communities including bacteria.)

57 Panchanathan, K. and Boyd, R.: **Indirect reciprocity can stabilize cooperation without the second-order free rider problem.** *Nature* 432 (Nov 25), 499-502 (2004). <> (See also perspective by E. Fehr, "Don't lose your reputation," with leader reading "collective action in large groups whose members are genetically unrelated is a distinguishing feature of the human species.")

58 Fehr, E. and Fischbacher, U.: **The nature of human altruism [Review].** *Nature* 425 (Oct 23), 785-791 (2003). <> (Covers recent thinking in this field. From opening paragraph: "human societies represent a huge anomaly in the animal world," based as they are on "cooperation between genetically unrelated [?] individuals in large groups. ... Even in other primate societies, cooperation is orders of magnitude [?] less developed then it is among humans, despite our close common ancestry [to each other? or to apes?]." Near the end, they charitably admit that "there are still some unsolved problems." Understatement of the day.)

59 Nowak, M. A. and Sigmund, K.: **Evolution of indirect reciprocity by image scor-ing.** *Nature* 393 (Jun 11), 573- (1998). <> (With perspective by R. Ferriere, p517; also discussed in *Science* 280:2070 by Claus Wedekind. "Image scoring" refers to how we perceive others based on our observations of them or their reputation for coop-eration or charity. According to Ferriere, this work "differs radically from classical approaches because they [Nowak and Sigmund] assume that any two organisms are unlikely to interact twice. This excludes direct reciprocity ...")

60 Nowak, M. A. and Sigmund, K.: **Evolutionary dynamics of biological games.** *Science* 303 (Feb 6), 793-799 (2004). <> (Overall survey for special issue on "Mathematics in biology.")

61 Nowak, M. A. and Sigmund, K.: **Evolution of indirect reciprocity (Review).** *Nature* 437 (Oct 27), 1291-1298 (2005). <> (Review article with 71 references. Emphasis is on mathematical models of human altruism based on indirect reciprocity and role of reputation.)

62 Wedekind, C. and Milinski, M.: **Cooperation through image scoring in humans.** *Science* 288 (May 5), 850-852 (2000). <> (With perspective by M.A. Nowak and K. Sigmund, p819-20. Reports results of experiments on volunteers intended to test results of game modeling concerning how the moves made by the players are affected by their perceptions of the other players, i.e. their reputations.)

63 Gintis, H., Bowles, S., Boyd, R. and Fehr, E., eds. **Moral Sentiments and Material Interests: The Foundations of Cooperation in Economic Life.** Cambridge, MA: MIT Press, (2005). <> (Reviewed in *Nature* 440:744, by A.M. Colman, who opens

quoting Robert May of late 2005. Cites examples of defectors in modern society such as those who fail to vaccinate their children on the assumption that enough others are vaccinated, and over-fishing, etc. Book covers the work of leaders in altruism theory from W.H. Hamilton, R. Trivers, and E. Fehr, but reviewer notes omission of R. Alexander's theory of 'indirect reciprocity' which he holds to be of major importance. Thus, book likely reflects the editors' allies, meaning that rivals would not be invited to contribute papers.)

64 Vogel, G.: **The evolution of the golden rule.** *Science* 303 (Feb 20), 1128-1131 (2004).

65 Bowles, S. **Microeconomics: Behavior, Institutions, and Evolution.** Princeton, N.J.: Princeton Univ. Press, (2004) <> (Reviewed in *Science* 306:1293, by E. Maskin. "A standard axiom in economic theory holds that humans are self-interested ... Economists recognize, of course, that the assumption is not literally true ... [Bowles] contends that self-interest is a poor approximation of reality. ... puts great weight on experimental results in 'public good' games." He argues that recognition of altruism is essential to understanding economics. Coincidentally, the next review is of book by G.J. Vermeij, *Nature: An Economic History,* which portrays human economics as co-extensive with the natural world.)

66 Seabright, P. **The Company of Strangers: A Natural History of Economic Life.** Princeton, N.J.: Princeton Univ. Press, (2004) <> (Reviewed in *Nature,* 431:245-6, Sep. 16, 2004. Book "exemplifies a new breed of economic analysis, seeking answers to fundamental questions ... The novelty is that he does so from a long-term evolutionary perspective." Touches on all the currently fashionable buzz-words: reciprocal altruism, gene-culture coevolution, etc.)

67 Camerer, C. F. and Fehr, E.: **When does "economic man" dominate social behavior?** *Science* 311 (Jan 6), 47-52 (2006). <> (Embraces psychological experiments concerning the rational basis of interacting players, such as with or without the opportunity to punish unfairness, etc., and is claimed to "provide better predictions of actual aggregate behavior than does traditional economic theory.")

68 Gurerk, O., Irlenbusch, B. and Rockenbach, B.: **The competitive advantage of sanctioning institutions.** *Science* 312 (Apr 7), 108-111 (2006). <> (With perspective under "Sociology" by J. Heinrich, p60-61, "Cooperation, punishment, and the evolution of human institutions," and widely noted elsewhere, e.g. *New Scientist.* Conclusion is that groups which punish non-cooperators fare better than those which do not. However, this seems to oppose traditional theory, that negative reinforcement doesn't work.)

69 Stevens, D. W., McLinn, C. M. and Stevens, J. R.: **Discounting and reciprocity in an iterated Prisoner's Dilemma.** *Science* 298 (Dec 13), 2216-2218 (2002). <> (With perspective p2146, but a clearer if less scholarly one is given in *Science News* 162:373, Dec. 14. Briefly, many previous workers had tried but failed to demonstrate

cooperation in birds – blue jays in this case – such as by placing two birds in adjacent cages equipped with levers [perches], some combinations of which would yield a larger food reward if the two birds cooperated. But these authors succeeded, chiefly by arranging the bigger reward from cooperating to accumulate in a transparent container so both could see it. There are several lessons here, first being that failure of an experiment does not necessarily falsify the premise since quite often, as in this case, a slightly different experiment will succeed. The second lesson is that birds are perfectly capable, if not always willing, to cooperate, depending on circumstances – as with us humans. The word "discounting" in the title refers to the delay in time that the birds are willing to put up with in order to reap the larger reward given by cooperating, a sensitive variable.)

70 Silk, J. B. and al, e.: **Chimpanzees are indifferent to the welfare of unrelated group members.** *Nature* 437 (Oct 27), 1357-1359 (2005). <> (This was widely reported, e.g. in *Science News*, "Chimps indifferent to others' welfare," 168:302, Nov. 5, 2005. However, this is very controversial as there is plenty of evidence, or at least opinion, in the other direction.)

71 Silk, J. B.: **Who are more helpful, humans or chimpanzees?** *Science* 311 (Mar 3), 1248-1249 (2006). <> (Perspective on two papers in this issue, by Melis et al, p1297, and by Warneken and Tomasello, p1301, both widely discussed, such as in *New Scientist* of Mar 11, p19. Dealing with capacity of apes for collaborative behavior, i.e. cooperation. Many related papers exist.)

72 Sober, E. and Wilson, D. S. **Unto Others: The Evolution and Psychology of Unselfish Behavior.** Cambridge, MA: Harvard Univ. Press, (1998) <> (Reviewed by John Maynard Smith in *Nature,* 393:639, 18 Jun 1998.)

73 Simon, H. A.: **A mechanism for social selection and successful altruism.** *Science* 250 (Dec 21), 1665-1668 (1990).

74 Burtsev, M. and Turchin, P.: **Evolution of cooperative strategies from first principles.** *Nature* 440 (Apr 20), 1041-1044 (2006). <> (Begins as usual: "One of the greatest challenges in the modern biological and social sciences is to understand cooperative behavior." Then explains that some progress has been made in previous modeling efforts – "general outlines of the answer to this puzzle are emerging" in light of kin selection, reciprocity, etc. – but indicates that previous assumptions were too simplistic. Therefore, they create a fancier game, adding to simple Hawk / Dove models imaginary Starlings and Ravens, and more strategic options, rather like a more realistic video arcade game. Cooperation emerges.)

75 Roughgarden, J. **Evolution's Rainbow: Diversity, Gender and Sexuality in Nature and People**: Univ. of California Press, (2004) <> (Reviewed in *Nature* 429:19-20, by Sarah Blaffer Hrdy.)

76 Roughgarden, J., Oishi, M. and Akcay, E.: **Reproductive social behavior: Cooperative games to replace sexual selection.** *Science* 311 (Feb 17), 965-969 (2006). <> (As I expected, this paper elicited strong comments in a dozen letters by leaders in the field, in the unusually lengthy letters section of *Science,* 312:689-697, issue of May 5.)

77 Forchhammer, M. C., Post, E. and Stenseth, N. C.: **Synchronized courtship in fiddler crabs.** *Nature* 391 (Jan 1), 31-32 (1998).

78 Buck, J. and Buck, E.: **Synchronous fireflies.** *Sci American* May, 74 (1976). <> (From the blurb: "Fireflies in temperate zones flash in unsynchronized fashion [independently] whereas certain species of Asia and the Pacific flash in unison."]

79 Hay, M.: **Synchronous spawning: when timing is everything.** *Science* 275 (Feb 21), 1080 (1997). <> (Perspective on report by K.E. Clifton, p1116, showing that numerous species ["up to nine species in five genera on a single morning") of coral reef-associated green alga spawn almost synchronously – "clouding the water" – though some "closely related species spawn at different times." This discovery comes 10 years after discovery of the synchronous spawning of the many species of the coral reel polyps.)

80 Kramarsky-Winter, E., Fine, M. and Loya, Y.: **Coral polyp expulsion.** *Nature* 387 (May 8), 137 (1997). <> (Reports that corals can also reproduce asexually, in contrast to the major discovery some years earlier of mass synchronous spawning of many coral species sexually.)

81 Sismondo, E.: **Synchronous, alternating, and phase-locked stridulations by a tropical katydid.** *Science* 249 (Jul 6), 55-58 (1990). <> (From the abstract: "Iteration of the Poincare map of the phase response predicts a variety of phase-locked synchronization regimes, including period-doubling bifurcations, in close agreement with experimental observations." Etc.)

82 Strogatz, S. **Synch. The Emerging Science of Spontaneous Order.** New York: Hyperion Press (Theia), (2003) <> (Reviewed in *Science,* 300:1878-9, by Nancy Kopell, under the caption, "A tree of fireflies, a flock of boson clouds," with a photo of a tree filled with Malaysian fireflies flashing in synchrony. Too complicated to encapsulate here, but the general approach of the book is to relate in mathematical terms the many biological phenomena which exhibit various kinds of periodic or synchronous oscillations under a larger umbrella that includes purely 'physical' systems. Review is lukewarm and perceptive, pointing out antecedents missed by Strogatz and inevitable shortcomings of the whole approach, chiefly, the fact that many unrelated phenomena obey similar math formulations.)

83 Greenfield, M. D. and Rolzen, I.: **Katydid synchronous chorusing is an evolutionarily stable outcome of female choice.** *Nature* 364 (Aug 12), 618-620 (1993). <> (They review prior hypotheses / explanations, including three which propose adaptive reasons, and finds all of them to be wrong by demonstrating that the choruses

are simply the result of each male trying to be the first and loudest to sing, since the females prefer him who sings first.)

84 Jones, D.: **The moral maze.** *New Scientist* (Nov 26), 34-36 (2005).

Ch 9 Refs & Notes

1 Lyons, E. J.: **Sex and synergism.** *Nature* 390 (Nov 6), 19 (1997). <> (Highlights from a symposium, *Summer School on the Evolution and Maintenance of Sex,* Seewiesen, Germany, said to be the first of its kind since 1988.)

1a Hurst, L. D. and McVean, G. T.: **...and scandalous symbionts.** *Nature* 381 (Jun 20), 650-651 (1996). <> (Reviews theories of the supposed need for sexual exchange to prevent extinction, in light of a bacterium adapted to the gut of aphids, without benefit of sexual exchange, which has nevertheless survived for about 100 million years, in conflict with theoretical expectation.)

2 Stebbins, G. L.: **The flowering of sex.** *The Sciences* May/June (May/Jun), 28-35 (1984). <> (Subtitle, "A botanist's search for the origin of life's strangest habit.")

3 Feist, M. and Feist, R.: **Oldest record of a bisexual plant.** *Nature* 385 (Jan 30), 401 (1997). <> (Beautiful fossil. The date assigned differs somewhat from estimates based on the mutational 'clock.')

4 Peck, J. R., Yearsley, J. M. and Waxman, D.: **Explaining the geographic distribution of sexual and asexual populations.** *Nature* 391 (Feb 26), 889-892 (1998).

5 Cherfas, J. and Gribben, J. **The Redundant Male.** N.Y.: Pantheon, (1985) <> (Reviewed in *The Sciences,* p56, Sep/Oct 1985, under caption, "Is sex irrelevant in the modern world?" Reviewer first speaks of the adaptive need hypothesis, then explains that Cherfas and Gribben argue that the main adaptive need of higher organisms is to enable rapid immunological defenses in the face of microbial pathogens.)

6 Elena, S. F. and Lenski, R. E.: **Test of synergistic interactions among deleterious mutations in bacteria.** *Nature* 390 (Nov 27), 395-398 (1997). <> (With perspective by S.P. Otto, p343.)

7 Otto, S. P.: **Unravelling gene interactions.** *Nature* 390 (Nov 27), 343 (1997). <> (Perspective on paper by Elena and Lenski, p395 this issue.)

8 Paland, S. and Lynch, M.: **Transitions to asexuality results in excess amino acid substitutions.** *Science* 311 (Feb 17), 990-993 (2006). <> (With background, "Why sex?" by Rasmus Nielson, p960-1, opening "This has been one of the most fundamental questions in evolutionary biology.")

9 Kimura, M.: **The neutral theory of molecular evolution.** *Sci Amer* 241 (Nov), 98-128 (1979).

10 Rifkin, S. A. et al: **A mutation accumulation assay reveals a broad capacity for rapid evolution of gene expression.** *Nature* 438 (Nov 10), 220-223 (2005). <> (They measured changes in patterns of gene expression as a proxy for mutations.)

11 Agosti, D., Grimaldi, D. and Carpenter, J. M.: **Oldest known ant fossils discovered.** *Nature* 391 (Jan 29), 447 (1997). <> (Perfectly preserved in amber for about 130 million years. Shows even a gland for producing antibiotic substances. Thousands of small creatures have been studied in amber and many are almost identical to those of today.)

12 Azevedo, R. B. R. and al, e.: **Sexual reproduction selects for robustness and negative epistasis in artificial gene networks.** *Nature* 440 (Mar 2), 87-90 (2006).

13 Michod, R. E.: **What's love got to do with it?** *The Sciences* (May/Jun), 22-28 (1989).

14 Michod, R. E. and Levin, B. R., eds. **The Evolution of Sex: An Examination of Current Ideas.** Sunderland, MA: Sinauer, (1987). <> (Reviewed *Amer Scientist* 76:393, Jul/Aug 1988, by Jon Lovett-Doust, who reports a "star-studded cast of contributors" but also absence of consensus, problems of definitions of sex – different authors talking about different things – and that "the explanation for its origin may be totally different from the reason for its persistence." Sex as a DNA repair mechanism is said to be one of the three main theories treated.)

15 Miller, S. and Agrawal, F.: **Sexual selection and the maintenance of sex / Sexual selection and the maintenance of sexual reproduction.** *Nature* 411 (Jun 7), 689-695 (2001). <> (It appears from a brief reading that both are building on prior ideas such as of A.S. Kondrashov, and seem to hold that if females exert sexual selection, and if males are under more intense natural selection pressure than females, then deleterious mutations are preferentially shed and beneficial mutations more readily fixed in the population than by asexual reproduction or if the females mated randomly with males. Both give similar math rationalization.)

16 Margulis, L. and Sagan, D. **Origins of Sex: Three Billion Years of Genetic Recombination.** New Haven, CT: Yale Univ Press, (1986) <> (Reviewed *Amer Scientist* 75:536, Sep/Oct 1987, by Ricard Michod, opening thus: "The problem of the origin and evolution of sex is considered by many to be one of the great unsolved problems in evolutionary biology." Review is harshly negative, saying that while the book is loaded with interesting material, it tells little of the origin of sex. He thinks that her thesis, that sex originated for DNA repair and no longer has any purpose or benefit, is poorly defended.)

17 Margulis, L. and Sagan, D. **What is Sex?** N.Y.: Simon & Schuster, (1998) <> (Briefly in *Nature,* 392:561: "Their book views sex simply as a consequence of thermodynamics. Organisms are open systems with orifices through which both energy and matter can flow. ... Sex exists, they argue, purely because the organisms [that do it] have survived; it has no inherent advantage over asexual reproduction.")

18 Lane, N. **Power, Sex, Suicide: Mitochondria and the Meaning of Life:** Oxford Univ. Press, (2005) <> (Reviewed in *Nature,* 437:1235-6, Sep 30, 2005, by John F. Allen, who is enthusiastic despite the "embarrassing" title. Includes a theory for why there are two sexes, based on the interplay between the mitochondrial and nuclear genomes. Reviewer calls it a must-read. Lane's previous book, *Oxygen,* was well acclaimed, reviewed in *Science,* 311:1869, Mar 31, 2006.)

19 Kondrashov, A. S.: **The asexual ploidy cycle and the origin of sex.** *Nature* 370 (Jul 21), 213-216 (1994).

20 West, S. A. and Peters, A. D.: **Paying for sex is not easy.** *Nature* 407 (Oct 26), 962 (2000). <> (Opening: "Explaining the maintenance of sexual reproduction remains one of the greatest challenges for biology, with more than 20 hypotheses ..." This short paper aims to show that a new hypothesis by Doncaster *et al, Nature* 404:281, actually reduces to a version of the Lotka-Volterra model – a.k.a. "the tangled bank" hypothesis – and is subject to the same objections. Nevertheless, West and Peters find the approach to contain useful insights.)

21 Gadagkar, R.: **Sex. Only if really necessary in a feminine monarchy.** *Science* 306 (Dec 3), 1694-1695 (2004). <> (Introduces article by Pearcy et al, p1780, "which uncovers a new dimension in the complexity of hymenopteran reproduction.")

22 Powwows, T. E. and Taylor, J. W.: **Organization of genetic variation in individuals of arbuscular mycorrhizal fungi.** *Nature* 427 (Feb 19), 733-737 (2004). <> (This group of organisms is of special interest because of its complex and fascinating relationships with the root systems of plants, but is challenging because each cell contains numerous nuclei.)

23 Lucentini, J.: **How sex may have started it all.** *The Scientist* (Mar 29), 26-28 (2004). <> (In addition to reporting on a novel theory for the origin of sex, offers short survey on controversies about whether lower forms such as bacteria truly engage in "sex" and the definition of this term.)

24 Bhattacharyya, R. P. et al: **The Ste5 scaffold allosterically modulates signaling output of the yeast mating pathway.** *Science* 311 (Feb 10), 822-826 (2006).

25 Qi, J. and al, E.: **Characterization of a *Phytophthora* mating hormone.** *Science* 309 (Sep 16), 1828 (2005). <> (Fine work isolating 1.2 mg of hormone from 1,830 liters of culture broth of *P. nicotianae,* of this genus of fungus-like organisms, many of which

are major plant pathogens such as caused the Irish potato famine and the recent destruction of oak forests in Europe and America. Their mating types and systems of reproduction are too complex to even briefly describe here, but it is noteworthy that the hormone they found is equally effective on all four species of the genus they tested it on.)

26 Fraser, J. A. and al, e.: **Same-sex mating and the origin of the Vancouver Island** *Cryptococcus gattii* **outbreak**. *Nature* 437 (Oct 27), 1360-1364 (2005).

27 Brownlee, C.: *Giardia* **bares all**. *Science News* 167 (Jan 29), 67 (2005). <> (Note on paper by J. Logsdon et al writing in Jan 26 issue of *Current Biology*. Shows that this primitive unicellular parasite *Giardia intestinalis* possesses genes for meiosis, used in higher forms for producing germ cells, implying they might reproduce in that way, though previously thought to reproduce only asexually.)

28 Staff: **Ancient sex**. *Nature* 437 (Oct 20), 1068 (2005). <> (Brief note on paper in *PNAS* concerning certain placozoans, the smallest and simplest known multicellular animal: "They are also completely asexual – or are they? Ana Signorovitch and colleagues [findings were] consistent with expectation for a sexual population ...")

29 Gratia, J. P. and Thiry, M.: **Spontaneous zygogenesis in Esceriacoli, a form of true sexuality in prokaryotes**. *Microbiology* 149, 2571-2584 (2003).

30 Camilli, A. and Bassier, B. L.: **Bacterial small-molecule signaling pathways [Review]**. *Science* 311 (Feb 24), 1113-1116 (2006). <> (Summarizes knowledge to date, which is considerable, but judging from the past, further complexity will emerge.)

31 Winans, S. C.: **Bacterial speech bubbles**. *Nature* 437 (Sep 5), 330 (2005). <> (Opens, "many bacteria socialize using diffusible signals." Supplies background for article on p422 identifying some of the chemicals used in messaging and showing that some are packaged in tiny lipid spheres. New insights come from messages "aimed at enemies rather than friends.")

32 Xavier, K. B. and Bassler, B. L.: **Interference with AI-2-mediated bacterial cell-cell communication**. *Nature* 437 (Sep 29), 750-753 (2005). <> (AI-2 isa group of chemicals produced by bacteria for communication, some of which are species-specific, others being more like a universal language. Some authors have spoken of similar chemical signals, such as in yeasts, as "pheromones" since they appear to be involved in sex-like conjugation. From the abstract: "Here we show that some species of bacteria can manipulate AI-2 signaling and interfere with other species' ability to assess and respond correctly.")

33 Morell, V.: **Sex frees virus from genetic 'ratchet'**. *Science* 278 (Nov 28), 1562-1563 (1997). <> (Indicates that Chao's experiments help settle another dispute in evolutionary theory, that between the stepwise-advance model of R.A. Fisher, and the

topographical landscape model of Sewell Wright. As might be expected, his experiments indicate that both conceptual metaphors have merit.)

34 Barinaga, M.: **Viruses launch their own 'Star Wars'.** *Science* 258 (Dec 11), 1730 (1992). <> (Notes that researchers "prided themselves" for a clever way of getting rid of the AIDS virus, namely, injecting the soluble form of the target CD4 molecule to "mop up" the virus. "Lately, however, they have been humbled on that point as well: It turns out that viruses had the idea first. ...")

35 McFadden, G.: **Even viruses can learn to cope with stress.** *Science* 279 (Jan 2), 40-41 (1998). <> (Introduces a report on p102 of a remarkable new strategy used by viruses, and supplies a table listing some of the more remarkable tricks devised by viruses; the list is now [2011] longer and more amazing.)

36 Vignuzzi, M. and al, e.: **Quasispecies diversity determines pathogenesis through cooperative interactions in a viral population.** *Nature* 439 (Jan 19), 344-348 (2006). <> ("An RNA virus population does not consist of a single genotype; rather, it is an ensemble of related sequences, termed quasispecies [which] arise from rapid genomic evolution powered by the high mutation rate of RNA viral replication. ... not simply a collection of diverse mutants but a group of interactive variants. .... selection indeed occurs at the population level rather than on individual variants." Experiments were on the poliovirus.)

37 Suttle, C. A.: **Viruses in the sea.** *Nature* 437 (Sep 15), 356-362 (2005). <> (One of five reviews in special issue, "Bio-oceanography," see p343-9, "Molecular diversity and ecology of microbial plankton," by Giovannoni and Sting.)

38 Thompson, J. R. and al, e.: **Genotypic diversity within a natural coastal bacterioplankton population.** *Science* 307 (Feb 25), 1311-1313 (2005). <> (They make the puzzling find that *Vibrio splendidus* consists of at least a thousand distinct genotypes and propose that these differences are adaptively neutral.)

39 Venter, J. C. and al, e.: **Environmental genome shotgun sequencing of the Sargasso Sea.** *Science* 304 (Apr 2), 66-74 (2004). <> (Startling technical feat by genomic guru Venter, who scooped up some mid-ocean water and sequenced about 1 billion base pairs to discover 1.2 million previously unknown genes, representing at least 1800 bacterial species. With commentary p58-60.)

40 Gewin, V.: **Discovery in the dirt.** *Nature* 439 (Jan 26), 384- (2006). <> (Survey of recent studies of soil bacteria using genomic tools, aiming towards new drug discoveries; and related topics. It is said that less than 1% of them can be cultured in the laboratory by classical methods.)

41 Gans, J., Wolinsky, M. and Dunbar, J.: **Computational improvements reveal great bacterial diversity and high metal toxicity in soil.** *Science* 309 (Aug 26), 1387-1390

(2005). <> (With perspective by Curtis and Sloan, p1331-2. Ingenious method devised to estimate bacterial species in soil from prior genomic studies. Previous estimates were 10,000 species in a gram of soil but this study finds at least 100-times more. It is said that only 1% to 5% can be cultured in the lab. Elsewhere it is shown that many soil bacteria have natural resistance to antibiotics, presumably for defense against their neighbors. A plot of number of species vs. their relative abundance is roughly a straight line on a log-by-log grid. They also find that soil polluted by heavy metals has the same number of cells but only 1% as much diversity.)

42 Arnqvist, G. and Rowe, L. **Sexual Conflict**. Princeton, N.J.: Princeton Univ. Press, (2006) <> (Reviewed in *Nature*, 439:537, by Tracy Chapman, who remarks that "the greatest strength of the book is in tackling the theory ... a first-rate job of dissecting models of sexual selection and sexual conflict, and in getting to grips with the distinctions between them.")

43 Arnqvist, G. and Rowe, L.: **Antagonistic coevolution between the sexes in a group of insects.** *Nature* 415 (Feb 14), 787-789 (2002).

44 Fedorka, K. M. and Mousseau, T. A.: **Female mating bias results in conflicting sex-specific offspring ratios**. *Nature* 429 (May 6), 65-67 (2004). <> (Study of the cricket, *Allonemobius socius*.)

45 Michiels, N. K. and Newman, L. J.: **Sex and violence in hermaphrodites**. *Nature* 391 (Feb 12), 647 (1998).

46 Anderson, A.: **The evolution of sexes**. *Science* 257 (Jul 17), 324 (1992). <> (Review of 'hot papers' on the origin of sex by L. Hurst and W.D. Hamilton in *Proceedings of the Royal Society-B*, with background on problems posed by Ronald Fisher in his "monumental 1958 work, *The Gentical Theory of Natural Selection*," said to be solved by the current work. Details are complex but reviewer A. Anderson assures us that the work is based on "exceptionally good evidence" including the curious sexual habits of numerous obscure species.)

47 Miller, P. M., Gavrilets, S. and Rice, W. R.: **Sexual conflict via maternal-effect genes in ZW species**. *Science* 312 (Apr 7), 73 (2006). <> (Math analysis of how "selfish elements" of the sex chromosomes, specifically in birds with ZW, could operate to favor the interests of daughters vs. sons.)

48 Gavrilets, S.: **Rapid evolution of reproductive barriers driven by sexual conflict.** *Nature* 403 (Feb 24), 886-889 (2000). <> (Followed by critical letter by Fayrer-Hosken et al in issue of Sep 14, p149-150, with response by Gavrilets, titled "Sexual conflict and speciation.")

49 Rice, W. R.: **Sexually antagonistic male adaptation triggered by experimental arrest of female evolution**. *Nature* 381 (May 16), 232-234 (1996).

50 Chapman, T. and Partridge, L.: **Sexual conflict as fuel for evolution.** *Nature* 381 (May 16), 189 (1996). < > (Perspective on paper by Rice, p232 this issue. Leader: "An experiment with fruit flies, in which females were not allowed to adapt genetically to male change, has unmasked the rapid evolutionary dynamics underlying an ostensibly stable system." Concludes by remarking that "males and females may be running frantically to stand still.")

51 Morell, V.: **Flies unmask evolutionary warfare between the sexes.** *Science* 272 (May 17), 953 (1996).

52 Borgia, G.: **Sexual warfare?** [Letter]. *Science* 272 (Jun 21), 1723 (1996).

53 Hamilton, W. D.: **Extraordinary sex ratios.** *Science* 156 (Apr 28), 477-488 (1967).

54 Gowaty, P. A.: **Birds face sexual discrimination.** *Nature* 385 (Feb 6), 486-487 (1997). < > (Background for report by Komdeur *et al,* p522, dealing with "adaptive modification" of sex ratio of a warbler's eggs, confirming theoretical expectation.)

55 Brown, J. L. **Helping and Communal Breeding in Birds: Ecology and Evolution.** Princeton, NJ: Princeton Univ. Press, (1987) < > (Reviewed in *Amer Sci* v76, Nov/Dec 1988, p609-10, by R.L. Curry, who remarks that "interest in cooperation and altruism has intensified since 1963, when Hamilton developed the concept of inclusive fitness, which predicts the evolution of caring for non-descendant relatives as well as offspring." Said to include a comprehensive review of the literature. "Brown's harshest criticisms concern perceived failures to consider alternative hypotheses, but he rarely questions [his own assumption] whether regularly occurring helping is adaptive. Recent critics argue that basic assumptions about influence of helping on fitness remains untested ...")

56 Greenwood, J. J. D.: **Pretty polly, polly, polly,** *Nature* 389 (Oct 2), 442 (1997). < > (Short report of work in then recent *Proc Roy Soc,* on variable hatchling sex ratios of the parrot, *E. roratus,* which uniquely among parrots breeds cooperatively – up to 10 males associated with a single female. One female produced 20 sons in succession. However, much of the data were from captive birds, and other caveats are mentioned.)

57 Ridley, M.: **A boy or a girl: is it possible to load the dice?** *Smithsonian* (Jun), 113-123 (1993). < > (Cites the theory of Trivers and Willard. Mentions that well-fed opossum parents in the wild made 40% more sons. Conversely, hamsters kept hungry made more females. Similar findings for nutria, white-tailed deer, and wood-rats. But Clutton-Brock found that among red-tailed deer on the island of Rhum, high social status of the female gave more males, as confirmed by Meg Symington in Peruvian spider monkeys: all 21 babies born to the lowest ranking female were female, while 75% of those born to high-status females were male.)

58 Crews, D.: **Animal sexuality.** *Sci American* (Jan), 108-114 (1994). < > (Subtitle: "Animals have evolved a range of mechanisms to determine whether an individual

takes on masculine or feminine traits." Emphasis is on role of hormones. Focus on the whiptail lizard.)

59 Ens, B. J.: **Love thine enemy?** *Nature* 391 (Feb 12), 635-636 (1998). <> (Background and summary of report by Heg and van Treuen, p687, on the social ecology of oystercatchers, which are normally socially and sexually monogamous but sometimes become polygynous. The explanation of the latter, arising in certain social situations, is that "if you can't beat 'em, join 'em.")

60 Davies, N. B. **Dunnock Behavior and Social Evolution**. New York: Oxford Univ. Press, (1992) <> (Reviewed in *Science*, 260:374, Apr 16, 1993, by Robert Gibson, under the wry caption, *Menage a Plusieurs*. As the reviewer points out, this bird raises the question of why so many other birds are more strictly confined to one particular mating practice.)

61 Pizzari, T. and Birkhead, T. R.: **Female feral fowl eject sperm of subdominant males**. *Nature* 405 (Jun 15), 787-789 (2000). <> (In these fowl, copulation is commonly coerced, almost rape, but the female has the option of expelling sperm from undesired males.)

62 Norman, M. D. and Lu, C. C.: **Sex in giant squid**. *Nature* 389 (Oct 16), 683-684 (1997). <> (Reports capture of female giant squid containing many packets of spermatophores. References are cited to the effect that many female cephalopods store sperm packets from multiple matings "sometimes for several months." By the way, it is known that cephalopods, too, play the mating game in an astonishing variety of ways.)

63 Eberhard, W. G. **Female Control: Sexual Selection by Cryptic Female Choice**. Princeton, NJ: Princeton Univ. Press, (1996) <> (Reviewed in *Science*, 275:1075, Feb 21, 1997, by R.H. Wiley, who calls this a "mesmerizing review of the intricacies of mating by diverse animals." A major theme of the book is said to be the many ways in which females affect fertilization after copulation.)

64 Roberts, K. A. and Thompson, M. B.: **Viviparous lizard selects sex of embryos**. *Nature* 412 (Aug 16), 698 (2001). <> (Temperature dependent sex determination, TSD, demonstrated in a skink, *Eulamprus tympanum*, unexpected because they bear live young. In a lab where they were free to choose their temperature, they preferred 32C and produced all males but in two seasons of field collecting, equal numbers of males and females were found. Their ref 9 reports another skink known "to balance" sex ratios of offspring in response to that of the adult population.)

65 Holden, C.: **Fish sex: all in the head?** *Science* 275 (Jan 10), 163 (1997). <> (Short comment on work by Godwin *et al* writing in *PNAS*, 22 Dec, on a Caribbean bluehead wrasse [fish] "distinguished by the ability of females to change into males when males are lacking." Shows that this is controlled by a non-gonadal hormone, and therefore

throws into doubt numerous other observations suggesting that male/female traits and behaviors are mechanistic consequences of sex hormones.)

66 Anderson, M. W.: **A sperm finds its egg.** *The Scientist* (Nov 3), 37 (2003). < > (Note on two papers identifying the biochemistry by which an egg resists fertilization by sperm from alien species of sea urchin, blocking hybridization.)

67 Olsson, M., Shine, R., Madsen, T., Gullberg, A. and Tagelstrom, H.: **Sperm selection by females.** *Nature* 383 (Oct 17), 585 (1996). < > (This is a comparatively early paper on what is now an accepted and much studied phenomenon, sperm selection and competition. Paper purports to show that sand lizard females, which seem to indiscriminately accept mating from males, store the sperm and exert control over which sperm cell are used to fertilize their eggs.)

68 Simmons, L. W. **Sperm Competition and its Evolutionary Consequences in the Insects.** Princeton, N.J.: Princeton Univ. Press, (2001) < > (Reviewed in *Nature*, 417:122, May 2, by M.T. Siva-Jothy, who notes that three other books surveying this field already exist, but that Simmons offers a novel perspective and is really good.)

69 Levitan, D. R. and Ferrell, D. L.: **Selection on gamete recognition proteins depends on sex, density, and genotype frequency.** *Science* 312 (Apr 14), 267-269 (2006). < > (Nice study examining the gene for binding, one of the key proteins determining compatibility between sperm and egg. Aim is to better understand speciation, viz., reproductive isolation.)

70 Milius, S.: **Anemone wars.** *Science News* 167 (Jun 4), 355-356 (2005). < > (Note on article in upcoming issue of *Animal Behavior* on battles between neighboring colonies of the sea anemone, *A. Ellegantissima.*)

71 DeTomaso, A. W. and al, e.: **Isolation and characterization of a protochordate histocompatibility locus.** *Nature* 438 (Nov 28), 454-459 (2005). < > (With perspective on p437-8 by G.W. Litman, "Colonial match and mismatch." The same sea squirt, *Botryllus schlosseri,* was the subject of another study showing that stem cells from one colony can invade and take over another, as noted in *Nature,* 439:120, referring to a paper in *Cell,* 123:1351.)

72 Ainsworth, C.: **The story of i.** *Nature* 440 (Apr 6), 730-733 (2006). < > (Subtitle, "Multicellular creatures can be battlegrounds for competing populations of cells," reviewing the tale of the tunicate, *Botryllus schlosseri,* studied by Irving Weissman for some 30 years, with implications for understanding cancer and immunology.)

73 Barratt, E. M. et al: **DNA answers the call of pipistrelle bat species.** *Nature* 387 (May 8), 138-139 (1997). < > (Demonstration that "Europe's most abundant and well-studied bat" actually consists of two distinct species, initially discovered by acoustic analysis of their cries, then supported by DNA typing. They coexist in most

of their range but maternal roosts harbor only one. The authors speculate that "it is more likely that acoustic divergence between [these] species occurs after genetic isolation ..." rather than being a cause of the divergence but their argument is unconvincing, running as follows: that in other species such as grasshoppers, birds, frogs, toads and crickets, divergence of calls is said to be due to whimsies of female preference, whereas in these bats it serves to echolocate prey of different size or form. But we have argued that these 'mechanisms' are related, in that hunting a different prey could have the same kind of 'sex appeal' as does singing a different song.)

74 Editor: **A tale of two salamanders**. *Science* 288 (May 5) (2000). < > (Note on paper in *PNAS USA* 97:4106 pertaining to 'the law of limiting similarity,' that two species which are identical except for non-mating cannot coexist because one is bound to displace the other, here shown to be upheld for the case of two similar America salamanders, since their jaws are slightly different to facilitate capturing prey. The implication is that they diverged into two different species because of this difference in adaptive specialization, however, we suggest a motivational reason behind the divergence.)

75 Singh, R. S. and Hale, L. R.: **Regional variation in fruit flies [Letter]**. *Nature* 369 (Jun 9), 450 (1994). < > (Critique of a paper, with reply, based on questionable methods, one analyzing mitochondrial DNA, the other nuclear. But the point here is that both detected unsuspected degrees of population structure, likely including many cryptic or incipient species.)

76 Brownlee, C.: **DNA bar codes**. *Science News* 166 (Dec 4), 360-361 (2004). < > (Reports status of big effort to devise simple DNA tests, analogous to bar codes, for identifying specimens. An unexpected spin-off was the finding that several actually consist of multiple species, each breeding to itself, although visually indistinguishable, e.g., a skipper butterfly, *Astraptes fulgerator*, actually consisted of 10 distinct species. Advocate Jenzen thinks this may be convergent mimicry, that is, many species having become mimics of one of them. But that strikes this writer as far-fetched.)

77 Pennisi, E.: **Speciation standing in place**. *Science* 311 (Mar 10), 1372-1374 (2006). < > (Survey of the controversies, with many new examples discussed.)

Ch 10, Refs & Notes

1 Aitken, J. R. and Graves, J. A. M.: **The future of sex**. *Nature* 415 (Feb 28), 963 (2002). < > (Short essay on the impending loss of the Y, "in about 10 million years.")

2 Morell, V.: **Rise and fall of the Y chromosome**. *Science* 263 (Jan 14), 171-172 (1994). < > (Background for paper by Rice, p230 this issue. Includes remarks on sexual selection, e.g. theory that a gaudy red spot on male guppies which appeals to females was

once possessed by females as well but was lost in females by natural selection because it is maladaptive, being attractive to predators. Not very persuasive.)

3 Charlesworth, B.: **The evolution of sex chromosomes**. *Science* 251 (Mar 1), 1030-1033 (1991).

4 Sykes, B. **Adam's Curse: A Future Without Men**. New York: Norton, (2004) <> (By the author of the eminently readable, *The Seven Daughters of Eve*. Argues for the eventual disappearance of males owing to the degenerate and shrinking Y chromosome.)

5 Staff: **Dawn of the Y**. *Science News* 165 (Jan 24), 52 (2004). <> (Reporting claim by R. Ming and colleagues in Jan. 22 *Nature* that the Y chromosome of the papaya tree is the most youthful yet found. By the way, only about 5% of plants have genders, and only some of them are sexually dimorphic, marijuana [*Cannabis sativa*] among them. The sex chromosomes of mammals are said to be ten times older, at around 300 millions years, than those of plants, but very different estimates have been given by others.)

6 Johnstone, N.: **And then there was Y**. *The Scientist* (Sep 26), 24-25 (2005). <> (Short review of "Hot Papers" concerning Y chromosome, in *Nature* [423:825, 423:873, with commentary, 423:810, all 2003; and 431:931, 2004; and 437:101, 2005], the main thrust being that, contrary to widely publicized reports, the Y is vigorous – but also rather mysterious.)

7 Ridley, M.: **Y chromosome exposed**. *Discover* (Jan), 56 (2004). <> (Listed with *Advances of the Year*. "Quashing that theory [that Y will soon vanish] is the discovery that the Y has devised a way to survive, says David Page of MIT.")

8 Repping, S. and al, e.: **High mutation rates have driven extensive structural polymorphism among human Y chromosomes**. *Nature Genetics* 38 (4), 463-467 (2006). <> (They examined a Y chromosome from one person of each of 47 human racial groups, and found a lot of differences. Contrary to earlier reports, they find evidence of competitively high mutational activity on Y, 10,000 times faster than the average rate of single nucleotide substitutions.)

9 Pennisi, E.: **Mutterings from the silenced X chromosome**. *Science* 307 (Mar 18), 1708 (2005). <> (Summary of a report in *Nature* that "genes once thought to be silenced in women are sometimes expressed – and that their degree of expression varies from woman to woman.")

10 Reik, W. and Ferguson-Smith, A. C.: **The X-inactivation yo-yo**. *Nature* 438 (Nov 17), 297-298 (2005). <> (Introduces report by Okamoto et al, p369 this issue, being the latest [2005] on the mechanisms of X inactivation, now involving imprinting; i.e. it is now far more complicated and subtle as compared with, for example, to a 2002 paper on this subject, *Science* 295:287.)

11 Kimura, M.: **The neutral theory of molecular evolution.** *Sci Amer* 241 (Nov), 98-128 (1979).

12 Ohta, T. and Aoki, K., eds. **Population Genetics and Molecular Evolution.** N.Y.: Springer-Verlag, (1985). < > (Reviewed *Amer Scientist,* 75:313, May/Jun 1987. Papers honoring Motoo Kimura, founder of the once-heretical neutral theory of molecular evolution, on his 60th birthday. For background, see "The neutral theory of molecular evolution," *Sci Amer,* Nov 1979, p98.)

13 Gillespie, J. H. **The Causes of Molecular Evolution.** N.Y.: Oxford Univ. Press, (1991) < > (Reviewed in *Science,* 257:420, 17 Jul 1992, under the canny caption, "Molecular Panselectionism." The reviewer begins by noting Kimura's 'neutral theory' of molecular evolution 25 years earlier, then heretical but now the basis of measuring evolutionary distances, i.e., time of branchings.)

14 Kondrashov, A. S.: **Fruit fly genome is not junk.** *Nature* 437 (Oct 20), 1106 (2005). < > (Commentary on paper by Andolfatto et al, p1149, comparing non-coding regions of two fruit fly genomes, with results "dealing a double blow to the neutral theory of molecular evolution.")

15 Silvertown, J. **Demons in Eden: The Paradox of Plant Diversity.** Chicago, IL: Univ. of Chicago Press, (2005) < > (Reviewed in *Nature,* 438:27, by P.D. Moore. The "demon" in the title refers to the ease with which invasive alien species sometimes take over. Sounds like a very interesting book, and review itself is informative.)

16 Nee, S.: **Thinking big in ecology.** *Nature* 417 (May 16), 229-230 (2002). < > (Highlights of symposium, *Macroecology: reconciling divergent perspectives on large-scale ecological patterns,* April 16-19, Birmingham, U.K. Sub-head speaks of the "new discipline known as macroecology" as a "bonny baby." Includes debates on why larger areas hold more species diversity, if they do. Neutral theory was a prominent topic.)

17 Hanski, I. **Metapopulation Ecology.** Princeton, NJ: Princeton Univ. Press, (1999) < > (Reviewed in *Nature* 399:747 by C. Parmesan, who opines that this book is a "tour de force, ... not for the faint-hearted ... thick enough to stand a spoon in ... requires a high degree of knowledge, as well as motivation" to read. Complains of poor print quality resulting in a "solid headache within 20 minutes." Nonetheless, reviewer "applauds Hanski for embracing what many ecologists emotionally reject: the incorporation of evolution" into this field, but which seems to be a growing trend. He asks "how much bigger can this bouncing baby [macroecology] become?" Reviewer predicts book will become a "classic". Headache and all?)

18 Hubbell, S. P. **The Unified Neutral Theory of Biodiversity and Biogeography.** Princeton, NJ: Princeton Univ. Press, (2001) < > (Called "the bible" of neutral theory in ecology.)

19   Whitfield, J.: **Neutrality versus the niche.** *Nature* 417 (May 30), 480-481 (2002).
<>

20 Ostling, A.: **Neutral theory tested by birds.** *Nature* 436 (Aug 4), 635-636 (2005).
<> (Perspective on work by Graves and Rahbek in *PNAS.* "Despite its simplicity, neutral theory has proven robust. Claims that it has been falsified have been followed by persuasive counter-arguments. Graves and Rahbek mount a new line of attack ...")

21 Dornelas, M., Connolly, S. R. and Hughes, T. P.: **Coral reef diversity refutes the neutral theory of biodiversity.** *Nature* 440 (Mar 2), 80-82 (2006). <> (With informative perspective by J.M. Pandolphi on p35-36, "Corals fail a test of neutrality.")

22 Wills, C. and al, e.: **Nonrandom processes maintain diversity in tropical forests.** *Science* 311 (Jan 27), 527-531 (2006). <> (Some 34 authors including Stephen Hubbell. Conclusion is not stated bluntly but is clearly at variance with neutral theory.)

23 Read, D.: **The ties that bind.** *Nature* 388 (Aug 7), 517 (1997). <> (Introduces report by Simard *et al,* p579 this issue, on the fungal symbionts linking the roots of forest trees.)

24 Giovannoni, S.: **Oceans of bacteria.** *Nature* 430 (Jul 29), 515-516 (2004). <> (Commentary on paper by Acinas et al, p551, "Fine-scale phylogenetic architecture of a complex bacterial community.")

25 Copley, J.: **Ecology goes underground.** *Nature* 406 (Aug 3), 452-455 (2000). <> (Survey of recent work on the importance of soil microbes on communities of plants.)

26 Callaway, R. M. and al, e.: **Positive interactions among alpine plants increase with stress.** *Nature* 417 (Jun 20), 844-848 (2002). <> (Thirteen authors study 115 species in 11 mountains ranges, the largest to date to investigate their interactions, finding that at low elevations where nutrients are limiting they are competitive but at high altitudes where wind, weather, etc. are the main stresses they interact cooperatively ["positively"]. This tests an assumption of species-independence, accepted in ecology since 1926, Lotka's era. From abstract: "Thus the notion that the distribution and abundances of plant species are independent of other species may be inadequate as a theoretical underpinning.")

27 Callaway, R. M. and al, e.: **Soil biota and exotic plant invasion.** *Nature* 427 (Feb 19), 1731-1733 (2004). <> (Incidentally, article following this one concerns the genetic of soil-dwelling arbuscular micorrhizal fungi.)

28 S, A. M.: **Plant wars.** *Science* 311 (Mar 17), 1525 (2006). <> (Short note on article by Weir and Vivanco in *Planta* showing that the spotted knapweed (*C. maculosa*) invading American grasslands produces a toxin from its roots, catechin, which eliminates competing native neighboring plants; but that some native plants, such as *Luppinus*

sericeus, resist the attack by the chemical countermeasure of releasing oxalate, which also protects its vulnerable neighbors.)

29 Gesteland, R. F., Cech, T. R. and Atkins, J. F. **The RNA World [3rd ed'n]:** Cold Spring Harbor Laboratory Press, (2006) <> (Big book with contributions from experts in various specialties, reviewed in *Nature Genetics* 38(4):393-4, by J.S. Mattick, who judges it to be a good reference but already outdated, and brings to attention several active and exciting topics not covered in the book.)

Ch. 11, References & Notes.

1 Stein, L. D.: **End of the beginning.** *Nature* 431 (Oct 21), 915 (2004).

2 Rennie, J.: **DNA's new twists.** *Sci American* (Mar), 122- (1993).

3 Lawrence, P. A.: **Science or alchemy?** *Nat Rev Gen* 2 (Feb), 139-142 (2001).

4 Pennisi, E.: **Genome data shake tree of life.** *Science* 280 (May 1), 672-673 (1998).

5 Whitfield, J.: **Born in a watery commune.** *Nature* 427 (Feb 9), 674-676 (2004). <> (Survey of efforts to identify LUCA – Last Universal Common Ancestor – and the problems that have come up, chiefly, rampant horizontal gene transfer.)

6 MacDonald, D. et al: **Structural basis for broad DNA-specificity in integron recombination.** *Nature* 440 (Apr 27), 1157-1162 (2006). <> (Work towards clarifying exactly how "single-stranded genetic material is recognized and exchanged between bacteria.")

7 Chen, I., Christie, P. J. and Dubnau, D.: **The ins and out of DNA transfer in bacteria.** *Science* 310 (Dec 2), 1456-1460 (2005). <> (Reviews complexities of topic, including descriptions of the several specialized gateways in the cell wall, named Types I, II, III and IV, each of which is a complicated apparatus for regulating the egress, entry or exchange of DNA.)

8 Houck, M. A., Clark, J. B., Peterson, K. R. and Kidwell, M. G.: **Possible horizontal transfer of *Drosophila* genes by the mite *Proctolaelaps regalis*.** *Science* 253 (6 Sep), 1125-1129 (1991). <> (See p1092 for commentary by J. Marx.)

9 Fedoroff, N. and Bottstein, D., eds. **The Dynamic Genome: Barbara McClintock's Ideas in the Century of Genetics.** Cold Springs Harbor, NY: Cold Springs Harbor Laboratory Press, (1992). <> (As reviewed in *Science* 259:1206-8, by A.C. Spradler.)

10 Doolittle, R. F.: **A bug with excess gastric avidity.** *Nature* 388 (Aug 7), 515-516 (1997). (Commentary on *H. pylori* genome sequence by Tombs *et al* (42 authors), p539, where related comments appear.)

11 Doolittle, R. F.: **Microbial genomes opened up.** *Nature* 392 (Mar 26), 339-342 (1998). < > (Discusses questions of phylogenetic relations among bacteria, archea, eucaryots, etc. in view of the complete genome of *Aquifex aeolicus* reported p353 of this issue; see also p351, "Horizontal gene transfer.")

12 Giovannoni, S.: **Oceans of bacteria.** *Nature* 430 (Jul 29), 515-516 (2004). < > (Commentary on paper by Acinas et al, p551, "Fine-scale phylogenetic architecture of a complex bacterial community.")

13 Mower, J. P. and al, e.: **Gene transfer from parasitic to host plants.** *Nature* 432 (Nov 11), 165-166 (2004). < > (Reports study of the mitochondrial gene *atp1* in 43 species of mostly weedy plants of the genus *Plantago*, finding that three of the species have acquired an additional variant pseudo-gene copy of it, apparently from a type of parasitic vine known as a dodder, of the genus *Cuscata*. Authors indicate that the mechanism was probably by way of the tiny hollow spines with which the parasite penetrates the flesh of the host, and suggest that such transfers between host and parasitic plants, in both directions, was common among the 4,000 species of parasitic plants.)

14 Hurtley, S.: **Geography of gene swapping.** *Science* 310 (Oct 14), 197 (2005). < > (Note on paper by Davis et al in *Proc. R. Soc. London Sec. B*, documenting "horizontal gene transfer between more distantly related plants: Part of the genome of the rattlesnake fern, *Botrychium virginianum,* appears to be derived from ... the parasitic sandalwoods and mistletoes." Geographical distributions suggest that the transfer was recent.)

15 Hooper, R.: **Hydra may keep genetic souvenir of lost partner.** *New Scientist* (Jun 4), 16 (2005). < > (Tells of finding alien plant gene in the simple animal, Hydra, not derived from its alga symbiont. Refers to *J Exp Biol* 208:2157.)

16 Pennesi, E.: **Researchers trade insights about gene swapping.** *Science* 305 (Jul 16), 334-335 (2004). < > (Highlights of a symposium, "Horizontal gene flow in microbial communities.")

17 Bushman, F. **Lateral DNA Transfer: Mechanisms and Consequences.** Cold Springs Harbor, N.Y.: Cold Springs Harbor Laboratory Press, (2002) < > (Reviewed in *Science* 295:2219-20.)

18 Novick, R. P.: **Plasmids.** *Sci Amer* (Dec), 102-109 (1980). < > (Subheading: "These accessory genetic elements in bacteria, best known as carriers of resistence to antibiotics, and as vehicles for genetic engineering, are actually subcellular organisms poised on the threshold of life." See also Nov 1987 p62. Although now routinely exploited in biotechnology, their evolution and precise role remains poorly understood.)

19 Clewell, D. B. **Bacterial Conjugation**. NY: Plenum Press, (1993) <> (Reviewed *Science,* 263:839, Feb 11, 1994, under heading "Plasmid Transfer." Said to be a "delightful smorgasborg.")

20 Baumberg, S., Young, J. P. W., Wellington, E. M. H. and Sanders, J. R. **Population Genetics of Bacteria**. N.Y.: Cambridge Univ. Press, (1995) <> (Reviewed *Science,* 270:1233, Nov 17, 1995, where it is noted that this is "only the third [book on this subject] to appear in the last two decades," citing the earlier two, then asking, "So what's new? A lot!.")

21 Jeljaszewicz, J. and Pulverer, G. **Antimicrobial Agents and Immunity**. N.Y.: Academic Press, (1986) <> (Reviewed *Amer Scientist,* 76:199, Mar/Apr 1988.)

22 Neu, H. C.: **The crisis in antibiotic resistence**. *Science* 257 (Aug 21), 1064-1073 (1992). <> (Special issue reviewing the topic. See esp. p1073 and p1078.)

23 Ferber, D.: **New hunt for the roots of resistance**. *Science* 280 (Apr 3), 27 (1998). <> (Wide use of antibiotics, as in animal feed, selects for resistant strains. "The resistant strains can then jump species lines into disease-causing bacteria ... For example, ampicillin has lost its potency against ... *Haemophilus influenzae* because the bacterium picked up an ampicillin-resistance gene from *Escheria coli* in the 1970's.")

24 Tomasz, A.: **Weapons of microbial drug resistance abound in soil flora**. *Science* 311 (Jan 20), 342-343 (2006). <> (Commentary on paper by V.M. D'Costa et al, p374, "Sampling the antibiotic resistome,' in which some 480 distinct bacterial types isolated from soil were tested for antibiotic resistance: all had natural resistance to several, and some were resistant to all 20 tested. But see also letter on this paper, 312:529.)

25 Shrag, S. J. and Perrot, V.: **Reducing antibiotic resistance**. *Nature* 381 (May 9), 120-121 (1996). <> (Effort to supply theory or rational strategy to circumvent problem, as well as the multi-drug resistence (MDR) seen in cancer patients. MDR was traced to a specific gene but further work revealed additional layers of complexity, poorly understood. One may ponder the relation of these phenomena to our own ability to quickly develop tolerance to drugs and some poisons.)

26 Veen, H. W. v. and al, e.: **A bacterial antibiotic-resistance gene that complements the human multidrug-resistence P-glycoprotein gene**. *Nature* 391 (Jan 15), 291-295 (1998).

27 Yin, Y. and al, e.: **Structure of the multidrug transporter EmrD from *Escheria coli***. *Science* 312 (May 5), 741-744 (2006).

28 Paulson, I. T. and al, e.: **Role of mobile DNA in the evolution of vancomycin resistent *Enterococcus faecalis***. *Science* 299 (Mar 28), 2071-2074 (2003). <> (Abstract of this paper, with 32 authors, states that "more than a quarter of the genome consists

of probable mobile or foreign DNA," and that the "propensity for the incorporation of mobile elements probably contributed to the rapid acquisition" of drug resistence in this group.)

29 Syvanen, M.: **In search of horizontal transfer.** *Nature Biotechnology* 17 (Sep), 833 (1999). <> (Survey of evidence for evaluating the risk that genes from genetically engineered food crops may escape to other species.)

30 Wheeler, D. A. and al, E.: **Molecular transfer of a species-specific behavior from** *Drosophilia simulans* **to** *Drosophilia melanogaster.* *Science* 251 (Mar 1), 1082-1085 (1991). <> (The gene *per* was introduced artificially to transfer the courtship song from one fruit fly species to another. They conclude that "further scrutiny ... may eventually determine whether *per* plays a role in drosophilid speciation.")

31 Systems, R. D.: **Viral cytokines.** *Cytokine Bulletin* (Summer), 4-5 (1998).

32 Amabile-Cuevas, C. F. and Chicurel, M. E.: **Horizontal gene transfer.** *Amer Scientist* 81 (Jul/Aug), 332-341 (1993).

33 Quiring, R., Walldorf, U., Kloter, U. and Gehring, W. J.: **Homology of the** *eyeless* **gene of** *Drosophila* **to the** *small eye* **gene in mice and** *anaridia* **in humans.** *Science* 265 (Aug 5), 785-789 (1994). <> (For perspective, see p742 by C.S. Zucker.)

34 Gould, S. J.: **Common pathways of illumination.** *Natural History* (Dec), 10-20 (1994). <> (Perspective on evolution of the eye *vis a vis* the recent discovery that eye development in many distantly related genera seem to be controlled in part by a homologous gene.)

35 Pennisi, E.: **Worm's light-sensing protein suggests eye's single origin.** *Science* 306 (Oct 21), 796 (2004). <> (Perspective on article by Arendt et al, p869.)

36 Arnhelter, H.: **Eyes viewed from the skin.** *Nature* 391 (Feb 12), 632-633 (1998). <> (Conjectures on why vertebrate melanophores [skin pigment cells] contain an invertebrate opsin, *vis a vis* the evolution of eyes, horizontal gene transfer, the purposes of melanophores, etc.)

37 Frigaard, N. U. and al, e.: **Proteorhodopsin lateral gene transfer between marine planktonic Bacteria and Archaea.** *Nature* 439 (Feb 16), 847-850 (2006). <>

38 Wehner, R.: **Brainless eyes.** *Nature* 435 (May 12), 157-159 (2005). <> (Perspective on report by Nilsson et al, p201 this issue, showing that a box jellyfish, a.k.a, a cubomedusa or 'cubozoan' of species *Tripedalia crystophora*, possesses small but sophisticated sets of image-forming eyes. It is an active swimmer and predator in near-shore environments.)

39 Shubin, N.: **Evolutionary cut and paste**. *Nature* 394 (Jul 2), 12-13 (1998). <> (Perspective on report by J. Clark, p66, bringing to attention a "melange of crown-group characters" in a Carboniferous tetrapod, and citing other such cases.)

40 Toner, M.: **When squid shine**. *International Wildlife* 24 (May/Jun) (1994).

41 Pepling, R. S.: **All that glows**. *Chem Engin News* (Apr 3), 24-25 (2006). <> (Brief survey of bioluminesence, including the recently discovered deep sea creature that features a red glowing lure to attract prey [*Science* 309:263].)

42 Johnsen, S., Balert, E. J. and Widder, E. A.: **Light-emitting suckers in an octopus**. *Nature* 398 (Mar 11), 113 (1999). <> (The light-emitting organs appear to be modified suckers.)

43 Pieribone, V. and Gruber, D. F. **Aglow in the dark: The revolutionary science of biofluorescence**. New York: Belknap Press, (2006) <> (Reviewed in *Nature* 440:280 by T.G. Oertner, who has a few "quibbles" but on the whole finds the main narrative "riveting.")

44 Albert, V. A., Williams, S. E. and Chaset, M. W.: **Carnivorous plants: phylogeny and structural evolution**. *Science* 257 (Sep 11), 1491-1495 (1992). <>

45 Barthlott, W., Porembski, S., Fischer, E. and Gemmel, B.: **First protozoa-trapping plant found**. *Nature* 392 (Apr 2) (1998). <>

46 Heslop-Harrison, Y.: **Carnivorous plants**. *Sci American* (Feb), 104 (1978). <>

47 Milius, S.: **Killer flatworm**. *Science News* 169 (Feb 18), 100 (2006). <> (Tells of report by Ritson-Williams et al in current *PNAS* showing that a newly discovered marine flatworm kills its molluscan prey by injecting the neurotoxin, tetrodotoxin, under the lip of the shell. This toxin is found in diverse genera such as in certain frogs, puffer fish, blue-ringed octopus, and in land-dwelling rough-skinned newts. Author thinks it is most likely obtained from bacteria; if so, the worm must be somehow protected against its tox effect. Other flatworms studied by the authors acquire toxins from sponges and sea squirts.)

48 Travis, L.: **Genetic pickup: did animals get brain genes from bacteria?** *Science News* 165 (Jun 12), 372-373 (2004). <> (On work by Klein et al, July *Trends in Genetics,* arguing that many key genes acting in the brain and elsewhere in higher animals were picked up from microbes, being absent in plants and other intermediates. Controversial. One view holds that the genes in question decayed and were lost in intermediate life forms. Is a puzzle.)

49 Roelofs, J. and VanHaastert, P. J. M.: **Genes lost during evolution**. *Nature* 411 (Jun 28), 1013 (2001). <> (He finds a gene for adenyl cyclase in a slime mold, which gene is also found in humans but is absent in certain worms, fruit flies, plants, yeasts.)

50 McClintock, J.: **This is your ancestor**. *Discover Magazine* (Nov), 64-70 (2004). <> (On work by Mitchell Sogin of the Marine Biological Laboratory, Woods Hole, MA, that we have many genes in common with sponges even though these genes are absent in many other forms thought to be intermediate.)

51 Holmes, B.: **Lowly sea animals boast world-class genetic armory**. *New Scientist* (Dec 3), 10 (2005). <> (Comments on article in *Trends in Genetics,* 21:633, on the genomes of lowly cnidarians, which includes sea anemones and corals, with the surprise finding that they have about 25,000 genes, "'the same ballpark humans are in,' says team member Eldon Ball." This means that the many organisms which arose later and which have much simpler genomes must have lost many genes. Thus, overlooking the simple and primitive marine creatures gave rise to the false impression that genomes have become steadily more complex.)

52 Kurland, C. G., Collins, L. J. and Penny, D.: **Genomics and the irreducible nature of eukaryote cells**. *Science* 312 (May 19), 1011-1014 (2006). <> (Defends the hypothesis that the major branches of living things have lost many genes that were present in the last universal common ancestor, LUCA. But if so, LUCA must have been one hell of a critter.)

53 Constans, A.: **Those mysterious noncoders**. *The Scientist* (Mar), 68 (2006). <> (Commentary on what is called a landmark paper, though it was originally rejected by *Science* because it was too radical, by S. Cawley et al, *Cell* 116:499, 2004, finding function in 'junk DNA' – noncoding DNA. Specifically, they mapped binding sites for three regulators of transcription to such regions.)

54 Schmitt, S. and Paro, R.: **A reason for reading nonsense**. *Nature* 429 (Jun 3), 510-511 (2004). <> (Background on paper by Martens et al, p571, to the effect that transcription of non-coding 'noise' upstream of the gene controls expression of that gene. Details are technical but the interesting conclusion they reach is that this is done by the mere act of the machinery reading through the sequence, not by the transcript produced, i.e. it seems to be a kind of allosteric effect.)

55 Travis, J.: **Possible evolutionary role explored for 'jumping genes'**. *Science* 257, 884-885 (1992). <> (Report on meeting, "Transposable Elements and Evolution." Term "jumping genes" refers to mobile DNA sequences that can jump from chromosome to chromosome, discovered by Barbara McClintock, Nobel laureate 1983.)

56 Federoff, N. V.: **Transposable genetic elements**. *Sci American* (Jun), 84-ff (1984).

57 Federoff, N. V.: **The restless gene: How the colors of Indian Corn have led to an understanding of wandering DNA**. *The Sciences* (Jan/Feb), 22-28 (1991).

58 Faulker, G. J. and Carninci, P.: **Altruistic functions for selfish DNA**. *Cell Cycle* 8 (18), 2895-2900 (2009).

59 Kondrashov, A. S.: **Fruit fly genome is not junk.** *Nature* 437 (Oct 20), 1106 (2005). < > (Commentary on paper by Andolfatto et al, p1149, comparing non-coding regions of two fruit fly genomes, with results "dealing a double blow to the neutral theory of molecular evolution.")

60 Bejerano, G. and al, e.: **A distal enhancer and an ultraconserved exon are derived from a novel retroposon.** *Nature* 441 (May 4), 87-90 (2006). < > (From abstract: "These add to a growing list of examples in which relics of transposable elements have acquired a function that serves their hosts, a process termed 'exaptation.'" They date this one back to a primitive fish, the coelecanth, about 410 million years ago.)

61 Bushman, F.: **Selfish elements make a mark.** *Nature* 429 (May 20), 253-255 (2004). < > (Subhead: "Transposons qualify as 'selfish' DNA elements, adding new copies of themselves into our genomes without regard for the consequences. This willful habit may, however, help in normal gene regulation." Concluding paragraph indicates that transposon insertions "provide a 'molecular rheostat'" governing gene activity, and proposes that "this could be a key feature of many speciation events.")

62 O'Neill, R. J. W., O'Neill, M. J., and Graves, J. A. M.: **Undermethylation associated with retroelement activation and chromosome remodelling in an interspecific mammalian hybrid [wallaby].** *Nature* 393 (7 May), 68-72 (1998). < > (For perspective by M.G. Kidwell and D.R. Lisch, see "Transposons Unbound," p22-23 this issue.)

63 Pennisi, E.: **Rogue fruit fly DNA offers protection from insecticides.** *Science* 309 (Jul 29), 681 (2005). < > (Begins by noting the reputation of transposons for wreaking genetic havoc in the course of jumping; "but sometimes these interlopers can do some good." Specifically, as detailed in the report on p764, one of them is now identified as having been recruited to supply crucial help in resisting insecticides, just within the last century or so. Author states, "they may play a much larger role in evolutionary novelty than is currently appreciated.")

64 Moffat, A. S.: **Transposons help sculpt a dynamic genome.** *Science* 289 (Sep 1), 1455-1457 (2000). < > (Surveys then-new developments on this type of 'junk DNA,' such as that transposons can help a genome to shrink as well as expand. The ends link up like an inchworm and the middle is cut out. Wide variations in rates of gain and loss are noted between fruit flies, mammals, crickets; see issue of Feb. 11, p1060.)

65 Editor: **An immune system that's so versatile it might kill you.** *New Scientist* (Jun 17) (2006). < > (Note on paper in *Genes and Development* 20:1575 explaining that leukocytes – white blood cells of the immune system – exploit transposons ['jumping genes'] as part of the mechanism for producing a diverse array of disease-fighting antibodies. This can cause genetic errors, but they are normally promptly corrected. However, this paper explains that in a certain fraction of cases such an error may go uncorrected, causing lymphoma, a cancer of these blood cells. Thus, rapid immune

adaptation is purchased at the cost of increased danger of cancer – according to the authors.)

66 Brownless, C.: **Trash to treasure**. *Science News* 166 (Oct 16), 243 (2004). <> (Highlights of paper by B. Knowles et al in Oct. issue of *Developmental Cell,* to the effect that certain retrotransposons, thought to be junk DNA and amounting to some 25% of the genome, are located near key genes involved with embryonic development and have important regulatory functions.)

67 Hammock, E. A. D. and Young, L. J.: **Microsatellite instability generates diversity in brain and sociobehavioral traits**. *Science* 308 (Jun 10), 1630-1634 (2005). <> (With perspective by E. Pennisi, p1538, titled "In voles, a little extra DNA makes for faithful mates." Also noted in *Science News*, 168:30, Jul 9, under title "More junk makes for better dads;" and in *New Scientist,* Jun 18, p21, under title, "Junk DNA keeps a vole devoted;" and more broadly in *The Scientist*, Feb 14, p20-21, by Jack Lucentini, entitled "Love is like an addiction," with photo captioned "Altering the expression of a single gene can switch meadow voles from a promiscuous to a monogamous lifestyle.")

68 Carter, S. C. and Getz, L. L.: **Monogamy and the prairie vole**. *Sci American* 6/93 (Jun), 100-106 (1993). <> (The opening blurb: "Studies of the prairie vole – a secretive mouse-like animal – have revealed hormones that may be responsible for monogamous behavior."

69 Fackelmann, K. A.: **Hormone of monogamy: prairie vole and the biology of mating**. *Science News* 144 (Nov 27), 360-365 (1993).

70 Pennisi, E.: **A genomic view of animal behavior**. *Science* 307 (Jan 7), 30-32 (2005). <> (Survey of recent studies seeking to identify gene sets for various behaviors such as maternal care, socialization, fidelity of pair bonds, etc. Thus, one study found 100 genes involved with male social status of a fish, and 2,200 of 5,500 bee genes differed between two kinds of bees with different behavior patterns.)

71 Reporter: **Vole love**. *Science* 265 (Sep 30), 2007 (1994). <> (Concludes with quotation, "'This work offers more evidence that social behavior has an important basis in biology.'")

72 Kaneko-Ishino, T. and Ishino, F.: **Retrotransposon silencing by DNA methylation contributed to the evolution of placentation and genome imprinting in mammals**. *Dev Growth Differ* 52 (6), 533-543 (2010).

73 Black, S. G., Arnaud, F., Palmarini, M. and Spencer, T. E.: **Endogenous retroviruses in trophoblast differentiation and placental development**. *Am J Reproduc Immunol* 64 (4), 255-264 (2010).

74 Flajnik, M. F. and Kasahara, M.: Origin and evolution of the adaptive immune system: genetic events and selective pressures. *Nat Rev Genet* 11 (1), 47-59 (2010). < >

75 Fugman, S. D.: The origins of the Rag genes – from transposition to V(D)K recombination. *Semin Immunol* 22 (1), 10-16 (2010).

76 Cordaux, R. and Batzer, M. A.: The impact of retrotransposons on human genome evolution. *Nat Rev Genet* 10 (10), 691-703 (2009).

77 Friedberg, E. C. Correcting the Blue Print of Life. An Historical Account of the Discovery of DNA Repair Mechanisms. Cold Spring Harbor, NY: Cold Spring Harbor Laboratory Press, (1997) < > (Reviewed *Science,* 277:1945, 26 Sep 1997, by L.A. Loeb; and in *Nature,* 390:573, 11 Dec 1997, by R.D. Wood.)

78 Jr, D. E. K.: Molecule of the year: the DNA repair enzyme. *Science* 266 (Dec 23), 1925 (1994). < > (Introduction to special issue on DNA repair; see esp. p1954-1959.)

79 Marx, J.: DNA repair comes into its own. *Science* 266 (Nov 4), 728-730 (1994).

80 Friedberg, E. C., Walker, G. C. and Siede, W. DNA Repair and Mutagenesis. Washington, DC: ASM Press, (1995) < > (Reviewed in *Science,* 270:1511, Dec 1, 1995, by B.S. Strauss, who remarks that this well-written book will remain useful even as new details come to light – but then cites three major developments missed, all in the year during the book's publication: *Science* 268:738, 268:1749, 269:909.)

81 Zhou, B. B. S. and Elledge, S. J.: The DNA damage response: putting checkpoints in perspective [Review]. *Nature* 408 (Nov 23), 433-439 (2000).

82 Sancar, A. and et al: Molecular mechanisms of mammalian DNA repair and the DNA damage checkpoints. *Annu Rev Biochem* 73, 39-85 (2004).

83 Workman, C. T. and al, e.: A systems approach to mapping DNA damage response pathways. *Science* 312 (May 19), 1054-1059 (2006). < > (Identifies 5,272 interactions among 30 DNA damage-related factors, assembled into "causal pathway models." Authors conclude optimistically that although complicated, it is not so extremely so as to prohibit future understanding. But they assume that all such pathways and mechanisms are known.)

84 Banerjee, A., Santos, W. L. and Verdine, G. L.: Structure of a DNA glycosylase searching for lesions. *Science* 311 (Feb 24), 1153-1157 (2006). < > (This particular type of repair molecule, part of a "lesion recognition complex," or LRC, searches out a very subtle kind of mutation, in this case, a G that has been chemically altered to OxoG, differing in just two atoms. Although the molecule studied, MutM, is bacterial, a similar function is performed in mammals by Ogg1.)

85 Weinert, T.: **A DNA damage checkpoint meets the cell cycle engine.** *Science* 277 (Sep 5), 1450-1451 (1997).

86 Pyle, A. M.: **Big engine finds small breaks.** *Nature* 432 (Nov 11), 187-188 (2004). <> (Introduces report by Wigley et al, p187, on the structure and functioning of the double-strand DNA break repair complex, RecBCD. Seeing the structure "helps us to visualize how the complex [RecBCD] careers along the DNA [at 1,000 bp/sec], applies the brakes at a designated site, and chops up unwound DNA in its wake." Remarks that "a double strand break in DNA is the equivalent of a 3-alarm fire," then tells of some of the equipment used for the repair.)

87 Tsukada, T. et al: **Chromatin remodeling at a DNA double-strand break site in** *Saccharomyces cerevisiae. Nature* 438 (Nov 17), 379-382 (2005).

88 Della, M. and al, e.: **Mycobacterial Ku and ligase proteins constitute a two-com-ponent NHEJ repair machine.** *Science* 306 (Oct 22), 643-646 (2004). <> (Concerns repair of double strand breaks; NHEJ stands for non-homologous end joining.)

89 Casci, T.: **One very tough bug.** *Nat Rev Gen* 2 (Jun), 402-403 (2001). <> (Comments on recent findings by Carlin and Mrazek, *PNAS USA* 98:5240, on the bacterium, *Deinococcus radiodurans,* known and so named for its exceptional power to resist and rebound from X-radiation doses thousands of times above those lethal to most.)

90 Hoyle, S. F. and Wickramasinge, N. C. **Evolution from Space: A Theory of Cosmic Creationism.** New York: Simon & Schuster, (1981).

91 Nemitz, E.: **Mutant of the month.** *Nature Genetics* 38 (4), 407 (2006). <> (After remarking on the many advances that have transformed the classical view of genet-ics, notes that "the *Arabidopsis hothead* mutant has the potential to push the laws of Mendelian inheritance to the breaking point ... because it has the amazing ability to revert, at relatively high frequency, back to an ancestral state." How this might work is a real puzzler but some conjectures are given. Refers to *Nature* 434:505. Note: As I recall, doubts about this report, and/or possibly the next cited, later surfaced, and may have been otherwise explained.)

92 Weigel, D. and Jurgens, G.: **Hotheaded healer.** *Nature* 434 (Mar 24), 443 (2005). <> (Comment on report by Lolle et al, p505, of the "spectacular discovery" that a forced mutation of a gene called *hothead* in the much-studied plant, *Arabadopsis thaliana,* which causes severe abnormalities, failed to have any obvious effects on some of the offspring, and that this mutation completely disappeared in most of later generations. "Another fascinating question is what template the cell uses to restore the original DNA sequence." Known mechanisms were ruled out.)

93 Brown, L. and McCarthy, N.: **A sense-abl response?** *Nature* 387 (May 29), 450-451 (1997). <> (Perspective on reports by Baskaran *et al,* p516, and by Shafman *et al,*

p520, this issue. See also Kharbanda *et al, Nature* 386:732, 17 Apr 1997. Reveals new complications in DNA damage response / repair, found in the first paper by study of the hereditary disease, ataxia telangiectasia, characterized by abnormal response to ionizing radiation, along with a variety of serious clinical manifestations.)

94 Bartek, J. and Lukas, J.: **Damage alert.** *Nature* 421 (Jan 30), 486-487 (2003). <> (Background on report by Bakkenist and Kastan on p499, explaining principles of gene repair, especially of double strand breaks. Paper deals with biochemical workings, with focus on the faulty repair system gene, *ATM*, in the hereditary disease, ataxia telangiectasia.)

95 Kastan, M. B.: **DNA damage responses: cancer and beyond.** *The Scientist* (Oct 10), 24-25 (2005). <> (Discusses prospects for cancer therapy based on new knowledge of DNA repair in relation to cancer.)

96 Editor: **Slug beats puma.** *Nature* 438 (Nov 24), 398 (2005). <> (Short note on paper in *Cell,* 123:641, on the latest findings concerning p53, a protein of great interest in cancer research, long believed to promote cancer but now known to "either initiate DNA repair or trigger cell suicide.")

97 Friend, S.: **p53: A glimpse at the puppet behind the shadow play.** *Science* 165 (Jul 15), 334-335 (1994). <> (Short review as of then, of cancer-associated mutations in p53 in light of structure determination.)

98 Service, R.: **Slow DNA repair implicated in mutations found in tumors.** *Science* 263 (Mar 11), 1374 (1994). <> (Perspective on report by Gao *et al,* p1438.)

99 Hausen, H. Z.: **Papillomavirus and p53.** *Nature* 393 (May 21), 217 (1998). <> (Perspective on report by Storey *et al,* pg 229 this issue, elucidating its role in this particular cancer.)

100 Rennie, J.: **Kissing cousins: a DNA repair system stops species from interbreeding.** *Sci American* (Feb), 22-23 (1990). <> (Survey of studies with bacteria reported in *Nature,* Nov 1989; in this theory, defects in the DNA mismatch repair system "may help create new species by erecting a reproductive barrier between groups ... with slightly different DNA even if there are no geographical obstacles.")

101 Bieniasz, P. D.: **Intrinsic immunity: a front-line defense against viral attack (Review).** *Nature Immunology* 5 (Nov), 1109-1115 (2004). <> (Describes a "third arm" of the immune system, additional to innate and adaptive, which seems to bridge the gap between traditional mechanisms and genomic defenses, dubbed *intrinsic* immunity. Author wisely notes that "it may be that present knowledge of intrinsic immunity represents only a small fraction of its true complexity.")

102 Peck, w. R. and Eyre-Walker, A.: **The muddle about mutations.** *Nature* 387 (May 8), 135-136 (1997). <> (Concerning uncertainties about variable rates of mutations

in various species; cites a *PNAS* report which has "shaken the foundations on which current estimates" of such rates rest. This is another headache for those attempting to establish phylogenies based on the mutation rate 'molecular clock.')

103 Moxon, E. R. and Thaler, D. S.: **The tinkerer's evolving toolbox.** *Nature* 387 (Jun 12), 659-660 (1997). <> (Perspective on reports by Taddei *et al.* and by Sniegowski *et al.*, p700 and p703 this issue, concerning the production of mutator alleles stimulated by environmental stresses.)

104 Doolittle, W. F.: **Phylogenetic classification and the universal tree.** *Science* 284 (25 Jun), 2124-2128 (1999). <> (On very early branchings of major life domains and kingdoms. Discusses difficulties in attempts to identify what is called the 'last universal common ancestor,' LUCA. Based on molecular phylogenies.)

105 Sugden, J. M. and et al: **Charting the evolutionary history of life.** *Science* 300 (Jun 13), 1691 (2003). <> (Introduces theme issue on subject, with six articles and editorial.)

106 Ridley, M. **Evolution and Classification: The Reformation of Cladism.** London: Longman, (1986) <> (Reviewed in *Amer Scientist,* 76:37, Jan/Feb 1988. Surveys the rise of "phylogenetic systematics, a school of taxonomic classification founded during the 1950's by Willi Hennig.")

107 Naylor, G. J. P. and Brown, W. M.: **Structural biology and phylogenetic estimation {Letter}.** *Nature* 388 (Aug 7), 527-528 (1997). <> (On controversy about whether DNA sequences yield reliable divergence time estimates and phylogenies similar to those based on biological [phenotypic, cladistic] criteria. The letter is not totally negative to the molecular approach but points out concerns. For more on this, see opening letter in *Science,* Jun 13, 1997.)

108 Roberts, J. P.: **New growth in phylogeny programs.** *The Scientist* (Dec 20), 22-24 (2004). <> (Subsequent articles survey other topics in computer-based bioinformatics.)

109 Rokas, A. and al, E.: **Genome-scale approaches to resolving incongruence in molecular phylogenies.** *Nature* 425 (Oct 23), 798-804 (2003). <> (See also perspective by H. Gee p782.)

110 Gibbons, A.: **Calibrating the mitochondrial clock.** *Science* 279 (Jan 2), 28-29 (1998). <> (Report on First International Workshop on Human Mitochondrial DNA, Oct 25-28 1997, Washington, D.C. Serious problems, discrepancies found.)

111 Newton, I. **The Speciation and Biogeography of Birds.** N.Y.: Academic Press (Elsevier), (2003) <> (Reviewed *Science*, 304:1249, 28 May 2004, by A. Townsend Peterson, with references, under caption, "A New Synthesis or just the [old] New Synthesis?" To us non-experts, this would probably be a very rewarding read but expert Peterson, who calls it "boldly titled," finds it disappointing for several reasons,

such as (i) phylogenetic methods outdated and assumed accuracy of molecular clock unreliable, (ii) relies on Mayr's biospecies concept (BSC) and "more or less rejects alternatives" which are proving more illuminating, (iii) "many exciting results from recent research have been dismissed, downplayed, or ignored, including the impressive avifaunal richness of Oceania before arrival of humans, pre-Pleistocene speciation ..." Yet, reviewer calls it "eminently readable" and probably the "best available ...", yet falls short, is "curiously dated" in being "firmly embedded in the new synthesis." But reviewer does not identify alternatives to the New Synthesis.)

112 Felsenstein, J. **Inferring Phylogenies**. Sunderland, MA: Sinauer, (2004) < > (Reviewed *Science* 303:767, Feb 6 2004, by F. Ronquist, under heading, "A broad look at tree building," noting that the author is an "icon in the field." Said to be excellent on some topics (tree building), weak on others (e.g. systematics), but this is a huge field, nobody could cover it all. Said to neglect Willi Hennig, regarded as the father of modern phylogenetics.)

113 Hafner, M. S. and *et al*: **Disparate rates of molecular evolution in cospeciating hosts and parasites**. *Science* 265 (Aug 19), 1087-1090 (1994). < > (Lice, being very host-specific, were examined for some mitochondrial DNA – for cytochrome C oxidase subunit I – in 14 species of pocket gopher and their corresponding lice, producing beautifully parallel sets of co-evolving phylogenetic trees. However, the mutation rate of the lice was 3-fold to 11-fold faster than the host rate, leading the authors to claim their findings support a neutral theory, and that mutation rates would be similar between louse and host if calculated per generation rather than uniform in time.)

114 Hillis, D. M., Huelsenbeck, J. P. and Cunningham, C. W.: **Application and accuracy of molecular phylogenies**. *Science* 264 (Apr 29), 671-677 (1994). < > (Defends this now-ubiquitous method of dating divergences among species and establishing their 'genetic distance' based on 'random changes' in amino acids in proteins. The authors emphasize in their conclusion that these methods are "powerful enough to reconstruct evolutionary histories with a high degree of accuracy, as long as the rates of change of the observed characters are appropriate for analysis.")

115 Peirce, J. L.: **Following phylogenetic footprints**. *The Scientist* (Sep 27), 34-37 (2004). < > (This computer-based approach to searching for regulatory elements is based on the assumption that if these regions of the genome are important, they will tend to be conserved – little changed – among related species and genera, compared to genomic regions that are not important. But reasons to doubt this come to mind, such as that some related species may be actively speciating, giving rise to higher mutation rates, at least in some regions of the genome.)

116 Rosenberg, S. M. and Hastings, P. J.: **Worming into genetic instability**. *Nature* 430 (Aug 5), 625-626 (2004). < > (Comments on paper by Denver et al, p679, measuring mutation rates in a worm, with results 10-fold higher than expected. They offer additional possible explanations, such as the SOS response, and revive an idea from

the 1980s of stress-induced mutations. They also remark on the ability of molecular chaperones to correct protein conformations whose gene has been damaged, citing Rutherford among others.)

117 Ayre-Walker, A.: **Size does not matter for mitochondrial DNA.** *Science* 312 (Apr 28), 537-538 (2006). <> (Commentary on paper by Bazin et al, p570, with brief note on p496 titled "Unreliable mitochondrial DNA," reading: "Thus, mtDNA is far from a neutral marker; its diversity is essentially unpredictable and may not reflect population history and demography.")

118 Wedemayer, G. J. and al, e.: **Structural insights into the evolution of an antibody combining site.** *Science* 276 (Jun 13), 1666-1669 (1997). <> (Study of the molecular structure of an antibody combined with its target antigen, before and after nine steps of hypermutation during which the antibody "matured" to tighter and tighter binding affinity. Details of exactly how this is accomplished remain obscure.)

119 Basu, U. and al, e.: **The AID antibody diversification enzyme is regulated by protein kinase A phosphorylation.** *Nature* 438 (Nov 24), 508-511 (2005). <> (Recent news on this recently discovered enzyme. For related key papers, see "Hot Papers" in *The Scientist*, Dec. 5, 2005, p26-27. For an earlier key paper on the mechanism, see *Nature*, 418:99, 2002.)

120 Alder, M. N. and al, e.: **Diversity and function of adaptive immune receptors in a jawless vertebrate.** *Science* 310 (Dec 23), 1970-1973 (2005). <> (Study of immunity in lamprey, a primitive fish, with perspective p1892-3. Similar immune system was earlier shown in the shark; see J. Rast and G. Litman in *PNAS*, Sep. 27, 1994, and as noted in *Science*, 265:2006.)

121 Pasquier, L. D.: **Insects diversify one molecule to serve two systems.** *Science* 309 (Sep 16), 1826-1827 (2005). <> (Perspective on paper by Watson et al, p1874, showing that a fruit fly protein called Dscam, which is homologous to a human protein involved with Down's syndrome, called DSCAM [two forms, encoded on chromosomes 11, 21], and functions to generate the equivalent of antibody diversity in the fly immune system. Table summarizing immune mechanisms in various creatures is given. Authors are intrigued by implicit connection between immune and nervous systems. Paper has been widely noted, e.g. in *The Scientist*, Sep. 24, p24, where the figure of "more than 18,000" variations is quoted.)

122 Nowak, R.: **How the parasite disguises itself.** *Science* 269 (Aug 11), 755 (1995). <> (Excerpt: "Work by three independent teams [named, some at the project for 10 years] shows that the malaria parasite avoids detection by the immune system by switching between as many as 150 genes, each encoding a different version of a protein known as EMP-1 ... which is made by the parasite after it infects red blood cells ...")

123 Brownlee, C.: **Codes for killers.** *Science News* 168 (Jul 16), 25 (2005). <> (Summary of report in July 15 issue of *Science* revealing more than 800 genes for surface proteins of *T. brucei*, the cause of African sleeping sickness, which it "mixes and matches" to evade the immune system of its host. *T. cruzi*, the agent of Chagas disease, which is transmitted by the bite of the kissing bug, *Triatoma infestans*, and the protozoan, *Leishmania major*, all "share a common core of 6,200 genes arranged in similar order.")

124 Boeke, J. D.: **A is for adaptation.** *Nature* 431 (Sep 23), 408-409 (2004). <> (Commentary on paper p476 this issue showing that a virus which infects bacteria of the genus *Bordatella*, which include the cause of whooping cough, has evolved a diversity-generating mechanism to quickly adapt to the equally fast-changing coat proteins of its target.)

125 Editor: **Quick costume change.** *Nature* 441 (Jun 21), 1030 (2006). <> (Short note on work in *PloS Biol.* 4, e185, finding that the bacterium *Neisseria gonnerhoeae* has from two to six copies of its genome per cell, said to help it to rapidly modify its outer coat proteins, called pilins, to thwart immune defenses, since the extra copies give it more resources for mixing variant genes or exons for these proteins.)

126 Hurtley, S.: **Not so spineless.** *Science* 310 (Dec 23), 1871 (2005). <> (Note on paper by Nair et al in *Physiol. Genomics* 22:33 on the immune system of a sea urchin, which has several modes including phagocytes and a complement system, and achieves the equivalent of specific antibodies by alternative splicing.)

127 Guttman, D. S. and Dykhuizen, D. E.: **Clonal divergence in *Escheria coli* as a result of recombination, not mutation.** *Science* 266 (Nov 25), 1380-1383 (1994).

128 Harris, R. S., Longerich, S. and Rosenberg, S. M.: **Recombination in adaptive mutation.** *Science* 264 (Apr 8), 258-260 (1994). <> (With perspective by D.S. Thaler, p258-260, called "The evolution of genetic intelligence" [sic], discussing the many theories to explain the apparent results, with 31 refs. The Harris paper confirms that recombination is essential for the effect at issue.)

129 Lunt, D. H. and Hyman, B. C.: **Animal mitochondrial DNA recombination.** *Nature* 387 (May 15), 247 (1997).

130 S, S. J.: **Rendered powerless by heme.** *Science* 312 (May 5), 661 (2006). <> (Note on paper in *J Biol* 5:5, showing that the malarial *Plasmodium* parasite causes general suppression of the host's immune system, evidently as a result of degraded hemoglobin.)

131 S, S. J.: **Helpful helminths.** *Science* 310 (Dec 2), 1393 (2005). <> (Note on paper by Wilson et al, in *J Exp Med*, 202:1199, and by Smith et al in the same journal, 202:1319, each showing that a parasite had effects blunting the host immune response. The first was a gut nematode worm which disabled the host's CD4+ T cell response. The second was a trematode which produced a protein which inactivated

cytokines. Many dozens of other such strategies are now known [2011], each devilishly clever and highly specific, not only by parasites but by viruses, bacteria, fungi.)

132 Shannon, M. and Weigert, M.: **Fixing mismatches**. *Science* 279 (Feb 20), 1159-1160 (1998). <> (Perspective on paper by Cascalho et al, p1207, showing that "a protein ordinarily involved in correcting mutations is involved in causing V gene mutations" for antibody variation. Opens by stating that "complex biological processes evolve by co-opting bits and pieces of pre-existing cellular machinery and using them for new purposes." Other speak of this sort of thing as "molecular cut-and-paste".)

133 Pal, C., Papp, B. and Lercher, M. J.: **An integrated view of protein evolution**. *Nat Rev Gen* 7 (May), 337-348 (2006). <> (This article is far more cautious than similar articles on this subject of ten years earlier, which are already badly outdated.)

134 Lee, J. T.: **Complicity of gene and pseudogene**. *Nature* 423 (May 1), 26-28 (2003). <> (Perspective on paper by Hirotsune et al, p91, giving evidence that at least one pseudogene is not "junk" but performs vital function. He notes that many would have missed this discovery and praises the author for perceptive and diligent work. Writer notes that "this study generates new and exciting questions.")

135 Duret, L. and al, e.: **The Xist RNA gene evolved in eutherians by pseudogenization of a protein-coding gene**. *Science* 312 (Jun 16), 1653-1655 (2006). <> (This gene functions to silence one of the two X chromosomes in females of higher mammals [eutherians] but remarkably, not in lesser mammals [marsupials] or in birds or fish. More remarkably, it arose from a protein-coding gene which lost its function [pseudogene] and now encodes only a long RNA transcript, which is not translated into protein. How this happened seems to be a mystery.)

136 Kafri, R., Bar-Even, A. and Pilpel, Y.: **Transcription control reprogramming in genetic backup circuits**. *Nature Genetics* 37 (3), 295-299 (2005). <> (With commentary by L.D. Hurst ad C Pal, p214-5. Addresses the question of "dispensable" DNA, specifically, why so many duplicate genes, known as paralogs, exist. Though we lack space for details, the blurb quoted is from *Nature*, Mar 24, 2005, pg xv. Of course, we wonder how the genome knows to call upon the backup when needed. Late note added [Dec 2010]: It is now very clear that copy number variations are very important, and are increasingly revealing differences between human individuals in, for example, susceptibility to certain diseases and other problems.)

137 Ohno, S. **Evolution by Gene Duplication**. London: Allen & Unwin, (1970) <> (Also published Heidelberg, Germany, by Springer-Verlag)

138 Lewis, R.: **Genome evolution: first a bang, then a shuffle**. *The Scientist* (Jan 27), 18-20 (2003). <> (Subtitle: "Did duplication fuel vertebrate genome evolution?" Survey of the theory. Article title is from Ken Wolfe's succinct statement: first a bang [duplication], then a shuffle [modification of the duplicated part].)

139 Le, T. N. M. and al, e.: **Gene duplication of coagulation factor V and origin of venom prothrombin activator in Pseudonaja testilis snake.** *Thromb Haemost* 93 (3), 420-429 (2005). <> (Beautiful work built on their prior findings showing that a key component of this snake venom is made from a duplicate of a gene for a normal blood clotting protein but modified to block its normal shutdown mechanism, and expressed specifically in the venom gland, at a level much higher than the normal liver site of synthesis.)

140 Dujon, B. and al, E.: **Genome evolution in yeasts.** *Nature* 430 (Jul 1), 35- (2004). <> (About 70 authors, French; hard for me to read.)

141 Kellis, M., Birren, B. W. and Lander, E. S.: **Proof and evolutionary analysis of ancient genome duplication in the yeast *Saccharomyces cerevisiae*.** *Nature* 428 (Apr 8), 617-624 (2004).

142 Gu, Z. and al, E.: **Role of duplicate genes in genetic robustness against null mutations.** *Nature* 421 (Jan 2), 63-66 (2003). <> (See also p31 for perspective by Axel Myer.)

143 Silverman, P. H.: **Rethinking genetic determinism.** *The Scientist* (May 24), 32 (2004).

144 Jaffe, S.: **Alternative splicing goes mainstream.** *The Scientist* (Dec 15), 28-30 (2003).

145 Jaffe, S.: **Humanizing protein splicing.** *The Scientist* (Jun 7) (2004). <> (Sub-head: "Human immunity uses a post-translational modification [type of splicing] previously shown in plants and yeasts." The paper reviewed is highly praised. "Most scientists would have given up ... much earlier, says Nabilh Shastri ... 'It was the clever thought processes and extremely hard work that made their paper so impressive.'" Caption of figure reads, "It slices, it dices, it even splices.")

146 Hansson, K. and Stenflo, J.: **Post-translational modifications in proteins involved in blood clotting.** *J Thromb Haemost* 3 (12), 2633-2648 (2005).

147 Kwong, P. D. and al, e.: **HIV-1 evades antibody mediated neutralization through conformational masking of receptor binding sites.** *Nature* 420 (Dec 12), 678-682 (2002). <> (Shows how the AIDS virus defeats the host immune system, in part by variable patterns of glycosylation. More recent work on immune response to HIV-1 is highly interesting but beyond this note.)

148 Davis, B. G.: **Mimicking post translational modifications of proteins.** *Science* 303 (Jan 23), 480-482 (2004). <> (Survey of efforts to mimic the natural process, with important practical applications, such as to protein drugs made by recombinant technology which lack natural glyosylation. Authors note that "the most widespread and complex" form is glycosylation, served by about 1% of the human genes, yet is

generally absent in lower forms such as bacteria. However, they note that at least one virus uses glycosylation to circumvent host immunity, citing P.D. Kwong et al, *Nature* 420:678.)

149 Rutherford, S. L. and Lindquist, S.: **Hsp90 as a capacitor for morphological evolution.** *Nature* 396 (Nov 26), 336-342 (1998). <> (Landmark paper, with commentary by A. Cossins, p309-10. Hsp stands for heat-shock protein, discovered by their enrichment in heat-stressed cells, but now known to have diverse roles, including chaperoning. These authors studied fruit flies with mutated or chemically inhibited Hsp90, resulting in numerous abnormalities. At risk of oversimplifying, the bottom line is that Hsp90 normally protects against defective protein from aberrant genes. However, the authors claim that when Hsp90 is inhibited, as by stress, natural variations in the genotype become overt, available for natural selection, hinting at new species. But the variants all look like monsters to me – not even hopeful ones – and are frankly described as "defects," so that interpretation is a reach, similar to appeal to random mutations.)

150 Ju, T. and Cummings, R. D.: **Chaperone mutation in Tn syndrome.** *Nature* 437 (Oct 27), 1252 (2005). <> (Rare autoimmune disorder due to defective glycosylation of blood cells by a certain enzyme. Authors trace this to absence of a chaperone needed for proper folding of the glycosylating enzyme,)

151 Kulathinal, R. J. and al, e.: **Compensated deleterious mutations in insect genomes.** *Science* 306 (Nov 26), 1553-1554 (2004). <> (Briefly, harmful mutations are often compensated – "cancelled out" – by corresponding changes elsewhere in the genome which nullify the effect; this is shown by comparing two closely related species of fruit fly, some having mutations that would be harmful in the other. In other words, as in the next reference, the organism often finds a way to defeat your efforts to limit its capabilities.)

152 Kruglinski, S.: **Bacteria find they can all get along.** *Discover Magazine* (Dec), 20 (2003). <> (Short note on remarkable work by G. Vilicer and wife, who genetically modified bacteria (*Myxococcus xanthus*) to lack their means of attaching to each other (pili) to form social colonies. But after a short time they developed alternative means for attaining the same end. "We watched novel forms of cooperation evolve before our eyes," they say.)

Ch. 12, References & Notes.

1 Pennisi, E.: **Searching for the genome's second code.** *Science* 306 (22 Oct), 632-635 (2004). <> (In special issue, "Genes in Action," p629-650. Of related interest, see p644, *Gene order and dynamic domains,* and p647, *Cis-acting regulatory variation in the human genome.*)

2 Wolfe, A. P.: **Transcriptional control.** *Nature* 387 (May 1), 16-17 (1997).

3 Blackwood, E. M. and Katonaga, J. T.: **Going the distance: A current view of enhancer action [Review].** *Science* 281 (Jul 3), 60-63 (1998).

4 Ghazi, A. and Raghaven, K. V.: **Control by combinatorial codes.** *Nature* 408 (Nov 23), 419-420 (2000). <> (Commentary on three papers in *Cell*, where a number of specific examples are considered in terms of the combinatoric model. General idea is that genes are regulated by logic circuits, and that enhancers and repressors act on genes somewhat like the voltages that control transistors.)

5 Levine, M. and Tjian, R.: **Transcription regulation and animal diversity (Review).** *Nature* 424 (10 Jul), 147-152 (2003). <> (Opens by noting the unexpectedly small number of human genes and other features of the genome compared to seemingly much simpler life; and that two explanations have been proposed to explain the complexity of higher forms, (i) alternative splicing, (ii) transcriptional regulation. This review surveys the latter.)

6 Posfai, G. et al: **Emergent properties of reduce genome** *Escherichia coli.* *Science* 312 (May 19), 1044-1046 (2006). <> (They knock out as much as 15% of the genome of this bacterium and find that it still functions, with some modified properties.)

7 Nobrega, M. A. et al: **Megabase deletions of gene deserts result in viable mice.** *Nature* 431 (21 Oct), 988 (2004).

8 Peplow, M.: **Junk no more?** *Nature* 431 (21 Oct), 923 (2004). <> (Note on paper by A.E. Peaston *et al,* in *Dev. Cell* 7:597, 2004, that at least one-third of human genome previously considered junk is in fact functionally important.)

9 Kenneally, C.: **The deepest cut.** *New Yorker* (Jul 3), 36-42 (2006).

10 Qiu, J.: **Unfinished symphony.** *Nature* 441 (May 11), 143-145 (2006). <> (Title likens the genome to a musical score which has an overlay of additional notations – the epigenetic code – such as for "key signatures, phrasing and dynamics, which tell how the notes should be played." Remarks that up to 70% of the cause of some diseases are epigenetic.)

11 Dutton, G. and Perkel, J. M.: **Shhh: silencing genes with RNA interference.** *The Scientist* (Apr 7), 42-44 (2003). <> (Sub-head: "Banking on RNAi's promise, academic and private researchers flock to the field." Focus is on short hairpin RNA, abbreviated shRNA, thus the title of the piece.)

12 Novina, C. D. and Sharp, P. A.: **The RNAi revolution.** *Nature* 430 (8 Jul), 161-164 (2004).

13 Lau, N. C. and Bartel, D. P.: **Censors of the genome.** *Sci Amer* (Aug), 35-41 (2003).

14 Hamilton, A. J. and Baulcombe, D. C.: **A species of small antisense RNA in post transcriptional gene silencing in plants.** *Science* 286 (Oct 29), 950-952 (1999). < > (Reprinted in booklet of four landmark papers in this area, included with the issue of March 25, 2006, sponsored by Sigma Chemicals. The others in the booklet concern role of Argonaute2 [2004], processing by dicer [2006], and action of the RITS complex on heterochromatin [2004].)

15 Gura, T.: **A silence that speaks volumes.** *Nature* 404 (Apr 20), 804-808 (2000). < > (Baulcombe, a leader in the field, is quoted to remark that "one of the amazing things about this to me is how many discoveries have come up by accident.")

16 Wagner, R. W. and Sun, L.: **Double-stranded RNA poses puzzle.** *Nature* 391, 744 (1998 19 Feb). < > (Perspective on report p806 by Fire et al.)

17 Dennis, C.: **The brave new world of RNA.** *Nature* 418 (Jul 11), 122-124 (2002). < > (Sub-head: "Most of the RNA transcribed from your genome doesn't make protein. Corina Dennis talks to the revolutionaries who believe that it functions in gene-regulatory networks that underlie the complexity of higher organisms.")

18 Sledz, C. A. and Williams, B. R. G.: **RNA interference.** *Blood* 106 (3), 787-794 (2005). < > (Review oriented to hematology.)

19 Natesan, S., Rivera, V. M., Molinari, E. and Gilman, M.: **Transcriptional squelching re-examined.** *Nature* 390 (Nov 27), 349-350 (1997). < > (Begins: "The introduction of a potent transcriptional activator into eukaryotic cells can paradoxically suppress the transcription of a co-introduced target gene.")

20 Maden, B. E. H.: **Guides to 95 new angles.** *Nature* 389 (Sep 11), 129 (1997).

21 Doudna, J. A.: **A molecular contortionist.** *Nature* 388 (Aug 28), 830-831 (1997). < > (Describes some of the interesting shapes assumed by various RNAs – hammerhead, hairpin, etc. – discussed at a symposium on RNA structure. Also gives table of 7 classes of small RNA.)

22 Storz, G.: **An expanding universe of noncoding RNAs.** *Science* 296 (May 17), 1260-1262 (2002). < > (Lead article in theme issue on RNA, with table of 18 classes of the new active RNAs.)

23 Editor: **New class of small RNAs found.** *C & EN* (Jun 17), 34 (2006). <> (Refers to work by T. Tuschl and G.J. Hannon in *Nature* online describing piRNA, so named because they bind to members of the Piwi family of proteins, related to the Argonauts, modulating gene expression, particularly in forming sperm. See *Nature*, July 13, 2006, p199, p363, also p203, and perspective p305.)

24 Adams, A.: **RNAi inches toward the clinic**. *The Scientist* (Mar 29), 32-35 (2004). <> (Mentions that there were 15 publications on this subject in 1998, and more than 1,000 in 2003.)

25 Riddihough, G.: **The other RNA world**. *Science* 296 (May 17), 1259 (2002). <> (Editorial introducing special issue on non-coding RNA, p1259-1273, with three essays and review article, plus related report, p1319-1321. Editor speaks of the 'RNome,' distinct from genome.)

26 Riddihough, G.: **In the forests of RNA dark matter**. *Science* 309 (Sep 2), 1507-1533 (2005). <> (Editorial introducing special issue on RNA: review articles, perspectives, wall poster, research article, p1534, and reports, p1559, 1564, 1584.)

27 Gesteland, R. F., Cech, T. R. and Atkins, J. F. **The RNA World {3rd edition}**: Cold Spring Harbor Laboratory Press, (2006) <> (Big book with contributions from experts in various specialties, reviewed in *Nature Genetics* 38(4):393-4, by J.S. Mattick, who judges it to be a good reference but already outdated.)

28 Fraser, A.: **Human genes hit the big screen**. *Nature* 428 (Mar 25), 375-376 (2004). <> (Background for two papers this issue, p427 and p431, developing techniques for applying RNAi to large-scale surveys − screens − of the genes of humans and other mammals by selectively silencing them with siRNAs.)

29 Kamath, R. S. and al, e.: **Systematic functional analysis of the *Caenorhabditis elegans* genome using RNAi.** *Nature* 421 (Jan 16), 231-235 (2003). <> (A related report by Ashrafi et al begins p268, using the same approach to find all the genes in the worm that affect fat storage, identifying 305 that reduced fat and 112 that increased it. For perspective, see p220 this issue.) STOPPED

30 Gwack, Y. et al: **A genome-wide *Drosophilia* RNAi screen identifies DYRK-family kinases as regulators of NFAT**. *Nature* 441 (Jun 1), 646-650 (2006). <> (Said to "break new ground.")

31 Arron, J. A. and al, e.: **NFAT dysregulation by increased dosage of *DSCR1* and *DYRK1A* on chromosome 21**. *Nature* 441 (Jun 1), 595-600 (2006). <> (With perspective by C.J. Epstein, p582-3, "Critical genes in a critical region." Down's syndrome was long known to be caused by an extra piece of chromosome 21, but the deeper nature of the problem was unclear. This work shows that the increased dosage of the two genes mentioned "destabilizes the regulation of signaling pathways involving the NFAT transcription factor.")

32 Ngo, V. N. and al, e.: **A loss-of-function RNA interference screen for molecular targets in cancer**. *Nature* 441 (May 4), 106-110 (2006). <> (With commentary by W.G. Kaelin, p32. Cited among "Papers to Watch" in *The Scientist*, July issue, p62.)

33 Loureiro, J. and al, e.: **Signal peptide peptidase is required for dislocation from the endoplasmic reticulum.** *Nature* 441 (Jun 15), 894-897 (2006). < > (Employs siRNA to learn how the human cytomegalovirus, HCMV, evades immune system. It was known that it does so by blocking MHC-I, a protein essential for certain immune detections, from appearing on the surface of the infected cell, but it was not known how it did this. The answer is that the virus encodes a protein which blocks the enzyme that releases MHC-I from the manufacturing machinery.)

34 Zimmerman, T. et al: **RNAi-mediated gene silencing in non-human primates.** *Nature* 441 (May 4), 111-114 (2006). < > (Demonstrates use of siRNA for the selective silencing or attenuation of specific genes, as a prelude to developing such agents for therapy of human diseases. Work extends the promise previously shown in tissue culture, as by Morris et al, to whole animals.)

35 Schmidt, C. W.: **Therapeutic interference.** *Modern Drug Discovery* (Jul), 37-42 (2003). < > (Surveys potential applications of siRNA in medical therapy. Much excitement.)

36 Pluvinet, S. et al: **RNAi-mediated silencing of CD40 prevents leukocyte adhesion on CD154-activated endothelial cells.** *Blood* 104 (Dec 1), 3642-3646 (2004).

37 Editor: **'Interfering' RNA helps monkeys recover from SARS.** *New Scientist* (Aug 27), 15 (2005). < > (Note on article in *Nature Medicine* showing that siRNA is an effective therapy, given as a nasal spray, against the lung-targeting SARS virus in monkeys.)

38 Rossi, J. J.: **A cholesterol connection in RNAi.** *Nature* 432 (Nov 11), 155-157 (2004).

39 Soutschek, J. et al: **Therapeutic silencing of an endogenous gene by systemic administration of modified siRNAs.** *Nature* 432 (Nov 11), 173-179 (2004). < > (With perspective by J.J. Rossi, p155-6. Aimed to silence a gene causing high cholesterol but problem was how to deliver the siRNA to the desired tissues. They succeeded by injecting the siRNA in combined with cholesterol. Also in this issue is new insight on the RNA microprocessor complex, see p231, p235.)

40 Grimm, D. et al: **Fatality in mice due to over saturation of cellular microRNA / short hairpin RNA pathways.** *Nature* 441 (May 25), 537-541 (2006).

41 Constans, A.: **RNAi's minor setback.** *The Scientist* (Jun 20), 22-23 (2005). < > (Explains why initial optimism has given way to concerns about the safety of RNAi, because of "off-target effects.")

42 Lim, L. P. and al, e.: **Microarray analysis shows that some microRNAs down regulate large numbers of target mRNAs.** *Nature* 433 (Feb 17), 769-771 (2005). < >

(Shows that some miRNAs suppress not just one or two genes but up to hundreds, and tend to be tissue-selective. For instance, one miRNA acted mainly on brain tissues, while another affected mainly genes active in muscle.)

43 Couzin, J.: **RNAi safety comes under scrutiny.** *Science* 312 (May 25), 1121 (2006). < > (Mentions that tests of RNAi on humans for diseases, including macular degeneration and a respiratory virus, are already underway. But experiments treating mice with hepatitis, initially promising, later caused most to die from liver injury.)

44 Nolan, G. P.: **Transcription and the broken heart.** *Nature* 392 (Mar 12), 129-130 (1998). < > (Introduces two reports this issue, p82 and p186, showing how an error of transcriptional regulation may lead to a common congenital heart valve defect.)

45 Bruneau, B. G.: **Tiny brakes for a growing heart.** *Nature* 436 (Jul 14), 181-182 (2006). < > (Perspective on paper by Zhao et al, p214-220, who used a clever method of their own devising to show that a specific miRNA plays a critical role in embryonic development of the heart in mice. Among other things, appears that miRNAs are regulated much like other genes.)

46 Meltzer, P. S.: **Small RNAs with big impacts.** *Nature* 435 (Jun 9), 145-146 (2005). < > (Introduces papers by Lu et al, p834, and by He et al, p828, showing that siRNAs "are now definitively linked to the development of cancer.")

47 Plasterk, R. H. A.: **RNA silencing: the genome's immune system.** *Science* 296 (May 17), 1263-1265 (2002). < > (Part of special issue, "The other RNA world.")

48 Wang, S. H. and al, e.: **RNA interference directs innate immunity against viruses in adult *Drosophilia*.** *Science* 312 (Apr 21), 452-454 (2006). < > (For related article, see online address given in *Science* 312:1435, June 9, 2006.)

49 S, S. J.: **A most discerning host.** *Science* 312 (May 26), 1108 (2006). < > (Short note on related papers in *Nature* 441:101 and in *PNAS USA*, identifying two "pattern recognition receptors" – RIG-1 and MDA-5 – each of which functions to recognize specific types of viral RNA. One is required for the type 1 interferon response, which sends the cell into emergency shut-down, and is needed to resist, for example, infection by a picorna virus. The other was found to be essential for resisting infection by influenza and paramyxoviruses. Related article in July 6 issue, v442:39.)

50 Brownlee, C.: **Herpes runs interference.** *Science News* 169 (Jun 3), 329-330 (2006). < > (Note on work by N. Fraser and colleagues in *Nature* 442:82, with comments p32, solving a mystery: It had been known that *Herpes* virus secreted a substance blocking the apoptosis immune response of the host's infected cells, but the protein responsible could not be found. The explanation is that it's not a protein from a normal gene, but is an miRNA coded in the virus that silences the host gene for the suicide of the infected cell.)

51 Jopling, C. L. and al, e.: **Modulation of hepatitis C virus RNA abundance by a liver-specific microRNA.** *Science* 309 (Sep 2), 1577-1581 (2005). <> (In theme issue on the new face of RNA.)

52 R, G.: **Mighty meaty miRNAs.** *Science* 312 (Jun 16), 1574-1575 (2006). <> (The genetic explanation for a particularly meaty breed of New Zealand sheep, known as Texels, lies in their miRNA, according to work by Ckop et al, in *Nature Genetics*.)

53 Schratt, G. M. and al, e.: **A brain-specific microRNA regulates dendritic spine development.** *Nature* 439 (Jan 19), 283-289 (2006). <> (This may be involved in higher cognitive skills since dendritic spines are linked to memory and learning.)

54 Holtmaat, A. et al: **Experience-dependent and cell-type-specific spine growth in the neocortex.** *Nature* 441 (Jun 22), 979-983 (2006). <> (Updates and refines prior discoveries concerning dendritic spines. See also related article in next issue, v441:1144.)

55 Krawetz, S. A.: **Paternal contribution: new insights and future challenges.** *Nat Rev Gen* 6 (Aug), 633-642 (2005). <> (Emphasis is on role of sperm, aside from the main genome, such as the several thousand mRNA and others now found to be carried in sperm. For shorter perspective, see survey in *Nature*, 436-770-1, "The secret life of sperm.")

56 Spehr, M. and al, e.: **Identification of a testicular odorant receptor mediating human sperm chemotaxes.** *Science* 299 (Mar 28), 2054-2058 (2003). <> (With perspective p1993-4 by D.F. Babcock, "Smelling the Roses" and widely reported elsewhere, as in *Science News*, 163:195, "By a nose?" with subtitle, "Human sperm may sniff out the path ...")

57 Constans, A.: **RNAs running the show.** *The Scientist* (May 24), 26-27 (2004).

58 Dennis, C.: **The genome's guiding hand?** *Nature* 420 (Dec 19/26), 732 (2002).

59 Dutton, G.: **Small worms, small RNAs, big questions.** *The Scientist* (Jul 22), 32-34 (2002).

60 Lim, L. P. and al, e.: **Vertebrate microRNA genes.** *Science* 299 (Mar 7), 1540 (2003). <> (Clever method of data analysis employed to identify 200-255 human genes which produce miRNA. However, they identify only those with high level of expression, so the relevant number may be much larger.)

61 Goodman, A. F., Bellato, C. M. and Khidr, L.: **The uncertain future for central dogma.** *The Scientist* (Jun 20), 20-21 (2005). <> (This article, dedicated to the memory of Paul Silverman, is a frank discussion of the crisis in our concept of how genetics – and biology in general – really works. No clear answers.)

62 Pearson, H.: **What is a gene?** *Nature* 441 (May 25), 399-401 (2006).

63 Covey, S. N., Al-Kaff, N. S., Langara, A. and Turner, D. S.: **Plants combat infection by gene silencing.** *Nature* 385 (Feb 27), 781 (1997). <> (Shows how the commercially important vegetable, kohlrabi, can recover from infection by CaMV virus, by PTGS.)

64 Ratcliff, F., Harrison, B. D. and Baulcombe, D. C.: **A similarity between viral defense and gene silencing in plants.** *Science* 276 (Jun 6), 1558 (1997).

65 Wickens, M. and Gonzalez, T. N.: **Knives, accomplices, and RNA.** *Science* 306 (Nov 19), 1299-1300 (2004).

66 Famulok, M.: **RNAs turn on in tandem.** *Science* 306 (Oct 8), 233-234 (2004). <> (Commentary on report by Mandal et al, p275, on new complexities of a riboswitch.)

67 Editors: **Copy editor stops presses.** *Nature* 439 (Jan 12), 121 (2006). <> (Short note on work by N. Nashikura et al in *Nature Struc. Mol. Biol.* showing that a family of enzymes, ADARs, known to edit RNA by altering base and thus to modulate the expression of several mammalian genes, actually works to suppress certain miRNAs: "This shows that RNA editing and RNA interference interact.")

68 Tollervey, D.: **RNA lost in translation.** *Nature* 440 (Mar 23), 425-426 (2006). <> (Perspective on paper by Doma and Parker, p561, discovering a 4th mechanism – pathway – of quality control of mRNA, causing destruction of faulty transcripts.)

69 Sontheimer, E. J. and Carthew, R. W.: **Argonaute journeys into the heart of RISC.** *Science* 305 (Sep 3), 1409 (2004). <> (Perspective on reports by Liu et al, p1437, and by Song et al, p1434, on mechanism.)

70 Farh, K. K. and al, e.: **The widespread impact of mammalian microRNAs on mRNA repression and evolution.** *Science* 310 (Dec 16), 1817-1821 (2005). <> (Concludes as follows: "It is hard to escape the conclusion that micro RNAs are influencing the expression or evolution of most mRNAs." Be aware that mRNA is ordinary classical messenger RNA.)

71 Cattanach, B. M. and Kirk, M.: **Differential activity of maternally and paternally derived chromosome regions in mice.** *Nature* 315 (Jun 6), 496-498 (1985).

72 Sapienza, C.: **Parental imprinting of genes.** *Sci American* 10/90 (Oct), 52-60 (1990). <> (Focus is on her own work, on the possible role of imprinting in cancers.)

73 Montminy, M.: **Something new to hang your HAT on.** *Nature* 387 (Jun 12), 654-655 (1997). <> (Perspective on reports p677 by Troche *et al,* and p733 by Heery *et al,* concerning histone acetyl transferase, HAT. See also R.J. Lin et al, *Nature* 391:811, 1998, on role of HAT in leukemia.)

74 Francis, D., Diorio, J., Liu, D. and Meaney, M. J.: **Nongenomic transmission across generations of maternal behavior and stress responses in the rat.** *Science* 286 (Nov 5), 1155-1158 (1999).

75 Constans, A.: **Parentage has effects outside the genome.** *The Scientist* (Mar), 69 (2006). <> (Commentary on two "hot papers" by I.C. Weaver et al, *Nat Neurosci* 7:847 and *J Neurosci* 25:11045, showing how degrees of mothering – licking, grooming, nursing – directly effect methylation state of the pups, effecting certain hormones, "triggering lasting changes in the expression of genes." This result was so radical that the paper was initially rejected. The effects could be modified, such as by a histone deacetylase inhibitor. With regard to the nature *vs.* nurture controversy, author remarks that "in this case, nature is nurture.")

76 Pennesi, E.: **Supplements restore gene function via methylation.** *Science* 310 (Dec 16), 1761 (2005). <> (Report of recent work showing that the vitamin, folic acid, promotes methylation to override the effect of a transposon in experimental mice otherwise causing a kinked tail and altered coat color. For similar work, see R.A. Waterland et al, *Mol Cell Biol* 23:5293, 2003, as noted by M. Fogarty in *The Scientist*, Sep. 8, 2003, p39.)

77 Shultz, L. D. and al, E.: **Thrombocytopenia and cardiomyopathy (trac): a new single-gene mutation resulting in severe platelet and heart abnormalities in mice.** *Blood* 106 (11), 217a #734 (2005). <> (Presented orally on 12/15 at the 47th convention of the *American Society of Hematology,* Atlanta, where it was remarked that the cardiac defect was not apparent until the food diet happened to be changed to a different mix, attributed by them to different sodium concentration.)

78 Hooper, R.: **Men inherit hidden costs of dad's vices.** *New Scientist* (Jan 7), 10 (2006).

79 Goll, M. G. et al: **Methylation of tRNA{Asp} by the DNA methyltransferase homolog Dnmt2.** *Science* 311 (Jan 20), 395-398 (2006).

80 Nicholls, R. D.: **The impact of genomic imprinting for neurobehavioral and developmental disorders.** *J Clin Invest* 105 (4), 413-418 (2000).

81 Hastie, N.: **Disomy and disease resolved?** *Nature* 389 (Oct 23), 785-786 (1997). <> (Perspective on report p805 by Sun et al., p809. The many defective organs and other problems in Beckwith Wiedermann syndrome occur in babies born with both copies of chromosome 11p15 from the father instead of one from each parent. Sun et al., marshall new evidence that the key factor is a gene for an insulin-like growth factor, IGF2, which is normally imprinted so that only one copy, from either parent, is active, implicating an imprinting defect in this congenital disorder.)

82 Reik, W. and Constanzia, M.: **Making sense or antisense?** *Nature* 389 (Oct 16), 669-671 (1997). <> (Perspective on work, correctly described as "exciting," by Wutz et

al, p745, opening: "Genetic imprinting is a mechanism by which genes are expressed from the maternal or paternal chromosomes, and loss of imprinting is involved in a variety of diseases and cancers." Explains that certain regions of DNA are methylated differently at C and G, and this pattern is suspected to be the signal from the parental germ cells that results in the allele-specific expression of imprinted genes, since blocking an enzyme that performs the methylation in mice results in abnormalities. It gets better but to say more is beyond the scope of this note.)

83 Baylin, S. P.: **Tying it all together: Epigenetics, genetics, cell cycle, and cancer.** *Science* 277 (Sep 26), 1948-1949 (1997). < > (Perspective, with 38 references, on paper by Chuang et al, p1996.)

84 Tycko, B.: **Epigenetic gene silencing in cancer.** *J Clin Invest* 105 (4), 401-408 (2000). < > (Good reading but already dated. With related article, "Gene silencing as a threat to the success of gene therapy," by T.H. Bestor, p409-12.)

85 Kling, J.: **Put the blame on methylation.** *The Scientist* (Jun 16), 27-28 (2003). < > (Subtitle: "This gene-silencing mechanism, not necessarily mutation, is often found culpable in creating cancerous cells." Accompanied by article p22, "MicroRNA shows macro potential," on "large new class of small RNAs.")

86 Costello, J. F.: **Comparative peignoirs of leukemia.** *Nat Genetics* 37 (3), 211-212 (2005). < > (Incidentally, this is followed by a survey by P.A. Wade, p212, on methylated DNA. See also *Science* 312:1435, June 9, 2006, with commentary p843, "Leukemia protein spurs gene silencing.")

87 Greener, M.: **Cancer epigenetics enters the mainstream.** *The Scientist* (Jun 20), 18-19 (2005).

88 Prey, L. A.: **Epigenetics: genome, meet your environment.** *The Scientist* (Jul 5), 14-20 (2004). < > (Survey recounts key experiments, interviews with leading figures. Topics include the new respect for Lamarck.)

89 Chen, D. and al, E.: **Regulation of transcription by a protein methyl transfer.** *Science* 284 (25 Jun), 2174-2177 (1999). < > (Methylation is here added to acetylation as a regulator of transcription. In this case, nuclear hormone receptors – for androgen, estrogen, thyroid – when activated by hormone increased gene transcription severalfold in the presence of methyl transferase, CARM1.)

90 Flintoff, L.: **Identical twins: Epigenetics makes the difference.** *Nat Rev Genetics* 6 (Sep), 667 (2005). < > (Note on recent work finding that monzygotic twins diverge in epigenetic marks as they grow older, being almost identical at age 3, quite different by age 50, attributed to "environmental factors.")

91 O'Neill, R. J. W., O'Neill, M. J. and Graves, J. A. M.: **Undermethylation associated with retroelement activation and chromosome remodelling in an interspecific**

mammalian hybrid {wallaby}. *Nature* 393 (7 May), 68-72 (1998). <> (For background by M.G. Kidwell and D.R. Lisch, see "Transposons Unbound," p22-23 this issue.)

92 Maher, B.: **The nucleosome untangled.** *The Scientist* (May), 34-41 (2006). <> (Includes debate between two leaders, B.M. Turner vs. S. Henikoff, on whether it is a "code" or not.)

93 Shogren-Knaak, M. and et al: **Histone H4-K16 acetylation controls chromatin structure and protein interactions.** *Science* 311 (Feb 10), 844-847 (2006). <> (The general idea seems to be that adding the acetyl group to the histone protein causes the DNA to be less tightly wound, presumably making it more available to interactions.)

94 Mohd-Sarip, A. and Verrijzer, C. P.: **A higher order of silence.** *Science* 306 (Nov 26), 1484-1485 (2004). <> (Perspective on two reports this issue, p1571 and p1574, dealing with how the regulated compaction of DNA into chromatin might be achieved. Provides background on nucleosomes, etc. as "key developmental regulators ..." Title refers to this as a "higher order" of gene silencing.)

95 Haince, J. F., Rouleau, M. and Poirier, G. G.: **Gene expression needs a break to unwind before carrying on.** *Science* 312 (Jun 23), 1752-1753 (2006). <> (Introduces paper by Ju et al, p1798, showing that the DNA must actually be broken before certain genes can be transcribed. Does not deal with epigenetics per se, but offers good explanation of the latest complexities.)

96 Zhang, Y.: **No exception to reversibility.** *Nature* 431 (Oct 7), 637-639 (2004). <> (Perspective on papers in *Cell* and in *Science* showing that histone methylation is reversible, contrary to previous indications. For comments on related work. see *Science* 306:2171, Dec. 24, 2004.)

97 Apone, L. M. and Green, M. R.: **Transcription *sans* TBP.** *Nature* 393 (May 14), 114-115 (1998).

98 Bestor, T. H.: **Methylation meets acetylation.** *Nature* 393 (May 28), 311-312 (1998). <> (Introduces paper by Nan et al, p386, with comments on another by Jones et al in *Nature Genetics*, on complex interplay between methylation, acetylation, and the reverse – the removal of those groups. Technically complex but with good historical background.)

99 Eissenberg, J. C. and Elgin, S. C. R.: **Antagonizing the neighbors.** *Nature* 438 (Dec 29), 1090-1091 (2005). <> (Introduces three papers this issue, p116, p1176, p1181. Key point for us here is the prior finding by Allis and colleagues that reversible marks "in the tail regions of histones could antagonize the binding of regulatory proteins to neighboring" regions. These papers confirm and extend such findings, and remark: "The monotonous beads-on-a-string [concept] conceals a rich variety of ornaments

that compete with one another to control gene expression." Others compare these 'ornaments' to "charms on a bracelet.")

100 Taghavi, P. and vanLohuizen, M.: **Two paths to silence merge.** *Nature* 439 (Feb 16), 794-795 (2006). <> (Introduces paper this issue, p871, basically showing that silencing by DNA methylation and histone modification overlap and interact. Main focus is 'cellular memory,' that is, how each kind of cell in the embryo 'remembers' what kind of job it has to do in the particular tissues of the developing offspring.)

101 Rassoulzadegan, M. and al, e.: **RNA-mediated non-mendelian inheritance of an epigenetic change in the mouse.** *Nature* 441 (May 25), 469-474 (2006). <> (Widely reported. Featured on cover, "Breaking Mendel's law, first plants, now animals." With perspective by P.D. Soloway, *Paramutable possibilities*, explaining "paramutations," a term coined in 1956.)

102 Kawasaki, H. and Taira, K.: **Induction of DNA methylation and gene silencing by short interfering RNAs in human cells.** *Nature* 431 (9 Sep), 211 (2004). <> (First finding of this in humans.)

103 Tsukamoto, Y., Kato, J. I. and Ikeda, H.: **Silencing factors participate in DNA repair and recombination in** *Saccharomyces cerevisiae*. *Nature* 388 (Aug 28), 900-903 (1997). <> (See also p829 for perspective by S.P. Jackson.)

104 Pennisi, E.: **Scientists find transcription by committee.** *Sci News* 145 (Apr 2), 213 (1994). <> (Note on paper in the March 31 issue of *Nature* describing how a large complex of several proteins, called a 'holoenzyme,' assembles like a committee to direct transcription. Dozens of such large complexes are known.)

105 Pennisi, E.: **How the nucleus gets it all together.** *Science* 276 (Jun 6), 1495-1497 (1997). <> (Overview of recent burst of work revealing new complexities of RNA, long neglected because it was thought to be understood.)

106 Maher, B. A.: **Histone methylation is making its mark.** *The Scientist* (Jan 27), 27-28 (2003). <> (In "Hot Papers" column, citing then recent key papers. Mentions "stacks of papers" written on these matters.)

107 Wang, Y. and et al: **Beyond the double helix: writing and reading the histone code.** *Novartis Foundation Symposium* 259, 3-21 (2004). <> (Also p163-9, "The histone code hypothesis," which the authors support. For update, see four papers in *Nature*, July 6, 2006, with comments p31 entitled "A finger on the mark," opening with the question, "Are we any closer to deciphering these encoded instructions?")

108 Flam, F.: **Hints of a language in junk DNA.** *Science* 266 (Nov 25), 1320 (1994). <> (Reports on work in *Phys Rev Ltrs*, Dec. 5, 1994, finding that junk DNA sequences have earmarks of ' language structure' by the Zipf criterion.)

109 Sauro, H. M.: **The next frontier in cellular networking**. *The Scientist* (Aug 1), 20-21 (2005). <> (From opening paragraph: "The new mantra, systems biology, is proclaiming a change in attitude, to convince us that this overwhelming complexity is, in fact, tractable to human understanding." The author, a self-proclaimed "card-carrying systems biologist," agrees.)

110 Wiley, H. S.: **Sytems biology: Beyond the buzz**. *The Scientist* 52-57 (2006). <> (With enthusiastic editorial, p13.)

111 Janes, K. A. and al, E.: **A systems model of signaling identifies a molecular basis set for cytokine-induced apoptosis**. *Science* 310 (Dec 9), 1646-1651 (2005).

112 Jimenez, J. J., Jy, W., Mauro, L. M., Horstman, L. L., Soderland, C. and Ahn, Y. S.: **Response to letter-to-editor by Laurence {Invited}**. *Br J Haematol* 125 (3), 416 (2004). <> (Response to letter, *Br J Haematol* 125(3):415, concerning questionable findings on apoptotic endothelial cells.)

113 Workman, C. T. and al, e.: **A systems approach to mapping DNA damage response pathways**. *Science* 312 (May 19), 1054-1059 (2006). <> (Identifies 5,272 interactions among 30 DNA damage-related factors, assembled into "causal pathway models." Authors conclude optimistically that although complicated, it is not so extremely so as to prohibit future understanding.)

114 Editor: **Querying alternative splicing in the brain**. *The Scientist* (Jan), 68 (2006). <> (Describes paper by J. Ule et al in *Nature Genetics* 37:844, indicating that alternative splicing, mediated by the protein called Nova, can explain much of the great variety of neuronal synapses in the brain. It is said that authors made good use of bioinformatics to develop powerful new methods of data analysis.)

115 Giot, L. and et al: **A protein interaction map of *Drosophilia Melanogaster***. *Science* 302 (Dec 5), 1727-1734 (2003). <> (With perspective by E. Pennisi, p1646-9, "Tracing life's circuitry," with subhead: "A new movement aims to integrate biology, mathematics, and engineering. Even if its objective is hard to define, it is all the rage in the academic world." Well, maybe in part of that world.)

116 Johnston, N.: **Comparative genomics on the rise**. *The Scientist* (Jun 6), 26-28 (2003). <> (Subtitle: "Yeast species comparison reveals annotional power in alignments." In the jargon of the trade, 'annotation' refers to assigning functions to genes, and 'alignments' refers to computer-assisted searches of gene sequences in different organisms to locate similar or related genes.)

117 Luscombe, M. M. and al, E.: **Genomic analysis of regulatory network dynamics reveals large topological changes**. *Nature* 431 (Sep 16), 308-312 (2004). <> (From abstract: "We uncover large changes in underlying network architecture that are unexpected given current viewpoints ... In response to diverse stimuli, transcription

factors alter their interactions to varying degrees, thereby rewiring the network. A few transcription factors act as permanent hubs, but most act transiently ...")

118 Crandall, K. A. and Buhay, J. E.: **Genomic databases and the tree of life.** *Science* 306 (Nov 12), 1144-1145 (2004). <> (Perspective on article beginning p1172 by A.C. Driskell et al, *Prospects for building the tree of life* ..., a pilot test of computational methods.)

119 Perkel, J. M.: **Validating the interactome.** *The Scientist* (Jun 21), 19-22 (2004). <> (Featured on cover: "Molecular cartographers map the cell.")

120 Tong, A. H. Y. and et al: **Global mapping of the yeast genetic interaction network.** *Science* 303 (Feb 6), 808-813 (2004). <> (Diagrams of dizzying complexity.)

121 Li, S. and et al: **A map of the interactome network of the metazoan *C. elegans*.** *Science* 303 (Jan 23), 540-543 (2004). <> (Diagrams of interactions; mind-boggling.)

122 Friedman, N.: **Inferring cellular networks using probabilistic graphical models.** *Science* 303 (Feb 6), 799-805 (2004).

123 Zhong, W. and Steinberg, P. W.: **Genome-wide prediction of *C. Elegans* genetic interactions.** *Science* 311 (Mar 10), 1481-1484 (2006). <> (With perspective by S.R. Eddy, p1381-2. A partial diagram, p1482, shows about 30 components of the proteosome complex of this worm, connected by so many lines that the figure is black with them.)

124 Jeong, H. and al, e.: **Lethality and centrality in protein networks.** *Nature* 411 (May 5), 41 (2001). <> (With commentary, p30-31. Diagram and conclusions resemble contemporaneous studies of the World Wide Web, www.)

125 Staff: **Genome data deluge.** *Science* 306 (Dec 17), 2012 (2004). <> ("Scorecard" of advances in 2003 lists many new -omics including the *regulome.*)

126 Secko, D.: **Computing gene regulation.** *The Scientist* (Jun 21), 28-29 (2004). <> (Survey of recent efforts taking "a statistical glimpse at how gene expression is controlled," combined with micro-array studies, in particular for the "multi-subunit complexes" known as TFIID and SAGA. Doubts and controversies are indicated at end.)

127 Remenyi, A., Schuler, H. R. and Wimanns, M.: **Combinatorial control of gene expression.** *Nat Struct Mol Biol* 11 (9), 812-815 (2004). <> (Asserts that "we demonstrate how protein and DNA regulators manage gene expression in a combinatorial fashion." But I doubt opening assumption, that there are only about 2,500 regulators for some 25,000 genes, in view of the blizzard of regulatory RNAs and epigenetic marks.)

128 Suthram, S., Sittler, T. and Ideker, T.: **The *Plasmodium* protein network diverges from those of other eukayrotes.** *Nature* 438 (Nov 3), 108-112 (2005). < > (Follows paper on the network itself, p103-7.)

129 Fogarty, M.: **Organellar proteomics.** *The Scientist* (Apr 12), 32-33 (2004). < > (Reports application to the study of proteins of organelles such as mitochondria; see comment on "Hot Paper" p30.)

130 DeFrancesco, L.: **Probing protein interactions.** *The Scientist* 16 (Apr 5), 28-30 (2002). < > (Subhead: "Researchers have found them by the thousands, but what do these interactions mean?" Also discusses some new techniques under heading, "New toys in the sandbox.")

131 Jones, R. B. and et al: **A quantitative protein interaction network for the ErbB receptors using protein microarrays.** *Nature* 439 (Jan 2), 168-174 (2006). < > (Abstract states that work involved "77,592 independent biochemical measurements" with results hopelessly complex, despite the smallness of the family of hormone receptors studied. But the spin claims useful insights.)

132 Borman, S.: **Any new proteomics techniques out there?** *C&EN* (Nov 26), 27 (2001). < > (Subhead: "Genome project head says many technologies for a human proteome project are not yet ready." Related article follows.)

133 Constans, A.: **Protein microarrays mature.** *The Scientist* (Aug 2), 42-44 (2004). < > (This favorable survey was written before news broke from several labs comparing arrays from four different manufacturers, all giving different and conflicting results.)

134 McCook, A.: **The human interactome falls into place.** *The Scientist* (Aug 1), 14-21 (2005). < > (Three related articles. Subhead, "Scientists race to map the totality of protein-protein interactions." Opens by noting that "some researchers are strongly opposed" to the project, explaining that these "pessimists" argue that it is bound to be "riddled with errors and pointless anyway." One database lists 30,000 interactions. Some partial diagrams are shown.)

135 Amato, I.: **Yeast cells brim with machinery.** *Chem / Eng News* (Jan 30) (2004). < > (Comment on work by 32 authors in *Nature,* "showing how 2,760 proteins in budding yeast cells assemble into 491 complexes." As I recall, that issue included a wall poster of the resulting map.)

136 Abbott, A.: **The society of proteins.** *Nature* 417 (Jun 27), 894-897 (2002). < > (Survey of recent work concerning the many complexes between and among various proteins, i.e. many or most proteins and enzymes work together in various combinations, or sequentially. This review gives new methods for identifying such complexes.)

137 Marx, V.: **Lipidomics fattens research on neglected molecules.** *Drug Discovery & Development* (Jan), 58-60 (2004). <> (Identifies about a dozen labs and pharms working on this.)

138 Stephan, M. W.: **Sugars get an 'ome of their own.** *The Scientist* (Aug 2), 45-47 (2004).

139 Perkel, J. M.: **Glycobiology goes to the ball.** *The Scientist* (Apr 29), 32-34 (2002). <> ("'What makes glycobiology so challenging,' says Jim Paulson ..., 'is that carbohydrate biosynthesis is so incredibly complex.'" He gives example of a genetic disease resulting from an error in a single enzyme, phospho-mannomutase, involving some 80 steps in normal people. Unlike proteins, which are non-branching chains of amino acids, carbohydrates are chains of sugars and can have many branches, vastly increasing the complexity of possible structures.)

140 Daviss, B.: **Growing pains for metabolomics.** *The Scientist* (Apr 25), 25-28 (2005). <> (Subhead: "The newest 'omic science is producing results – and more data than researchers know what to do with." Aim is to map out all metabolic pathways, to enable, for example, predicting bad effects of new drugs. But that hope is seen as a vision for the distant future.)

141 Valigra, L.: **Metabolic profiling: meet the latest 'omics'.** *Drug Discovery & Development* (Dec), 39-42 (2004).

142 Corporate: **The cancer kinome.** *Advertisement by Abgent* (2004). <> (Nice diagram of relations among a group of several hundred enzymes called protein kinases, about 187 of which have been implicated in cancers.)

143 Przybprski, J. and Lanzer, M.: **The malarial secretome.** *Science* 306 (Dec 10), 1897 (2004). <> (Perspective on two reports this issue concerning the many substances secreted by the malaria parasite – thus "secretome" – which enable it to evade immune defenses and otherwise fulfill its needs in the host, mosquito or human.)

144 Hall, N. and et al: **A comprehensive survey of the *Plasmodium* {malaria agent} life cycle by genomic, transcriptomic, and proteomic analysis.** *Science* 307 (Jan 7), 82-86 (2005).

145 Editor: **Mapping the epigenome.** *New Scientist* (Jan 7), 10 (2006). <> (Short note on plans to undertake mapping the epigenome, as set forth in *Cancer Research*, 65:11241. Cost is estimated at hundreds of megabucks.)

146 Vidal, M.: **Time for a human interactome project?** *The Scientist* (Mar), 46-51 (2006). <> (Subhead, ""An investment of $100 million should be enough ..." Shows a sample map from Vidal's group.)

147 Ray, L. B. and Gough, N. R.: **Orienteering strategies for a signaling maze.** *Science* 295 (May 31), 1632-1633 (2002). <> (Editorial for theme issue on signaling pathways, with 15 articles. The signaling systems within the cell, such as those that determine the effects of hormones and combinations of them, have become extremely complicated, as shown by the papers in this issue. For update, see special issue *Science*, Oct. 7, 2005, "Signaling: From stem cells to dead cells.")

148 Zamora, P. D. and Haley, B.: **Ribo-gnome: The big world of small RNAs.** *Science* 309 (Sep 2), 1519-1524 (2005). <> (In theme issue on the new face of RNA.)

149 Editor: **Connectome.** *New Scientist* (Nov 12), 62 (2005). <> (Note on initiative by neuroscientist Olaf Sporns to start mapping all brain connections in four dimensions. It is expected to be more costly than the billions spent on the human genome. Data will come from various imaging methods.)

150 Macer, D.: **The behaviorome mental map project.** *The Scientist* (Apr 20), 19 (2003). <> (Appears to be an invitation to contribute to this project of Macer's, based on his "nine classes of ideas.")

151 Editor: **Size matters.** *Nature* 441 (Jun 8), 670 (2006). <> (Note on paper by Hartl and Hartl in *Cell* 125:903, revealing the detailed structure of a chaperone protein, GroEL, resembling a tall knitted cap or beehive or bird cage, within which proper folding of the client protein takes place.)

152 Pennisi, E.: **Genes commute to factories before they start work.** *Science* 312 (Jun 2), 1304 (2006). <> (Pennisi's usual excellent reporting on breaking news of RNA factories, so called because they are local regions on the inner surface of the cell nucleus where the genes are actually transcribed to mRNA, which is then transported to the cytoplasm for translation into proteins. The truly amazing part is that widely separated genes evidently somehow move together to produce specific outputs.)

Ch. 13, References & Notes.

1 Sherman, D. C. and ET AL: **Intravenous ancrod for treatment of acute ischemic stroke.** *JAMA* 283 (18), 2395-2403 (2000). <> (with editorial, p2440)

2 Castro, H. C. and al, E.: **Snake venom thrombin-like enzymes: from reptilase to now.** *Cell Mol Life Sci* 61 (7-8), 843-856 (2004).

3 Hutton, R. A. and Warrell, D. A.: **Action of snake venom components on the haemostatic system.** *Blood Reviews* 7, 176-189 (1993).

4 Ouyang, C., Teng, C. M. and Huang, T. F.: **Characterization of snake venom components acting on blood coagulation and platelet function.** *Toxicon* 30 (9), 945-966 (1992).

5 Carsi-Gabrenas, J.: **Purification of toxins from green mamba venom with distinct receptor selectivities.** (PhD Thesis) (1997). <> (Dissertation; Univ. of Miami, Coral Gables campus, FL. Among the components she isolated, one was of interest to our laboratory but which we never found time to investigate. By the way, the name she gave to the toxin of her interest, thrombostatin, has since been usurped for another agent natural in humans.)

6 Li, Z.-Y., Wu, X.-W., Yu, T.-F. and Lian, E. C.-Y.: **Purification and characterization of lupus anticoagulant like protein from *Agkistrodon Halys Brevicaudus* venom.** *Thromb Haemost* 76 (6), 993-997 (1996). <> (Eric Lian is a heamtologist at the Univ. of Miami.)

7 Kinjoh, K. and et al: **Involvement of THR(7)-PHE(8) on the sequence of rabbit fibrinopeptide A in the specific action of habutobin.** *Thromb Haemost* Supplement (Abstract #40), 14 (1999). <> (XVII Congress of *International Society on Thrombosis and Haemostasis*, (ISTH) Washington, DC, Aug 14-21. I counted 27 abstracts on venoms at this meeting, mostly snakes. This one investigates why this enzyme from *Trimeresurus flavoridis* acts only on rabbit fibrinogen, not human, bovine or dog, to induce rapid clotting. By the way, barbourin, from another snake, is also "particularly potent for rabbits.")

8 Kroll, M. H., Andrews, R. K., Lopez, J. A. and Berndt, M. C.: **A 50-kDa alboaggregin from *Trimeresurus albolabris* viper venom initiates signals for platelet activation by binding to the glycoprotein Ib/IX/V complex.** *Blood* 86 (supp 1) (10), 613a (A#2438) (1995).

9 Andrews, R. K. and al, E.: **Structure-activity relationships of snake toxins targeting platelet receptors, glycoprotein 1b-IX-V, and glycoprotein VI.** *Curr Med Chem Cardiovasc Hematol Agents* 1 (2), 143-149 (2003).

10 Liu, C. Z., Peng, H. C. and Huang, T. F.: **Crotavarin, a potent platelet aggregation inhibitor purified from the venom of the snake *Crotalus viridis*.** *Toxicon* 33, 1289-1298 (1995).

11 Markland(Jr), F. S.: **Snake venom fibrinogenolytic and fibrinolytic enzymes: an updated inventory.** *Thromb Haemost* 79, 668-674 (1998).

12 Morita, T.: **Use of snake venom inhibitors in studies of the function and tertiary structure of coagulation factors.** *Int J Hematol* 79 (2), 123-129 (2004).

13 Blatteis, C. M. and al, E.: **Signalling the brain in systemic inflammation: role of complement.** *Front Biosci* 9, 915-931 (2004).

14 Vogel, C. W. and al, E.: **Recombinant cobra venom vector.** *Mol Immunol* 41 (2-3), 191-199 (2004).

15 Hamako, J. and EtAl: **Purification and characterization of Kaouthiagin, a von Wilebrand factor-binding and -cleaving metalloproteinase from Naja kaouthia cobra venom.** *Thromb Haemost* 80, 490-505 (1998).

16 Lizano, S. and al, E.: **Natural phospholipase A(2) myotoxin inhibitor proteins from snakes, mammals and plants.** *Toxicon* 42 (8), 963-977 (2003).

17 Verstraete, M.: **Recombinant hirudin (desirudin): Advantages, drawbacks, and initial clinical experience.** *Vessels* 1 (4-14) (1995).

18 Fenton, J. W.: **Leeches to hirulogs and other thrombin-directed antithrombotics.** *Hematol Oncol Clin North Amer* 6 (5), 1121-1129 (1992).

19 Nutt, E. M. and al, E.: **Purification and characterization of recombinant antistasin: a leech-derived inhibitor of coagulation factor Xa.** *Arch Biochem Biophys* 285, 37-44 (1991).

20 Krezel, A. M., Wagner, G., Seymour-Ulmer, J. and Lazarus, R. A.: **Structure of the RGD protein decorsin: Conserved motif and distinct function in leech proteins that affect blood clotting.** *Science* 264 (Jun 24), 1944-1947 (1994).

21 Connolly, T. M., Karczewski, J., Condra, C., Singh, S. B., Goetz, M., Friedman, P. A., Shafer, J. and Gould, R. J.: **The leech *Hirudo medicinalis* produces a prostaglandin which selectively inhibits platelet activation mediated by the thrombin receptor.** *Blood* 86 (supp 1), 547a (A#2178) (1995).

22 Faria, F. and al, E.: **A new factor Xa inhibitor (Lefaxin) from the *Haementeria depressa* leech.** *Thromb Haemost* 82, 1469-1473 (1999).

23 Rigbi, M., Orevi, M. and Eldor, A.: **Platelet aggregation and coagulation inhibitors in leech saliva and their roles in leech therapy.** *Sem Thromb Hemost* 22 (3), 273 (1996).

24 writer, Staff.: **Leechmania: turning back the clot.** *J NIH Res* 3 (Jan), 22 (1991). <> (News item on revival of use of live leeches in medical practice, mainly for improved wound healing.)

25 Sun, J. and al, E.: **Purification, characterization and cDNA cloning of a novel anticoagulant of the intrinsic pathway (Prolixin-S) from salivary glands of the blood sucking bug, *Rhodnius prolixus*.** *Thromb Haemost* 75 (4), 573-577 (1996).

26 Staff: **New drugs from the dreaded deerfly?** *Science News* 144 (Oct 9), 235 (1993). <> (News item on agent, chrysoptin, from saliva of locally caught deerflies, which

blocks platelet aggregation. Refers to report by S A Gravelink and E A Lerner, PNAS, Oct 1.)

27 Bark, N., Blomback, B. and Fatah, K.: On the occurrence of thrombin-like enzymes in mosquitoes. *Thromb Res* 81 (6), 623-634 (1996).

28 Ribeiro, J. M. C., Rossignol, P. A. and Spielman, A.: Salivary gland apyrase determines probing time in anophaline mosquitoes. *J Insect Physiol* 31, 689-692 (1985).

29 Stewart, R. J., Fredenburgh, J. C. and Weitz, J. L.: Characterization of the interactions of LYS-plasminogen and tissue and vampire bat plasminogen activators with fibrin and (DD)E. *Blood* 86 (supp 1) (10), 287a (A#1133) (1995).

30 Gardell, S. J., Duong, L. T., et al: Isolation, characterization, and cDNA cloning of a vampire bat salivary plasminogen activator. *J Biol Chem* 264, 17947-17952 (1989).

31 Apitz-Castro, R., Beguin, S., Tablanti, A., Bartoli, F., Holt, J. C. and Hemker, H. C.: Draculin, a newly discovered anticoagulant from vampire bat saliva, selectively inhibits human coagulation factors IXa and Xa. *FASEB J* 9 (6), A1500 (A#1409) (1995).

32 Ørvim, U., Barstd-Barsted? R. M., Vlasuk, G. P. and Sakariassen, K. S.: Effect of selective factor Xa inhibition [by tick anticoagulant peptide] on arterial thrombus formation triggered by tissue factor / factor VIIa or collagen in an ex vivo model of shear-dependent human thrombogenesis. *Arterioscl Thromb Vasc Biol* 15, 2188-2194 (1995).

33 Waxman, L., Smith, D., Arcuri, K. E. and Vlasuk, G. P.: Tick anticoagulant peptide (TAP) is a novel inhibitor of blood coagulation factor Xa. *Science* 248 (May 4), 593-596 (1990).

34 Francischetti, I. M. B. and al, E.: Penthalaris, a novel recombinant five-Kunitz tissue factor pathway inhibitor (TFPI) from the salivary gland of the tick vector of Lyme disease, *Ixodes scapularis*. *Thromb Haemost* 91, 886-898 (2003).

35 Pavao, M. S. G., Mourao, P. A. S., Molloy, B. and Tollefsen, D. M.: A unique dermatan sulfate-like glycosaminoglycan from ascidian *[Ascidia nigra]*: its structure and the effect of its unusual sulfation pattern on anticoagulant activity. *Blood* 86 (supp 1) (10), 884a (A#3523) (1995).

36 Bergum, P. W., Maki, S. L., Huang, S., Benham, P. P. and Vlasuk, G. P.: The anticoagulant strategy of the hookworm *A. caninum*: potent inhibition of human factors Xa (FXa) and VIIa/TF by a unique mechanism using FX and FXa. *Blood* 86 (supp 1) (10), 865a (A#3448) (1995).

37 writer, S.: **Patent dispute: New anticoagulant prompts bad blood between partners.** *Science* 271 (Mar 29), 1800 (1996). < >

38 Chow, G. and Kini, R. M.: **Exogenous factors from animal sources that induce platelet aggregation.** *Thromb Haemost* 85, 177-181 (2001).

39 Merchant, M., Hinton, J. and Geren, C.: **Observations of phospholipase activity in the venom of the brown recluse spider (*Lioncels reclusa*) by phosphorus-31 NMR.** *FASEB J* 9 (6), A#334 (1995).

40 Fuly, A. L. and al, E.: **Mechanism of inhibitory action on plately activation of a phospholipase A2 isolated from *Lachesis muta* (Bushmaster) venom.** *Thromb Haemost* 78, 1372-1380 (1997).

41 Donato, J. L. and EtAl: ***Lonomia obliqua* caterpillar spicules trigger human blood coagulation via activation of factor X and prothrombin.** *Thromb Haemost* 79, 539-542 (1998).

42 Nagase, H., Enjyoji, K. I., Kamikubo, Y. I., Kitazato, K. T., Kitazato, K., Saito, H. and Kato, H.: **Effect of depolymerized holothurian {sea cucumber} glycosaminoglycan (DHG) on tissue factor pathway inhibitor: *In vitro* and *In vivo* studies.** *Thromb Haemost* 78, 864-870 (1997).

43 Nagase, H., Enjyoi, K., Minamiguchi, K., Kitazato, K. T., Saito, H. and HKato: **Depolymerized holothurian {*Stichopus japonicus* Selenka, a sea-cucumber} with novel anticoagulant actions: {listed}.** *Blood* 85 (6), 1527-1534 (1995). < > (It may employ this agent for defensive purposes.)

44 Lewis, R. L. and Gutmann, L.: **Snake venoms and the neuromuscular junction.** *Seminars in Neurology* 24 (2), 175-179 (2004).

45 Rees, B. and al, E.: **Crystal structure of a snake venom cardiotoxin.** *PNAS USA* 84 (May), 3132-3136 (1987). < > (Main aim of this interesting study was to figure out why certain classes of neurotoxins vs. cardiotoxins could have such totally different actions despite being so very similar in amino acid sequence. Snake was *Naja mossambica mossambica*.)

46 Altamirano, M. M. and al, E.: **Oxidative refolding chromatography: folding the scorpion toxin Cd5.** *Nature Biotechnology* 17, 187-191 (1999).

47 Kini, R. M.: **Animal toxins of Asia and Australia. Molecular moulds with mutliple missions: functional sites in three-finger toxins.** *Clin Exp Pharm Physiol* 29 (9), 815-830 (2002). < > (Similar structures are found in natural physiology, e.g. the blood protein, urokinase plasminogen activator, see his ref. 17. The ancestral genes were duplicated several times to give rise to these weapons, via "accelerated evolution of their exons," see his refs 61-63. "As is the case with other super-families of toxins,

snakes have used the robust 3-finger protein mould to construct a group of toxins with wide variation in function involving just a few subtle changes in the functional sites.")

48 Olivera, B. M. and al, E.: **Diversity of *Conus* neuropeptides.** *Science* 249 (Jul 20), 257-263 (1990). <> (For background, see p250 this issue, "Science Digests the Secrets of Voracious Killer Snails.")

49 Tsetlin, V. I. and Hucho, F.: **Snake and snail toxins acting on nicotinic acetylcho-line receptors: fundamental aspects and medical applications.** *FEBS Letters* 557 (1-3), 9-13 (2004).

50 Caporale, L. H.: **Lessons from the most innovative genetic engineer.** *Nature Biotechnology* 16 (Oct), 908-909 (1998). <> (Highlights of symposium, *Molecular strategies in biological evolution,* which included talk by B. Olivera on cone snail venoms.)

51 Ohno, M. and al, E.: **Molecular evolution of myotoxic phospholipases A2 from snake venom.** *Toxicon* 42 (8), 841-854 (2003).

52 Pentapharm: **Advertisement [back cover].** *Thromb Haemost* 75 (1) (1996). <> (Full page ad by this supplier of venom products for hematology, with drawings of several snakes, appearing widely in hemtology journals.)

53 Clardy, J. and Walsg, C.: **Lessons from natural molecules.** *Nature* 432 (Dec 16), 829-837 (2004).

54 Finkelstein, J.: **Ringing the changes with vancomycin.** *Nature* 432 (Dec 23/30), 967 (2004). <> (Note on article by K Zerbe et al in *Angew. Chem. Int. Ed.,* 43:6709, announcing improved synthesis but still too complicated to be commercially viable.)

55 Sharon, N. and Lis, H.: **Lectins as cell recognition molecules.** *Science* 246 (Oct 13), 227-234 (1989).

56 Farmer, E. E.: **New fatty acid-based signals: a lesson from the plant world.** *Science* 276 (May 9), 912-913 (1997). <> (Supplies references showing how plants under attack by insect herbivores release airborne signals which specifically attract species to prey on the attacker. The interactions are amazing complex and specific. This piece introduces the technical report on p945 of this issue by Alborn *et al.,* "An elicitor of plant volatiles from beet armyworm oral secretion.")

57 Lewin, D. L.: **Plants under attack: no shrinking violets.** *J NIH Research* 7 (Jul), 31-34 (1995). <> (Supplies additional dimensions and references on the complexity and subtlety of plant chemical signals and defense, including electrical modes.)

58 Wolfe, G. W., Steinke, M. and Kirst, G. O.: **Grazing-activated chemical defense in a unicellular marine alga.** *Nature* 387 (Jun 26), 894-897 (1997). < > (Describes chemical defense of a marine plankton which deters attack by a protozoan.)

59 Ligterink, W., Kroj, T., Nieden, U. Z., Hirt, H. and Scheel, D.: **Receptor-mediated activation of a MAP kinase in pathogen defense of plants.** *Science* 276 (Jun 27), 2054-2057 (1997). < > (Describes activation of defense of parsley plant against an invading fungal plant pathogen.)

60 Marx, J.: **Plants, like animals, may make use of peptide signals.** *Science* 273 (Sep 6), 338-339 (1996). < > (Review of growing body of evidence for peptide signals in plants; refers to Becraft *et al.,* on p1406 of this issue, and to *Science* 19 July p370, others.)

61 Pare, J. W. and Tumlinson, J. H.: **Induced synthesis of plant volatiles.** *Nature* 385 (Jan 2), 30-31 (1997). < > (Advances on chemicals released by cotton plant in response to beet armyworm attack.)

62 O'Donnell, P. J., Calvert, C., Atzorn, R., Wasternack, C., Leyser, H. M. O. and Bowles, D. J.: **Ethylene as a signal mediating the wound response of tomato plants.** *Science* 274 (Dec 13), 1914- (1996).

63 Voinnet, O. and Baulcombe, D. C.: **Systemic signaling in gene silencing.** *Nature* 389 (Oct 9), 553 (1997). < > (Shows how a plant silences genes of a pathogen introduced to a leaf and, more remarkably, that the silencing signal eventually spreads throughout the whole plant. "The signal could move through the plant ahead of the inducing virus ..." They speculate that the signal may be a nucleic acid.)

64 Shulaev, V., Silverman, P. and Raskin, I.: **Airborne signaling by methyl salicylate in plant pathogen resistence.** 385 (Feb 20), 718-721 (1997). < > (This chemical is released by plants under viral attack [by tobacco mosaic virus in these experiments] and "may function as an airborne signal which activates disease resistence ... in neighboring plants and in the healthy tissues of the infected plant.")

65 Dyer, B. D. and Obar, R. A. **Tracing the History of Eukaryotic Cells: The Enigmatic Smile.** NY: Columbia Univ. Press, (1994) < > (Reviewed in *Nature,* 369:451, 9 Jun 1994, by L.D. Hurst, under caption, "A penchant for protists." By the way, the reviewer concurs with Dyer and Obar in his admiration for the 1991 book edited by Margulis *et al,* "Handbook of Protoctista," which he describes as "possibly her second great contribution to science." We note this by way of defending her status as a major biologist, opposing the ridicule to which she has often been subjected because of her 'radical' views – meaning her rejection of many orthodox evolutionary dogmas, defense of *Gaia,* etc.)

66 Parkinson, J. S. and Blair, D. F.: **Does *E. coli* have a nose?** *Science* 259 (19 Mar), 1701 (1993). <> (Evidence is put forth that this bacterium has a specialized sense organ.)

67 Breidenbach, M. A. and Brunger, A. T.: **Substrate recognition strategy for botulinum neurotoxin serotype A.** *Nature* 432 (Dec 16) (2004).

68 Pressman, B. C. and Fahim, M.: **Pharmacology and toxicology of the monovalent carboxylic ionosphores.** *Ann Rev Pharmacol Toxicol* 22, 465-490 (1982).

69 Reed, P. W.: **Ionophores.** *Methods in Enzymology* LV, 435-457 (1979).

70 Pressman, B. C.: **Biological applications of ionophores.** *Ann Rev Biochem* 45, 501-530 (1976).

71 Staff, N.: **Top ten breakthroughs of the year.** *Science* 306 (Dec 17), 2017 (2004).

72 Pray, L.: **Microbial multicellularity.** *The Scientist* (Dec 1), 20-23 (2003). <> (Notes several eminent biologists who now regard bacteria as "social creatures" but, as the title implies, they may also be viewed as "multicellular organisms." As usual, quorum sensing – meaning the ability of bacteria to detect and respond cooperatively to a certain number of others nearby – was originally thought to be rare and specific but is now known to be widespread.)

73 Kruglinski, S.: **Bacteria find they can all get along.** *Discover Magazine* (Dec), 20 (2003). <> (Short note on remarkable work by G. Vilicer and wife, who genetically modified bacteria (*Myxococcus xanthus*) to lack their means of attaching to each other (pili) to form social colonies. But astonishingly, after a short time they developed alternative means for attaining the same end. "We watched novel forms of cooperation evolve before our eyes," they say.)

74 Chicurel, M.: **Slimebusters.** *Nature* 408 (Nov 16), 284-285 (2000).

Ch. 14, References & Notes

1 Thomson, K. S.: **Ontogeny and phylogeny recapitulated.** *Amer Scientist* 76 (May/Jun), 273-275 (1988).

2 Pennisi, E.: **Haeckel's embryos: fraud rediscovered.** *Science* 277 (5 Sep), 1435 (1997). <> (Reports Richardson's finding that Haeckel's influential drawings of embryos were badly in error, causing a big stir, including among creationists. For comments, see Letters in *Science,* 279:1287, 27 Feb 1998, and 280:983, 15 May 1988.)

3 Duboule, D.: **How to make a limb?** *Science* 266 (28 Oct), 575-576 (1994). <> (From a theme issue on development.)

4 Pennisi, E. and Roush, W.: **Developing a new view of evolution**. *Science* 277 (Jul 4), 34-39 (1997). < > (Summary of advances in Evo-Devo.)

5 Raff, R. A. **The Shape of Life**. Chicago: Univ. Chicago Press, (1996) < > (Subtitle, "Genes, Development, and the Evolution of Animal Form." Reviewed in *Science* 273:1347, by ML McKinney, under the headline, "Understanding evolution: the next step," with high praise, calling it "one of the most important books of the decade in evolutionary biology.")

6 Gerhart, J. and Kirschner, M. **Cells, Embryos, and Evolution**. Cambridge, MA: Blackwell Scientific, (1997) < > (Subtitle: "Towards a Cellular and Developmental Understanding of Phenotypic Variation and Evolutionary Adaptability." Reviewed in *Nature,* 387:866, 26 Jun 1997, and in *Science,* 277:772, 8 Aug 1997. *Science* review says discovery of homeobox genes in 1980's ushered in a "new paradigm" and "set the stage fir major advances." Book "breaks now ground." Began some 15 years prior as effort to write development book but discovery of "amazing conservation" of basic mechanisms kept pushing authors toward evolution. Includes chapter on "exploratory behavior" to account for evolutionary diversity. Said to be "long and sometimes wordy" but worth the effort.)

7 DiBerardino, M. A. **Genomic Potential of Differentiated Cells**. NY: Columbia Univ. Press, (1997) < > (Reviewed in *Science,* 277:1056, 22 Aug 1997, which begins, "This book is both a monograph on the author's major research focus ... and a memoir of her work and associations ... over the last 50 years." The review is informative in pointing out areas of controversy.)

8 Kirschner, M. W. and Gerhart, J. C. **The Plausibility of Life: Resolving Darwin's Dilemma**. New Haven, CT: Yale Univ. Press, (2005) < > (Reviewed in *Science,* 310:1619-1620, Dec, 9, 2005, by Brian Charlesworth, who notes that the authors are eminent professors writing for a general readership. The theme is that developmental biology can supply the missing explanation. Also in *New Scientist,* Dec. 10, p57, by Michael LePage, under the caption, "No Creator Required," who speaks of a "new theory of evolution called 'facilitated variation,'" which "at first glance smacks of Lamarckian heresy, of a hidden hand guiding evolution." LePage finds it interesting but "more hypothesis than theory," whatever that means.)

9 Gehring, W. J. **Master Control Genes in Development and Evolution: The Homeobox Story**. New Haven, CT: Yale Univ. Press, (1999) < > (Reviewed in *Science,* 283:799, Feb 5, 1999, by Claude Desplan. "One feels the excitement in the lab where much of this happened ... this is science with passion.")

10 Meyer, A.: **Hox gene variation and evolution**. *Nature* 391 (Jan 15), 225-229 (1998).

11 Quiring, R., Walldorf, U., Kloter, U. and Gehring, W. J.: Homology of the *eyeless* gene of *Drosophila* to the *small eye* gene in mice and *anaridia* in humans. *Science*

265 (Aug 5), 785-789 (1994). <> (For perspective, see p742 by C.S. Zucker, "On the evolution of eyes: would you like them simple or compound?")

12 Glinka, A., Wu, W., Delius, H., Monaghan, A. P., Blumenstock, C. and Niehrs, C.: **Dickkopf-1 is a member of a new family of secreted proteins and functions in head inductions.** *Nature* 391 (22 Jan), 357-362 (1998). <> (Article identifies gene involved with producing the head of a developing emryo; hailed as a major discovery.)

13 Ahlberg, P. E.: **How to keep a head in order.** *Nature* 385 (Feb 6), 489-490 (1997). <> (Comment on progress reported in *Development* on formation of the head, especially the sources of the several kinds of bone in the head, face, and jaw, in the quail chick. Some features are common to all vertebrates, back to the hagfish, the earliest living example.)

14 Shubin, N., Tabin, C. and Carroll, S.: **Fossils, genes and the evolution of animal limbs.** *Nature* 388 (Aug 14), 639 (1997).

15 Averof, M. and Patel, N. H.: **Crustacean appendage evolution associated with changes in Hox gene expression.** *Nature* 388, 682-686 (1997 14 Aug).

16 Keller, R.: **Shaping the vertebrate body plan by polarized embryonic cell movements.** *Science* 298 (Dec 6), 1950-1954 (2002). <> (One of five review papers, with related research papers in special issue on polarity in development; and see perspective, p1901.)

17 Nash, J. M.: **Where do toes come from?** *Time Magazine* (Jul 31), 56-57 (1995). <> (Brief account of the ubiquity of the *Hox* box, based on a then-recent paper in *Nature* by Duboule *et al.*)

18 Brooke, N. M., Garcia-Fernandez, J. and Holland, P. W. H.: **The ParaHox gene cluster is an evolutionary sister of the Hox gene cluster.** *Nature* 392 (Apr 30), 920-922 (1998).

19 Chourrout, D. et al: **Minimal ProtoHox cluster inferred from bilaterian and cnidarian Hox complements.** *Nature* 442 (Aug 10), 684-688 (2006). <> (The authors consider various other hypotheses, such as "that cnidarians are in fact simplified bilaterians," but reject them as unlikely.)

20 Editor: **Not all animals have the animal gene.** *New Scientist* (Apr 8), 21 (2006). <> (Note on paper by D. Miller in *Current Biology* 16:4682, that Hox genes, said to be a defining feature of animals, are absent in some lower forms.)

21 Myerowitz, E. M.: **Plants compared to animals: the broadest comparative study of development [Review].** *Science* 295 (Feb 22), 1482-1485 (2002). <> (A key question addressed is whether plant and animal development show signs of a common ancestor. Answer seems to be no.)

22 Bunk, S.: **Big genes are back.** *The Scientist* (Mar 18), 24-25 (2002). < > (Subheading, "Recent 'evo-devo' findings revive the large-effect gene hypothesis.")

23 Davidson, E. H. and Erwin, D. H.: **Gene regulatory networks and the evolution of animal body plans.** *Science* 311 (Feb 10), 796-800 (2006). < > (Jerry Coyne replied to this article with a critique [Science 313:761, Aug 11, more online] which he says "proposed that phyla arise from ... large mutations acting on conserved core pathways of development. I question some of these assumptions and show that natural selection adequately explains the origin of new phyla.")

24 Editor: **A wrinkly mutant.** *Nature* 442 (Aug 24), 851 (2006). < > (Note on study in *Nature Genetics* led by Christopher Knight showing that a single small mutation in a bacterium alters some 52 different proteins involved with its ability to form colonial mats. Also noted in *Science* 313:1203, under caption "Pleiotropic tensegrity.")

25 Cowen, L. E. and Lindquist, S.: **Hsp90 potentiates the rapid evolution of new traits: Drug resistance in diverse fungi.** *Science* 309 (Sep 30), 2185-2189 (2005).

26 O'Malley, B. W.: **Little molecules with big goals.** *Science* 313 (Sep 22), 1749-1750 (2006). < > (Short review of "nuclear co-regulators," said to have opened up in 1995 [*Sci* 270:1354] but the big-gene hypothesis is dated from 1969 [*Sci* 165:349]. By the way, this paper is followed by one on the genetics of cleft palate, a developmental birth defect, showing how a small change can eventuate in a large effect.)

27 K, P. A.: **An interfering ego?** *Science* 287 (Mar 10), 1715 (2000). < > (Note on work by Smardon et al writing in *Curr. Biol.* 10:169, indicating that a gene called ego-1, essential for development in the worm *C. elegans*, encodes a protein, QDE-1, and an RNA-dependent RNA polymerase [RdRP] indicative of siRNA involvement.)

28 Perkel, J. M.: **MicroRNAs assume a developmental role.** *The Scientist* (Apr), 69 (2006). < > (Highlights of paper by S. Yetka et al in *Science* 304:594, 2004, with David Bartel, showing that miR-196 plays a role in regulating the HOXB8 gene expression affecting development of the hindquarters of developing mice, as they later clarified [*Nature* 438:671, 2005]. In 'Hot Papers' column because it was cited 129 times. Authors note that many other genes also involved.)

29 Giraldez, A. J. et al: **Zebrafish MiR-430 promotes deadenylation and clearance of maternal mRNAs.** *Science* 312 (Apr 7), 75-79 (2006). < > (With commentary by Cohen and Brennecke, p65. The switch from using the mother's genes to the embryo's own genes is accomplished in part by miR-430, which inactivates some 200 maternal mRNAs, in the first 4-8 hours of development. miR-430 is coded by a family of some 90 genes.)

30 Vastenhouw, N. L. et al: **Long-term gene silencing by RNAi.** *Nature* 442 (Aug 24), 882 (2006).

31 Sanchez-Elsner, E. et al: **Noncoding RNAs of Trithorax response elements recruit *Drosophilia* Ash1 to Ultrabithorax.** *Science* 311 (Feb 24), 1118-1123 (2006). <> (Concerns research on the development of the embryo of fruit fly *Drosophilia*, the key point here being that noncoding transcripts "play an important role in epigenetic activation of gene expression.")

32 Tromans, A.: **Competitive edge.** *Nature* 428 (Apr 8), 609 (2004). <> (Note on two papers in *Cell,* 117:107 and 117:117, looking at what happens to a developing fly wing if some cells are made to express unusually high levels of dMyc, a growth regulator. The result was that the cells got much bigger, but surrounding cells with normal dMyc got smaller, or some died, resulting in a wing of normal size. They interpret this in alternative ways.)

33 Sato, K. and Siomi, H.: **Is canalization more than just a beautiful idea? (Epub preprint].** *Genome Biol* 11 (3), 109 (2010).

34 Palmer, A. R.: **Symmetry breaking and the evolution of development (Review).** *Science* 306 (29 Oct), 828-833 (2004).

35 Aubret, F. and al, E.: **Adaptive developmental plasticity in snakes.** *Nature* 431 (16 Sep), 261 (2004). <> (Reports that certain snake jaw size varies depending on size of prey available. Experiment well-controlled for genetics because he finds two natural populations, one on island, one mainland, which normally eat different size prey, resulting in different jaw size due to genetics. The implication is that prey size can result in heritable modification independent of natural selection, i.e. Lamarckian.)

36 Pigliucci, M. **Phenotypic Plasticity: Beyond Nature and Nurture.** Baltimore, MD: Johns Hopkins Univ. Press, (2002) <> (Reviewed *Nature* 416:584, Apr 11 2002, by Ralph Tollrian, who says "phenotypic plasticity describes the ability of a genotype to form different phenotypes in distinct environmental conditions." Speaks of "renaissance of the 'reaction norm,'" a term coined by Woltereck in 1909." Book is said to offer a "balanced presentation" of a hotly debated field. Refers to his own book on related matters (R. Tollrian and C.D. Harvell, *The Ecology and Evolution of Inducible Defenses*, Princeton Univ. Press, 1999) and an earlier one by C.D. Schlicting and Pigliucci, *Phenotypic Evolution: A Reaction Norm Perspective* (Sinauer Press, 1999).)

37 West-Eberhard, M. J. **Developmental Plasticity and Evolution:** Oxford Univ. Press, (2003) <> (Reviewed *Nature,* 424:16, Jul 3 2003, by G. de Jong and R.H. Crozier. Reviewer finds much good in this book but also many flaws, concluding: "West-Eberhad asserts a vision but presents little analysis. A major new synthesis and research programme this book is not.")

38 Montague, A. **Growing Young**. N.Y.: McGraw Hill, (1983) <> (Discussed at length in *The Sciences*, May/Jun 1983, p50-54, by Ray P. Coppinger and C. Kay Smith. By the way, *The Sciences,* was a real delight through the 1980's, not least, because of the stunning visual arts.)

39 Dinsmore, C. E., ed. **A History of Regeneration Research**. N.Y.: Cambridge Univ. Press, (1992). <> (Reviewed in *Science,* 256:865, 8 May 1992. Topic is regeneration of lost or damaged limbs or organs.)

40 Shubin, N. H. and Dahn, R. D.: **Lost and found**. *Nature* 428 (Apr 15), 703-704 (2004). <> (Perspective on report by Shapiro et al, p717, who crossed a female stickleback from a region where they lack a pelvic spine with a male that has it, then looked at the expression of 231 genes in the 375 offspring of two generations, some of which had the spine, others not or partial. They mapped control of the spine to a gene called *Pitx1*, also found in mice, and when deleted in mice causes reduced hindlimbs and sometimes abnormalities of head, face, certain glands. But the key finding was that this gene was identical in all of the fish, with or without the spine, showing that the regulation of *Pitx1*, not the gene itself, was responsible.)

41 Carroll, S. B. et al: **Pattern formation and eyespot determination in butterfly wings**. *Science* 265 (Jul 1), 109-112 (1994). <> (This oft-cited paper shows that wing spots are controlled first by genes previously identified in fruit flies, which govern the location of the spot on the wing, and on top of that, by other genes that elaborate the patterning details.)

42 Prud'homme, B. et al: **Repeated morphological evolution through cis-regulatory changes in a pleiotropic gene**. *Nature* 440 (Apr 20), 1050-1053 (2005). <> (With perspective by G.A. Wray, p1001. Authors examine genetic basis of wing spots across some 77 fruit fly species of known genealogy, showing that the spots have appeared and disappeared on multiple occasions, nearly always by altered regulation of the same gene, yellow, supporting the emerging generality that modified regulation of genes, not the genes themselves, is the cause of anatomical variation. At another level, the changes are said to be caused by sexual selection acting on natural variation.)

43 McKinney, M. J. and McNamara, K. J. **Heterochrony: The Evolution of Ontogeny**. N.Y.: Plenum Press, (1991) <> (Reviewed in *Amer Scientist,* 81:292, by A.C. Burke who, after listing the virtues and appeal of the book, devotes about half the review to raising questions about adequacy of definitions, inconsistency of usage, etc.)

44 Minugh-Purvis, N. and McNamare, K. J. **Human Evolution through Developmental Change**. Baltimore, MD: Johns Hopkins Univ. Press, (2002) <> (Reviewed in *Nature* 420:609, Dec. 12, by Bernard Wood. Offers evo-devo perspective. Reviewer seems to find it interesting but rather inconclusive. Among the conclusions is rejection of the neoteny hypothesis, that we arose "because of some relatively simple retention of

juvenile characteristics." A major theme is said to be "heterochrony," changes to the relative timing and rate of development.)

45 Schlosser, G. and Wagner, G. P. **Modularity in Development and Evolution.** Chicago: Univ. Chicago Press, (2004) <> (Edited, with 23 chapters from leaders on this topic. Reviewed in *Science* 306:814, 29 Oct, by L.A. Myers. The review itself is informative, summarizing and classifying different approaches and definitions of terms, especially modularity itself.)

46 Pennisi, E.: **Searching for the genome's second code.** *Science* 306 (22 Oct), 632-635 (2004). <> (We cited this in an earlier chapter, with comment.)

Ch. 15, References & Notes

1 Sloman, S. A. and Rips, L. J. **Similarity and Symbols in Human Thinking.** Cambridge, MA: MIT Press, (1998) <> (Critically reviewed in *Nature* 396:325 by Jerry Fodor, opening with the statement quoted. Book is a collection of papers about how we conceptualize and/or classify things. Review is itself lively and informative, as expected from Fodor.)

2 Scheibe, K. E. **The Drama of Everyday Life.** Cambridge, MA: Harvard Univ. Press, (2000) <> (Reviewed in *Science* 289:1303 by Rom Harre. "Karl Scheibe's entertaining book is a reaffirmation of one of the most powerful alternative ways of looking at human affairs: from the dramaturgical point of view." It is interesting to note that a certain cadre of anthropologists and sociologists are also taking this route of analysis, trending towards convergence with the best of literary criticism.)

3 Jaffe, S.: **Fake method for research impartiality (fMRI).** *The Scientist* (Jul 19), 64 (2004). <> (Sound-bite: "dozens of studies are pouring out of the humanities aisle of academia" based on fMRI. Holds up to ridicule studies claiming to show why some Americans vote republican or democrat [difference in amygdala brain region], why some golfers score poorly [excess brain activity in cerebellar vermis], and why we ignore flaws in those we love [part of brain switched off], with references. The general contempt expressed toward such 'sciences' is widely shared by those in hard science.)

4 Mann, C. C.: **Behavioral genetics in transition.** *Science* (Jun 17), 1686-1689 (1994). <> (Special section runs p1685-1740. This piece opens citing studies back to 1923 in Germany claiming to link genes with defective abilities, followed by the disrepute of such studies due to the Holocaust, misguided practice of eugenics, and evidence that culture was the major determinant. The field resurged in the 1980s led by psychologists, then geneticists got into the game.)

5 Buller, D. J. **Adapting Mind: Evolutionary Psychology and the Quest for Human Nature**. Cambridge, MA: MIT Press, (2005) <> (Reviewed in *Science* 309:706, July 29, 2005, by Johan J, Bolhuis, under the caption, "We're not Fred or Wilma," meaning the Flintstones.)

6 C, G. J.: **Close encounters**. *Science* 313 (Jul 14), 148 (2006). <> (Note on paper in *J. Pers. Soc. Psychol.* 90:751, by Pettigrew and Tropp, giving the latest findings on "the effect of contact between groups on prejudice," listing major factors determining attitudes.)

7 Dennett, D. C. **Breaking the Spell: Religion as a Natural Phenomenon**. New York: Viking / Allen Lane, (2006) <> (Lead book review in *Nature* 439:535, Feb. 2, 2005, by Michael Ruse, under the leader, "A Darwinian philosopher turns his attention to the strength of religion." Says "a major plank in Dennett's discussion is that religion is all smoke and mirrors." The same could be said of some of Dennett's books. Ruse concludes that this is a worthy topic – the relation of science to religion – "but we need better books than this to address the issues." Also reviewed in *Science* 311:471, Jan 27, by Michael Shermer; and in *New Scientist*, March 4, p48, by evolutionist, Francisco Ayala.)

8 Wolpert, L. **Six Impossible Things Before Breakfast: The Evolutionary Origins of Belief**. New York: Faber and Faber, (2006) <> (Reviewed in *Nature* 442:137, July 13, 2006, by Crispin Tickell, under the headline, "Religious beliefs can be viewed as an adaptation that was most favored as the human race evolved." Reviewer observes, as if with surprise, that "the Will of God is still seen as critical in many societies." He judges this book to be "more succinct and much better argued" than Dennett's, published at the same time.)

9 Tremlin, T. **Minds and Gods: The Cognitive Foundations of Religion**: Oxford Univ. Press, (2006) <> (Reviewed in *Nature* 444:39, Nov. 2, 2006, by A.C. Kruger and M. Konner. Book by psychologist offers evolutionary cognitive psychology explanation for religion. Reviewers note that Tremlin largely omits emotional components.)

10 Orr, H. A.: **The God Project**. *New Yorker* (Apr 3), 80-83 (2006). <> (Review and commentary on Dennett's book, *Breaking the Spell*.)

11 Shermer, M.: **Vox populi: ... why evolution remains controversial**. *Sci Amer* (Jul), 37 (2002). <> (Shermer, publisher of *Skeptic* magazine and author of a biography of Alfred Wallace, has a regular column, *Skeptic,* and is a perfect heir to myth-buster Martin Gardner, long-time math columnist at *Sci Amer* and author of the genre-setting *Fads & Fallacies*, a best-seller of the 1960's, bashing so-called pseudo-science but backfiring due to Gardner's excesses and smugness. Here, Shermer ponders why he received hundreds of letters on his February column on evolution, compared to the usual dozen or so.)

12 Shermer, M. **How We Believe: Science, Skepticism, and the Search for God**. New York: Henry Holt, (2000)

13 Shermer, M. **The Science of Good and Evil**. New York: Henry Holt, (2004)

14 Collins, F. S. **The Language of God: A Scientist Presents Evidence for Belief**. New York: Free Press / Simon & Schuster, (2006) <> (Reviewed in *Science* 313:1890 by Robert Pollack, who begins by noting that Collins, as head of the *NIH Human Genome Project*, replacing James Watson, clearly has lofty scientific credentials, yet writes as an evangelical Christian. The reviewer sees no contradiction here, being himself a man of faith at the *Center for the Study of Science and Religion* at Columbia University. But the reviewer finds inconsistencies in Collins' effort to reconcile religion and evolution. By the "moral law" is meant that a fundamental concern of all humans (and I think of all life) is the question, "What ought I do?" Reviewer remarks that some of the ideas derive from C.S. Lewis. It may be added that Hauser's *Moral Law* is a central theme also of certain existentialists. Also reviewed in *New Scientist*, Aug. 26, 2006, by Steve Fuller. Both reviews are polite, in deference to Collins' scientific stature.)

15 Hauser, M. D. **Moral Minds: How Nature Designed Our Universal Sense of Right and Wrong**. New York: Ecco / Harper Collins, (2006) <> (Reviewed in *Science* 314:57 by Michael R. Waldmann, who explains that Hauser, a psychologist, develops a theory analogous to Chomsky's 'universal grammar' but a grammar of morality rather than language. Waldmann notes that a similar theory was developed by John Mikhail in a forthcoming book, *Rowl's Linguistic Analogy*, Cambridge Univ. Press. Different cultures modulate this moral grammar, as they do language. Reviewer is skeptical. Other theories of morality are discussed.)

16 Lloyd, E. A. **The Case of the Female Orgasm: Bias in the Science of Evolution**. Cambridge, MA: Harvard Univ. Press, (2005) <> (Reviewed in *New Scientist*, May 14, 2005, by Gail Vines, under the caption, "The Accidental Orgasm," who notes that the author has dug up 21 theoretical explanations in terms of evolutionary theory and gives reasons why all are wrong or doubtful, and then seems to conclude that that female orgasm has no purpose or function at all, and therefore must be an accidental carryover from males, much as the useless nipples on males is a carryover from female function.)

17 Konner, M.: **Is orgasm essential?** *The Sciences* (Mar/Apr), 4-7 (1988). <>

18 Wade, N. **Before the Dawn: Recovering the Lost History of Our Ancestors**. New York: Penguin, (2006) <> (Reviewed in *New Scientist*, Apr. 8, 2006, p53, by Steven Mithen, under the headline, "Ancestors Behaving Badly"; and in *Nature*, June 15, p813, by K.M. Weiss and A.V. Buchanan.)

19 Russo, E.: **Mind, the adaptive gap**. *The Scientist* (Mar 1), 26-28 (2004). <> (Subtitle, "Evolutionary psychologists try to shed the just-so story stigma." Attributes the

origin of EP to George Williams, W.D. Hamilton, Robert Trivers and E.O. Wilson. Wilson defines it as "the study of the biological basis of all forms of social behavior in human beings." Notes that the journal, *Evolution and Human Behavior*, is the leading forum of EP studies, under Martin Daly, editor, and that a 1992 anthropology, *The Adapted Mind*, is considered a landmark in the field.)

20 Macrakis, K. **Surviving the Swastika: Scientific Research in Nazi Germany**. New York: Oxford Univ. Press, (1993) <> (Reviewed in Science 265:124, Jul 1, 1994. "Eugenics and other 'racial sciences' flourished during the period, too, of course, but Macrakis argues that 'eugenics was not a product of National Socialist Germany. German eugenics was part of a worldwide movement that had started before 1933; the Nazis simply used it to legitimize their racial policies.")

21 Badcock, C.: **More danger in doctrine than in genetics**. *Nature* 388 (3 Jul), 13 (1997). <> (Although addressing a different topic – "imagined political and social dangers of a belief in biological determinism" [*Nature* 387:743, 1997] – this letter reminds us of the genocides and war crimes "of the Soviet Union, Cambodia, and elsewhere" attributable to "sociological theories" and that the genetic theories of Nazi Germany were only one such instance. That is to say, when science is made a handmaiden – pimp – of social and political theories, the dangers are real and great.)

22 Deichmann, U.: **An unholy alliance**. *Nature* 405 (Jun 15), 739 (2000). <> (Essay, subtitled "The Nazis showed that 'politically responsible' science risks losing its soul.")

23 Stein, G. J.: **Biological science and the roots of Nazism**. *Amer Scientist* 76 (Jan/Feb), 50 (1988). <>

24 deWaal, F. **Our Inner Ape: A Leading Primatologist Explains Why we Are Who We Are**. New York: Riverhead Books / Penguin, (2005) <> (Reviewed in *Sci Amer*, Dec., 2005, p124-5, by Henry Gee, Editor of *Nature*, along with a related book. "*Good Natured* is written for a general audience but with strong scientific foundation." Also in *Nature* 437:33, by Robert Sapolsky.)

25 deWaal, F. **Good Natured: The Origin of Right and Wrong in Humans and Other Animals**. Cambridge, MA: Harvard Univ. Press, (1996) <> (Reviewed in *Science* 276:1088, by Charles T. Snowdon.)

26 Holden, C.: **Long-ago people may have been long in the tooth**. *Science* 312 (Jun 30), 1867 (2006). <> (Reporting highlights of a meeting of the *Human Behavior and Evolution Society*, including a study showing that simple hunting-gathering people who survive to age 15 have life expectancies not shorter than more 'civilized' peoples, commonly living into their 70's and beyond, contrary to conclusions of previous studies from less direct evidence.)

27 Holden, C.: **Teeth of civilization.** *Science* 312 (Jun 9), 1449 (2006). < > (Note on paper by Clark Larsen in June issue of *Quarternary International* showing that the shift to agriculture was accompanied by a large increase in tooth decay, reduction in size of face and jaw, iron deficiency anemia, abnormal crowding of teeth, pathological bone porosity, and increased frequency of diseases like leprosy, syphilis, and tuberculosis. Other sources not at hand report decline in stature.)

28 Russell, R. J. **The Lemur's Legacy: Evolution of Power, Sex and Love.** New York: Tarcher / Putnam, (1993) < > (Reviewed in *Nature* 364:585, by Adrienne Zihlman, along with a related book, *Female Choices*, on the "sexual behavior of female primates," by Meredith F. Small [Cornell Univ. Press]. Russell's studies indicate that all the ingredients of higher primate social behavior – those in the title – are present in the lemurs of Madagascar, "a hypothetical first stage in the evolution of human behavior." Interesting hypotheses, such as that lemurs were female-dominated, a habit supplanted by the rise of male-dominated apes – along the lines of Graves' *The White Goddess* but different in time scale. Russell puts great weight on testosterone. Attributes the origin of EP to Darwin's 1872 *Expression of the Emotions in Man and Animals*. The two books diverge widely in their interpretation of the implications for humans. Reviewer judges Small's book to be more measured, fact-based, and less speculative, but perhaps for that reason, maybe less interesting.)

29 Sapolsky, R. and Cape, J. **Monkeyluv: And Other Lessons on Our Lives as Animals.** New York: Scribner, (2005) < > (Reviewed in *New Scientist*, Oct. 15, 2005, p51, by Kate Douglas.)

30 Forbes, S. **A Natural History of Families.** Princeton, N.J.: Princeton Univ. Press, (2005) < > (Reviewed in *Nature* 437:195, Sept. 8, 2005, by Jonathan Wright, under headline, "families behaving badly.")

31 Mock, D. W. **More than Kin and Less than Kind: the Evolution of Family Conflict.** Cambridge, MA: Harvard Univ. Press, (2004) < > (Reviewed in *Nature*, 429:23, by H.C. Godfray.)

32 Andreason, N. C. **Brave New Brain: Conquering Mental Illness in the Era** of the Genome: Oxford Univ. Press, (2001) < > (Reviewed in *Nature* 411:740, by Robert Plomin, who notes that the author is editor-in-chief of the *American Journal of Psychiatry* and winner of prestigious awards. One of her main concerns is practical: weak spending on mental health and our purely economic 'non-humanistic' approach to public health care.)

33 Andreason, N. C. **The Broken Brain.** New York: Harper & Row, (1984)

34 Feldman, D. E. and Brecht, M.: **Map plasticity in somatosensory cortex [Review].** *Science* 310 (Nov 4), 810-815 (2005). < > (Many lines of experimental evidence overturn the classical theory in favor of a "distributed model in which plasticity occurs at

multiple sites in the cortical circuit with multiple cellular / synaptic mechanisms and multiple likely learning rules for plasticity." Article is part of a special issue on brain. See also special issue of *Nature*, Oct 14, 2004, p759-803, "Plasticity and neuronal computation.")

35 Buller, D. J.: **Making modern minds**. *New Scientist* (Sep 10), 48-49 (2005). <> (Thumbnail sketch of Buller's latest book, coming across as an attack on EP and its proponents such as Stephen Pinker's *The Blank Slate,* and David Buss', *The Evolution of Desire* and *The Murderer Next Door*. Buller picks on two related fallacies, (i) the idea that the human mind is "massively modular" with thousands of specialized modules for solving various kinds of problems, and (ii) that "we are living fossils of our Stone Age ancestors" in the sense of being blindly driven or programmed by instincts from long ago.)

36 Gotschall, J. and Wilson, S. **The Literary Animal: Evolution and the Nature of Narrative**: Northwestern Univ. Press, (2005) <> (Reviewed in *Nature* 441:150, May 11, 2006, by Rebecca Goldstein. Also in *Science* 311:612, by Harold Fromm, from a different perspective, seeing it as another shot fired in the Science Wars which, "from where I'm sitting, are looking pretty alive." Cites related works, e.g. Joseph Carroll's *Evolution and Literary Theory* [Univ. of Missouri Press, 1995], but we lack space for sketch of this informative review.)

37 Lai, C. S. L., Fisher, S. E., Hurst, J. A., Vargha-Khadem, F. and Monaco, A. P.: **A forkhead domain gene is mutated in a severe speech and language disorder**. *Nature* 413 (Oct 4), 529-523 (2001). <> (Announcing discovery of the putative language gene, *FOXP2*, on chromosome 7, region 7q31, by group led by Anthony P. Monaco, based on study of three generations of affected family. With perspective, p465)

38 Shute, N.: **We the mutants: On the trail of the genes that made us human**. *U.S. News & World Report* (Sep 16), 64 (2002). <> (Mentions language gene FOXP2; head size gene citing recent issue of *Proceedings of the National Academy of Sciences* on sialic acid; all widely reported.)

39 Newbury, D. F. and Monaco, A. P.: **Genetic advances in the study of speech and language disorders**. *Neuron* 68 (2), 309-320 (2010). <> (Article discusses also FOXP1 and several other genetic associations with various types of speech and other language disorders. Many reviews on this subject turned up.)

Ch. 16, References & Notes

1 Merrell, D. S. and al, E.: **Host-induced epidemic spread of the cholera bacterium**. *Nature* 417 (Jun 6), 642-645 (2002). <> (The bacterium is widely present but not generally very harmful, and it is difficult to cause disease in human volunteers who ingest the bacteria grown in the laboratory, yet severe outbreaks often occur, and the

authors want to know why. They found that passage through the gut is required for the bacteria to shift to a more highly infectious state causing severe diarrhea.)

2 Morell, V.: **How the malaria parasite manipulates its hosts**. *Science* 278 (Oct 10), 223 (1997). <> (Reporting highlights from a meeting of the *European Society of Evolutionary Biologists*.)

3 Merali, Z.: **Mozzies home in on delicious smell of malaria**. *New Scientist* (Aug 13), 13 (2005). <> (Note on work by J. Koella carried out in Kenya, with commentary by other authorities, showing that malarial parasites "actually make their human hosts smell more attractive to mosquitoes when the parasites are ready to be transmitted, although its not yet clear how they do it." Quoting Koella: "the malarial parasite manipulates the transmission from within." Also noted in *Science News,* Sept 3, 2005, p157, citing the Sept. issue of *PLoS Biology* as source.)

4 Smith, D. L. et al.: **The entomological inoculation rate and *Plasmodium falciparum* infection in African children**. *Nature* 438 (Nov 24), 492-496 (2005).

5 Zimmer, C.: **Parasites make scaredy-rats foolhardy**. *Science* 289 (Jul 28), 525-526 (2000). <> (Note on paper in Aug. 7 issue of *Proc Roy Soc Lond B,* showing that the protozoan parasite, *Toxoplasma gondii,* causes rats to lose their fear of cats. By the way, it is noted that Zimmer has written a book, *Parasite Rex,* "to be published in Sept.")

6 Brownless, C.: **Danger mouse**. *Science News* 168 (Nov 26), 341 (2005). <> (Note on article by G. Shumyatsky and colleagues in the Nov. 18 issue of *Cell.*)

7 Culotta, E.: **The parasite X-files**. *Science* 267 (Jan 20), 331 (1995). <> (Reports on a flatworm parasite whose eggs are shed by the bird host in feces into a salt marsh. "The parasite then invades a marine snail and castrates it by devouring the snail's sex organs. Next, it swims from the snail to the killifish ... where it forms cysts on the fish's brain. The cycle is closed" when a bird eats the fish. The work shows that the cysts on the brain cause the fish to jerk and surface, making it a tempting target. The article doesn't explain why the parasite castrates the snail.)

8 Moore, J.: **Parasites that change the behavior of their host**. *Sci American* (May), 108-115 (1984).

9 B, T. M.: **Alien influence: parasitic intruders manipulate host behavior**. *Sci American* (Jul), 26 (1989). <> (Explains how larva of a parasitic wasp alters behavior of its aphid host to suit its purposes. "The new twist is that *A. nigripes* is selective, making its host hide when hiding is adaptive ... but not otherwise.")

10 Moore, J. **Parasites and the Behavior of Animals**. New York: Oxford Univ. Press, (2002) <> (Reviewed in *Nature* 417:592, Jun 6, 2002, by P. Schmid-Hempel, who describes it as "a gripping account" and wittily written.)

11 Ikeda, N., Chijiwa, T., Matsubara, K. et al: **Unique structural characteristics and evolution of a cluster of venom phospholipase A2 isozyme genes of Protobothrops flavoviridis snake.** *Gene* 461 (1-2), 15-25 (2010).

12 Nei, M. and Rooney, A. P.: **Concerted and birth-and-death evolution of multigene families.** *Annu Rev Genet* 39, 121-152 (2005).

13 Froy, O., Sagiv, T., Poreh, M., Urbach, D., Zilberberg, N. and Gurevitz, M.: **Dynamic diversification from a putative common ancestor of scorpion toxins affecting sodium, potassium, and chloride channels.** *J Molec Evol* 48 (2), 187-195 (1999).

14 Rodriguez, R. C., Schwartz, E. F. and Posani, L. D.: **Mining on scorpion venom bio-diversity.** *Toxicon* 56 (7), 1155-1161 (2010).

15 Ma, Y., Zhao, Y., Zhao, R., Zhang, W., He, Y., Wu, Y., Cao, Z., Guo, L. and Li, W.: **Molecular diversity of toxic components from the scorpion Heterometrus petersii venom revealed by proteomic and transcriptome analysis.** *Proteomics* 10 (13), 2471-2485 (2010).

16 Moffat, A. S.: **A challenge to evolutionary biology.** *American Scientist* 77 (May/Jun), 224-226 (1989).

17 Rainey, P., Moxon, R. et al: **Unusual mutational mechanisms and evolution (Technical comments).** *Science* 260 (Jun 25), 1958-1960 (1993). <> (Comments on paper by Lenski and Mittler in Jan. 8 issue, p188, with other comments by D.A. Watson and by L.D. Hurst, and response by Lenski and Mittler. Very technical.)

18 Lenski, R. E. and Mittler, J. F.: **The directed mutation controversy and neo-Darwinism.** *Science* 259 (Jan 8), 188-194 (1993). <> (Article with 64 ref's. Abstract concludes that "further experiments that address these criticisms [of Cairns' thesis] do not support the existence of directed mutations.")

19 Foster, P. L.: **Whither directed mutation? (Letter).** *Science* 262 (Oct 15), 317-319 (1993). <> (Letter on letter by Lenski and Mittler in the May 28 issue, p1222, with response. Basically, Foster is defends but they stand fast.)

20 Shapiro, J. L., Foster, P. L. and Thaler, D.: **Adaptive mutation (Letters).** *Science* 265 (Sep 30), 1994-1996 (1994). <> (Letters on paper by R.N. Harris et al and the perspective on it by Thaler, *Science* 264:258, 254, with response by Thaler. The exchange is inconclusive. One prescient pearl considered is whether the SOS response is involved.)

21 Harris, R. S., Longerich, S. and Rosenberg, S. M.: **Recombination in adaptive muta-tion.** *Science* 264 (Apr 8), 258-260 (1994). <> (With perspective by D.S. Thaler,

p258-260, "The evolution of genetic intelligence," discussing the many theories to explain the apparent results, with 31 ref's. The Harris paper confirms that recombination is essential for the effect at issue.)

22 Rosenberg, S. M., Longerich, S., Gee, P. and Harris, R. S.: **Adaptive mutation by deletions in small mononucleotide repeats.** *Science* 265 (Jul 15), 405-407 (1994).

23 Foster, P. L. and Trimarchi, J. M.: **Adaptive reversion of a frameshift mutation in** *Escheria coli* **by simple base deletions in homopolymeric runs.** *Science* 265 (Jul 15), 407-409 (1994).

24 Culotta, E.: **A boost for "adaptive" mutation.** *Science* 265 (Jul 15), 318-319 (1994). <> (Reporter introduces two articles this issue, by Foster and Trimarchi, p407, and by Rosenberg, p405, which appear to support Cairn's thesis by showing that "a distinct type of mutation is indeed at work under certain selective pressures – but neither paper proves that useful mutations arise faster than non-useful ones." Slightly simplistic.)

25 Bridges, B. A.: **Hypermutation under stress.** *Nature* 387 (Jun 5), 557-558 (1997). <> (Commentary on two then-recent papers, by Foster [*J. Bact.* 179:1550] and by Torkelson et al [*EMBO J.,* 16:3303], relevant to the directed mutation controversy.)

26 Shapiro, J. A.: **Adaptive mutation: Who's really in the garden?** *Science* 268 (Apr 21), 373-374 (1995). <> (Perspective on two reports in this issue, by Radicella et al, p418, and by Galitski and Roth, p421. His use of "the unicorn in the garden" refers to that simile employed by F. Stahl in *Nature* 335:112, 1988, wondering if something mystical is involved with "adaptive mutations.")

27 Jackson, D. E., Holcombe, M. and Ratnieks, F. L. W.: **Trail geometry gives polarity to ant foraging networks.** *Nature* 432 (Dec 16), (2004). <> (Scent marked trails do not give information about the home direction. The authors show that forks in the trail networks are set at certain angles making directions to and away from the nest clear.)

28 Wohlgemuth, S., Ronacher, B. and Wehner, R.: **Ant odometry in the third dimension.** *Nature* 411 (Jun 14), 795=798 (2001). <> (It was known that ants keep track of the distance and direction from home by continuous calculation, called "path integration." This work adds another dimension by interposing sawtooth-like ramps to make the ants increase the distance walked. The conclusion is that they maintain accurate knowledge of the horizontal component of distance. The ants used are a desert species inhabiting flat terrain. See also commentary p792. "Homing in on ant navigation.")

29 Hemmi, J. M. and Zeil, J.: **Robust judgment of inter-object distance by an arthropod.** *Nature* 421 (Jan 9), 160-163 (2003). <> (Fiddler crabs, despite supposedly poor eyesight, diligently defend their burrows against conspecifics (neighbors) and here it

is shown that they can accurately judge distance of a potential rival from their home burrow at considerable remove and at varying angles, by geometric relations. Some might say "duh!" but this paper makes explicit another "intellectual ability" in a lower form.)

30 Hsu, C.-Y. and Li, C.-W.: **Magnetoreception in honeybees.** *Science* 265 (Jul 1), 95-96 (1994). <> (See also Technical Comment on this article by H Nichol and M Locke in *Science,* 269:1888, 19 Sept 1995.)

31 Boles, L. C. and Lohmann, K. J.: **True navigation and magnetic maps in spiny lob-sters.** *Nature* 421 (Jan 2) (2003). <> (Lobster is first invertebrate shown to perform "true navigation" as distinct from simpler means of path finding. The animals were moved distances of 12-37 km in the Bahamas in covered boxes, with lots of turning, yet could orient towards home, but artificial magnetic fields caused wrong headings. See also p27 for commentary and background by Thomas Alerstam, "The lobster navigators." Also noted in *Science News,* 1/4/03, p4, with interview of authors.)

32 Seachrist, L.: **Sea turtles master migration with magnetic memories.** *Science* 264 (Apr 29), 661 (1994).

33 Lohmann, K. J. and al, E.: **Geomagnetic map used in sea-turtle navigation.** *Nature* 428 (Apr 29), 909-910 (2004). <> (This takes the magnetic sense beyond mere direc-tion finding since the animals seem to recognize local geomagnetic parameters.)

34 Editors: **How blind mole rats find their way home.** *Science News* 165 (Feb 14), 110 (2004). <> (Describes findings by Kimchi and Terkel in Jan 27 issue of *PNAS,* build-ing on their earlier work of 2001, establishing magnetic compass orientation in these animals, additional to dead-reckoning.)

35 Jerome, R.: **Animals and magnetism.** *The Sciences* (Jan/Feb), 7 (1993). <> (Recounts work appearing in *Nature* by J.B. Phillips and S.C. Borland showing that newts do indeed possess magnetic compass orientation but it requires light of certain wave-lengths entering the eye.)

36 Barinaga, M.: **Giving personal magnetism a whole new meaning.** *Science* 256 (May 15), 967 (1992). <> (Interviews J. Kirschvink on his work in-press in *PNAS,* on his group's finding of magnetite crystals in the human brain similar to those found in bacteria. They speculate that it is "a vestigial remnant from our migratory ancestors." Explains the dark spots seen in MRI brain images. Cites prior articles on controversy about possible harmful effects of electromagnetic fields: *Science,* 1990, Sep 7 p106 and Sep 21 p1378.)

37 Becker, R. O. and Selden, G. **The Body Electric: Electromagnetism and the Foundation of Life.** New York: William Morrow and Co., (1986) <> (Reviewed in *The Sciences,* Jan/Feb 1986, p53-56, by distinguished W. Ross Adey, along with

N.H. Steneck's *The Microwave Debate,* from MIT Press. Review finds some merit in both but strongly objects to their hysterical tone, implying some sort of government conspiracy of silence on the then-contentious issue of health dangers posed by electromagnetic fields of powerlines, etc. One of the authors, a surgeon, is noted for his discovery that a weak electric current can speed the healing of fractured bones.)

38 Walker, M. M., Diebel, C. E., Haugh, C. V., Pankhurst, P. M., Montgomery, J. C. and Green, C. R.: **Structure and function of the vertebrate magnetic sense.** *Nature* 390 (Sep 27), 371-376 (1997).

39 Blakemore, R. P. and Frenkel, R. B.: **Magnetic navigation in bacteria.** *Sci Amer* (Dec), 58-65 (1981). <> (Manetite crystals aligned in chains, of size calculated to be optimal for the purpose, found in dozens of species inhabiting anoxic sediments in fresh and salt water, enabling them to avoid noxious environments. In northern hemisphere they are north-seeking, and opposite in the southern. They are synthesized by special enzymes from soluble iron and are sheathed in multi-layered capsules.)

40 Wiltschko, W., Munro, U., Ford, H. and Wiltschko, W.: **Red light disrupts magnetic orientation of migratory birds.** *Nature* 364 (Aug 5), 525 (1993).

41 Ritz, T. and al, E.: **Resonance effects indicate a radical-pair mechanism for avian magnetic compass.** *Nature* 429 (May 13), 177-180 (2004). <> (They conclude that evidence "suggest that the magnetite discovered is not involved in magnetic compass orientation but instead forms the basis of a magnetic intensity sensor, potentially used in a magnetic "map sense" for determining geographical position.")

42 Able, K. P. and Able, M. A.: **Daytime calibration of the magnetic orientation in a migratory bird requires a view of skylight polarization.** *Nature* 364 (Aug 5), 523 (1993).

43 Sauman, I. and al, E.: **Connecting the navigational clock to sun compass input in Monarch butterfly brain.** *Neuron* 46 (May 5), 457-467 (2005). <> (As referenced in *The Scientist,* June 6, 2005, p28. Deals with connection between polarized light, UV light (detected separately), and two genes involved with time keeping, *Period* and *Timeless.*)

44 Matthews, G. V. T. **Bird Navigation.** London, U.K: Cambridge Univ. Press; 2nd ed., (1968) <> (See especially Chs. 2, 3, and 10 on astronavigation. This is vol. 3 of Cambridge Monographs in Experimental Biology, with over 500 references and a nice table of the common and scientific names of some 100 species discussed.)

45 Alerstam, T. **Bird Migration.** London, U.K.: Cambridge Univ. Press, (1991) <> (Three other books on aspects of bird migration are reviewed in *Science,* 272:1896, Jun 28, 1996.)

46 Gould, J. L.: **Constant compass calibration.** *Nature* 375 (May 18), 184 (1995). < > (Background on report by Able and Able, p230 this issue, "Interaction in the flexible orientation system of a migratory bird.")

47 Gould, J. L.: **Fly (almost) south young bird.** *Nature* 383 (Sep 12), 123 (1996). < > (Background for report by Weindler *et al*, p158 this issue, "Magnetic information affects the stellar orientation of young bird migrants.")

48 Qui, J.: **Flight of the navigators.** *Nature* 437 (Oct 6), 804-806 (2005). < > (Sketch of work by T. Alerstam and colleagues aboard the research ice-breaker *Oden,* attempting to figure out how the many bird species which migrate to and across north polar regions with unerring accuracy do so.)

49 Gunnarsson, T. G. et al: **Arrival synchrony in migratory birds.** *Nature* 431 (Oct 7), 646 (2004). < > (The Icelandic black-tailed godwit winters between Britain and Iberia but breeds in Iceland. Pairs mate for life but winter separately, hundreds of kilometers apart, yet return to meet within a couple of days of each other, "avoiding costly divorce." Also noted in *Science News,* 10/9/04, p228.)

50 Berthold, P., Helbig, A. J., Mohr, G. and Querner, U.: **Rapid microevolution of migratory behavior in a wild bird species.** *Nature* 360 (Dec 17), 668-670 (1992). < > (Demonstrates genetic basis of recently altered migration habit of a portion of the European blackcap population. For commentary by W.J. Sutherland, see p625, "Genes mark the migratory route.")

51 Yoon, C. K.: **Bird-watching biologists see evolution on the wing.** *New York Times* (Dec 22) (1992).

52 Nussey, D. H. and al, E.: **Selection on heritable phenotypic plasticity in a wild bird population.** *Science* 310 (Oct 14), 304-307 (2005). < > ((The factual observation is that some of the population of migratory great tits in the Netherlands have recently altered their timing of arrival and reproduction to better coincide with caterpillar food availably in response to climate change. For digest by reporter E. Pennisi, see p215 this issue: "Better habits sometimes heritable.")

53 Editors: **Smart settlers.** *Nature* 436 (Jul 7), 5 (2005). < > (Note on paper in *Proc. Roy. Soc. Lond.* claiming that birds with bigger brains are less likely to fly south for the winter, supporting "the theory that migration evolved in birds that weren't smart enough to survive cold weather.")

54 Stettner, L. J. and Matyniak, K. A.: **The brain of birds.** *Sci Amer,* 64 (1968).

55 Milius, S.: **Where'd I put that?** *Science News* 165 (Feb 14), 103-105 (2004). < > (Short review of recent studies of bird memory.)

56 PBS: **Ravens.** *web site, www.pbs.org/wnet/nature/ravens* (posted Nov 4) (2005).

57 Emery, N. J. and Clayton, N. S.: **The mentality of crows: convergent evolution of intelligence in corvids and apes (Review).** *Science* 306 (Dec 10), 1903-1907 (2004).

58 Kenward, B. and al, E.: **Tool manufacture by juvenile crows.** *Nature* 433 (Jan 13), 121 (2005). < > (Appears to decisively establish that the manufacture and use of certain tools by crows is now heritable, i.e. genetic, inborn, instinctive. Only one other bird, the woodpecker finch, is known to use tools in the wild. However, in commentary on this article in *Science News*, 1/14/05 p24, another authority remarks that the burrowing owl also does so. "Kenward speculates that people's ancestors may have had a crow-like propensity for toolmaking." Wow.)

59 Milius, S.: **Techno crow.** *Science News* 163 (Mar 22), 182 (2003). < > (Sketch of paper in *Proc Roy Soc Lond*, by Hunt and Gray, studying local and regional variations in styles of tools made by crows, clearly demonstrating a component of cultural learning.)

60 Hawkins, T. H.: **Opening of milk bottles by birds.** *Nature* 165 (Mar 18), 435-436 (1950). < > (The first recorded instance was in 1921 after which this habit of blue tits spread through much of England. Other birds, "at least eleven species," have copied this habit. Author gives evidence that stealing milk from doorsteps was invented more than once, then copied by the local population. Further comment appears in *Nature* 169:1006, June 14, 1952, and articles in *British Birds*, v42, Nov. 1949, and v44, Dec. 1952.)

61 Neumann, P. et al: **Home-site fidelity in migratory honeybees.** *Nature* 406 (Aug 3), 474-475 (2001). < > (See also related article following, by J. Parr and colleagues, "Giant honeybees return to their nest sites," p475. Quote: "Although colonies regularly migrate over many kilometers [up to 200 km], we find that they often return to their original nest site even after an absence of up to two years. How the bees do this is unknown, as workers live for only a few weeks." This is despite the fact that "returning swarms usually have hundreds of alternative nest sites.")

62 Milius, S.: **Thoroughly modern migrants.** *Science News* 165 (Jun 26), 408-410 (2004). < > (Surveys recent work on migratory moths and butterflies. Includes work by Dingle, author of 1996 book on migrations, *The Biology of Life on the Move*, Oxford. Notes that Monarchs can also use polarized light for orientation, citing paper in *Current Biology*, Jan. 20.)

63 Gilman, A. G., Goodman, L. S., Rall, T. W. and Murad, F. **The Pharmacological Basis of Therapeutics.** New York: Macmillan Co., (1985) < > (A standard textbook. Regarding antibiotics, see Ch. 49, sulfonamides; Ch. 50, beta-lactams (penicillins, cephalosporins, etc.); Ch. 51, aminoglycosides (gentamycin, streptomycin, kanamycin, etc.); Ch. 52, tetracyclines, chloramphenicols, erytrhromycin, etc.; Ch. 53, 54, miscellaneous others. Several classes are not covered, e.g. the ionophores.)

64 Ives, A. R.: **Evolution of insect resistance to *Bacillus thuringiensis*-transformed plants.** *Science* 273 (Sep 6), 1412-1413 (1996). <> (More than 30 agricultural crop species have been engineered to contain genes from this bacterius, which produces a toxin conferring resistance to many pests. "However, several insect species have [already] evolved resistance to" these toxins. This report finds fault with a strategy reported in a prior issue to prevent or minimize acquired resistance.)

65 Taylor, M. F. J., Shen, Y. and Kreitman, M. E.: **A population genetic test of selection at the molecular level.** *Science* 270 (Dec 1), 1497-1499 (1995). <> (This article confirms what might be expected: that insects [tobacco budworms] in and near regions where certain insecticides [pyrethroids] are extensively applied show adaptive resistance to it, acquired within about 10 years. This particular insecticide acts to disrupt function of a voltage-gated sodium channel of the cell membrane.)

66 Baker, M.: **The hunt for new antibiotics.** *The Scientist* (Oct 10), 16-21 (2005).

67 Cowen, L. E. and Lindquist, S.: **Hsp90 potentiates the rapid evolution of new traits: Drug resistance in diverse fungi.** *Science* 309 (Sep 30), 2185-2189 (2005).

68 Kirschner, M. W. and Gerhart, J. C. **The Plausibility of Life: Resolving Darwin's Dilemma.** New Haven, CT: Yale Univ. Press, (2005) <> (See above, ref. 8 of Ch. 14, for comments.)

69 Editor: **New activity, old enzyme.** *Science* 310 (Nov 11), 945 (2005). <> (Note on paper by Horvat and Wolfenden, *PNAS USA,* 102:16199, 2005.)

70 Editor: **Songs of civilization.** *Science* 293 (Jul 6), 45 (2001). <> (Lyrebirds in Australia mimic almost perfectly the ringing of cell phones, vehicle back-up beeps, etc.)

71 Marshall, A. J.: **Bowerbirds.** *Sci Amer* June (Jun), 48-53 (1956).

72 Gillard, E. T.: **The evolution of bowerbirds.** *Sci Amer* (Aug), 38- (1963).

73 Borgia, G.: **Sexual selection in bowerbirds.** *Sci Amer* (Jun), 92 (1986).

74 Borgia, G.: **Why do bowerbirds build bowers?** *Amer Scientist* 83 (Nov/Dec), 542 (1995). <> (Subtitle: "Females prefer to visit courtship areas that provide easy avenues of escape, thereby protecting them from forced copulation." The author gives evidence that this is one criteria of the female's choice, but surely other criteria figure in the balance. It should be added that there is extensive and fascinating literature on these birds, supplying many different views.)

75 Milius, S.: **Cheap taste?** *Science News* 165 (Jan 17), 38 (2004). <> (Note on a study in *Behavioral Ecology and Sociobiology* of the spotted bowerbird, whose males construct an arched passageway of grasses decorated with various ornaments; found that the

ornaments are not selected on the basis of rarity, contrary to anecdotal reports, nor do rare items improve courtship success. However, another authority consulted for this article doubts the generality of the authors' conclusions.)

Ch. 17, References & Notes

1 Editor: "**The state of physics today ...**" *New Scientist* (Dec 10), 6-7 (2005). <> (See also editorial, p5, "Ideas needed: The hunt for a theory of everything is going nowhere fast.")

Appendix: References & Notes

1 Darwin, C. **On The Origin of Species, or the Preservation of Favored Races in the Struggle for Life**. London: John Murray, Albemarle Street, (1859) <> (The library copy I am now using [Random House / Modern Library, 1993] does not specify which edition it is but is probably the 6th of 1872. later editions became increasingly complicated and confusing as Darwin sought to reconcile his theory to various critical attacks, such as Kelvin's calculation of the age of the earth, which turned out to be wrong, and Jenkins' genetic calculations, also wrong.)

2 Harris, M. **The Rise of Anthropological Theory**. N.Y.: Thomas Y. Crowell & Co. (1st edn), (1968)

3 Secord, J. A. **Victorian Sensation: The Extraordinary Publication, Reception and Secret Authorship of** *Vestiges of the Natural History of Creation*. Chicago, IL: Univ. of Chicago Press, (2001) <> (Reviewed at length in *Nature* 409:285, Jan 18, 2001, by David Oldroyd. Robert Chamber's 1844 *Vestiges*, was written anonymously. Main theme was to doubt the Biblical account in view of emerging knowledge of geological time, the parallels between embryonic development and animal classification, etc., all pointing towards the evolution of species.)

4 Hegel, G. W. F. **Philosophy of History**. New York: Collier and Sons (1902, Engl. trans.), (1837)

5 Eiseley, L. **Darwin's Century (1st ed'n.)**. Garden City, N.Y.: Doubleday Inc. / Anchor Books; 1st ed'n., (1958)

6 Slotten, R. **The Heretic in Darwin's Court: The Life of Alfred Rissel Wallace**. N.Y.: Columbia Univ. Press, (2004) <> (Reviewed in *Nature,* 431:630, by G. Beccaloni, along with another on Wallace, *An Elusive Victorian ...* by Martin Fichman. The reviewer remarks that Fichman's "is more challenging to read and is likely to appeal to serious Wallace scholars," then adds that in order to understand why a book like Fichman's is needed, you should first read Slotten's, "the most detailed yet published"

and a "vivid account." Reviewer reminds us that Wallace was a heroic naturalist, first spending four years collecting in the Amazon but losing most of the collection and his notes in a shipwreck, then eight years collecting tens of thousands of specimens, including 1,000 new to science, in Southeast Asia, enduring all kinds of hardships.)

7 Raby, P. **Alfred Russel Wallace: A Life**. Princeton, N.J., and London: Princeton Univ. Press, (2001) <> (Reviewed in *Nature*, 413:357, Sept 27, 2001, by Jane R. Camerini, who praises the book, revealing this very interesting life, noting that several works of fiction have been based on his life. Also praises previous work by Ruby, *Bright Paradise,* an account of Victorian scientific travelers and expeditions. However, the reviewer finds the book a bit thin by the new high standards set by other scientific biographies such as those by Janet Browne on Darwin, Adrian Desmond on Huxley, and Desmond and James Moore on Darwin. But she says it's the best yet on Wallace.)

8 Berry, A., ed. **Infinite Tropics: An Alfred Russell Wallace Anthology**. N.Y.?: Verso, (2002). <> (Reviewed *Nature*, 416:790, 25 Apr 2002, along with another Wallace anthology edited by Jane Camerini, from Johns Hopkins Univ. Press, 2002.)

9 Desmond, A. **The Politics of Evolution: Morphology, Medicine, and Reform in Radical London**. Chicago, IL: Univ. Chicago Press (503 pages), (1990) <> (Reviewed in *The Sciences,* Nov/Dec 1990, p45-49, by Steven Shapin, professor of sociology and science studies, under the caption, "Revolutionary Biology." Desmond is now regarded as among the very best of science historians. Reviewer informs us that Desmond previously examined the impact of evolutionary ideas on working-class politics in the 1987 issue of *Osiris,* the present book focusing on the key role of the medical establishment, which at that time had recently won a broader middle class membership, in bringing to popular attention the relevant issues.)

10 Ritvo, H. **The Platypus and the Mermaid and Other Figments of the Classifying Imagination**. Cambridge, MA: Harvard Univ. Press, (1997) <> (Reviewed by Sherries Lyons in *Science,* 279:38, 2 Jan 1998, under headline "Taxonomy Recapitulates Society.")

11 Ruse, M. **Monad to Man: The Concept of Progress in Evolutionary Biology**. Cambridge MA: Harvard University Press, (1997) <> (Reviewed by David L. Hull in *Nature,* 385:497, Feb 6 1997; also by F. Ayala in *Science,* 275:495, Jan 24 1997; and widely elsewhere.)

12 Badash, L.: **The age of the earth debate**. *Sci Amer* (Aug), 90 (1989). <> (More accurately, Archbishop Ussher's calculation of the date of creation based on scripture was improved by John Lightfoot in 1654, who gave the date of creation as Oct. 26, 4004 B.C., at 9:00 AM, by Julian calendar.)

13 Danielson, D.: **Scientist's birthright**. *Nature* 410 (Apr 26), 1031 (2001). <> (Short essay on the coining of the word "scientist" in 1834 by William Whewell, including

the less well known fact that it was inspired by Mary Somerville's book, *On the Connexions of the Physical Sciences*. That forgotten book apparently inspired many others as well, such as Alexander von Humboldt.)

14 Malthus, T. R. **An Essay on the Principle of Population as it Affects the Future Improvement of Society, with Remarks on the Speculations of Mr. Godwin, M. Condorcet, and Other Writers**. London, (1798) <> (The original booklet was repeatedly expanded, finally to the large 6th edition of 1826. According to the *Encyclopedia Britannica* (1968, 14:717), "it was Malthus who first set Charles Darwin and A.R. Wallace, quite independently, on the train of reasoning which led to the principle of natural selection." The writer summarizes: "He briefly, crudely, yet strikingly argued that infinite human hopes for happiness must be in vain, for population will always tend to outrun the growth of production.")

15 Carlson, E. A. **The Unfit: A History of a Bad Idea**: Cold Spring Harbor Laboratory Press, (2001) <> (Reviewed *Nature*, 414:583, 6 Dec 2001, by Nick Martin, along with *Eugenics: A Reassessment*, by Richard Lynn (Praeger, 2001). Includes comments on alternative ways to achieve social benefits, such as evidence that legalized abortion resulted in reduced crime 18 years later. Reviewer appears to have some sympathy for limited eugenics, as do most people, except for wariness of the slippery slope.)

16 Deichmann, U.: **An unholy alliance**. *Nature* 405 (Jun 15), 739 (2000). <> (Essay, subtitled "The Nazis showed that 'politically responsible' science risks losing its soul.")

17 Pernick, M. S. **The Black Stork: Eugenics and the Death of 'Defective' Babies in American Medicine and Motion Pictures since 1915**. NY: Oxford Univ Press, (1996) <> (Reviewed in *Nature,* 382:217, 18 Jul 1996.)

18 Løvtrup, S. **Darwinism: Refutation of a Myth**: Croom Helm, (1987) <> (Reviewed in *Amer Scientist,* 76:394, by Keith Stewart Thomson. This book by a distinguished developmental biologist is apparently highly iconoclastic. But the reviewer may be right in remarking that "the reader will find a lot of this rhetoric irritating," even while admitting that a lot of it is on the money.)

19 Sulloway, F. J. **Born to Rebel: Birth Order, Family Dynamics, and Creative Lives**. N.Y.: Pantheon Press, (1996) <> (Reviewed in *Nature,* 384:126, 14 Nov 1996. Attempts to explain Darwin's achievement in terms of his birth order; see also *New Yorker* article, cited adjacent. The thesis clearly has some limited statistical merit, but the evidence suggests that Darwin was not at all a 'rebel' but a strait-laced conformist. Numerous other questions arise about his general thesis, such as the fact this it is limited to western cultures. Author charged with misconduct about dubious data delayed book by 4 years litigation; see *Nature,* 431:889, 21 Oct 2004.)

20 Boynton, R. S.: **The birth of an idea**. *New Yorker* (Oct 7), 72-81 (1996). <> (Profile of Harvard whizz-kid Sulloway and his observations about the importance of birth order in worldly achievement. "'The engine of history, he writes, is 'sibling rivalry.'" We cite this

because it is more than a comment on Darwin; it is a theory bearing on E.O. Wilson's Sociobiology – another black-eye for Harvard – *vis a vis* the theme of sibling rivalry.)

21 Landau, M. **Narratives of Human Evolution**. New Haven, CT: Yale Univ. Press, (1991) <> (Reviewed in *Science,* 252:993, 17 May 1991, by D.H. Hull. At first glance, this is a book on the rise of Darwinian thinking in the 19th century but reviewer sees it as an effort to view the theory as a kind of myth, in the style of "that proposed by Vladimir Propp in his classic [1928] *Morphology of the Folktale.*" However, the reviewer feels that she bungles the job with a too-heavy hand and too-simplistic mind,)

22 Richards, R. J. **The Meaning of Evolution: The Morphological Construction and Ideological Reconstruction of Darwin's Theory**. Chicago, IL: Univ. Chicago Press, (1992) <> (Lead review in *Science,* 256:1223, 22 May 1992, by P.R. Sloan, who notes that the title alludes to G.G. Simpson's classic, which "many of us cut our evolutionary teeth on," by way of saying that "in spite of the wide agreement I find with [Richards], the more polemical point ... requires comment. ... Richards' book is more broadly an attack upon the historiography adopted by neo-selectionist evolutionary theory – especially Stephen J. Gould and Peter Bowler – and behind them Simpson and Ernst Mayr ..." The reviewer tries to mediate, but appears to be won over. This again highlights the many sharp disputes within the field.)

23 White, M. and Gribbin, J. **Darwin: A Life in Science**. N.Y.: Dutton, (1995) <> (Reviewed in *Science,* 271:455, Jan 26, 1996, by W. Montgomery, who says that "in recent years ... Darwin has been the subject of a wave of biographies," e.g., by Desmond and Moore, and another major one in progress by J. Browne. This book is apparently for general readers, not Darwin experts. Reviewer feels that authors unjustly downplay the rationality of opponents of Darwin's theory, a persisting fallacy.)

24 Bowler, P. J. **Charles Darwin: The Man and His Influence**. Cambridge, MA: Blackwell, (1990) <> (Reviewed, along with the biography by Bowlby, in *Science,* 252:992, 17 May 1991: "Bowlby's biography is a newsy, intimate account of private life among comfortably situated Victorians. ... He makes good use of [the recently published] *Correspondences* ... In contrast to Bowlby, Peter Bowler concentrates specifically on the science ..." giving Darwin's personal life the same short shrift that Bowlby gives to his science.)

25 Bowlby, J. **Charles Darwin: A Biography**. N.Y.: Norton, (1991) <> (Reviewed with the biography by Peter J. Bowler; see next.)

26 Bowler, P. J. **Life's Splendid Drama: Evolutionary Biology and the Reconstruction of Life's Ancestry**. Chicago, IL: Univ. Chicago Press, (1996) <> (Reviewed in *Nature,* 385:127, 9 Jan 1997, by K. Padian. A History of bioevolutionary thought.)

27 Kohn, M. **A Reason For Everything: Natural Selection and the British Imagination**. N.Y.: Faber & Faber, (2004) <> (Reviewed *Nature,* 431:21, Sep 2 2004, by Steve Jones, under headline, "When giants walked the earth: a pedigree of

Darwin's well-bred English bulldogs." Profiles and relationships of 6 major figures, from old to recent, from Darwin himself to Alfred R. Wallace, R.A. Fisher, J.B.S. Haldane, John Maynard Smith, and Bill Hamilton. All except Wallace were high-born to privilege. Well-written review conveys a fascinating book. Touches on J.R.R. Tolkien, author of *Lord of the Rings*, which is compared to Fisher's *Genetical Theory of Natural Selection*, also full of gnomic and portentous truths, and with a nasty social agenda lurking beneath. As Kohn points out, Fisher's followers, like those of composer Wagner (*The Ring*), are "obsessed with the fine detail of what the great begetter meant ...")

28 Kardiner, A. and Preble, E. **They Studied Man**. New York: World Publishing, (1961) < > (Brief biographies include Charles Darwin, Herbert Spencer, Edward Tylor, James Frazer, Emile Durkheim, Franz Boaz, B. Malinowski, Alfred Kroeber, Ruth Benedict and Sigmund Freud. Kardiner was a leader of the psychological school of anthropology, which may explain why Freud and Frazer are included but not Morgan.)

29 Keynes, R. **Annie's Box: Charles Darwin, His Daughter and Human Evolution**: Fourth Estate Press, (2001) < > (Reviewed in *Nature,* 411:739, Jun 14, 2001, by Bruce Weber, who notes that Keynes, although not an historian, descended from Charles Darwin and so has access to rich sources. Reviewer feels that this book nicely complements the two recent major biographies, one by Desmond and Moore, *Darwin, the Life of a Tormented Evolutionist* [1994], the other by Janet Browne, *Charles Darwin: Voyaging* [1996]. The Annie of the title is Darwin's daughter, whose death agonized him – as did several other deaths besetting him. Mentions Darwin's shift in temperament as revealed from his notebooks of 1839 *vs.* 1859.)

30 Desmond, A. and Moore, J. **Darwin**. NY: Warner, (1992) < > (Reviewed in *Science,* 257:419, 17 Jul 1992, by historian F.M. Turner, who appears to accept the book as a major biography, despite its "radical ... revisionist" character. But he has some caveats, such as lack of attention to Darwin's relation with his wife, excessive credulity for the usual account of the Darwin-Wallace collision (letters on which are missing from his published correspondence), a "whiggish bias" in discussing Darwin's opponents, and "perhaps too much anti-clericalism." The review itself is informative and supplies balance to the book.)

31 Richards, R. J. **Darwin and the Emergence of Evolutionary Theories of Mind and Behavior**. Chicago, IL: Univ. Chicago Press, (1988) < > (Lead review in *Science,* 239:198, by historian J.C. Greene, under caption, 'Darwinism and Moral Purpose." We object only to the reviewer's perception that this is the first book to recognize the importance of Spencer, long appreciated in anthropological circles.)

32 Plotkin, H. **The Nature of Knowledge: Concerning Adaptations, Instinct and the Evolution of Intelligence**. Cambridge, MA: Allen / Lane / Harvard Univ. Press, (1994) < > (Reviewed in *Nature,* 368:701, 21 Apr 1994, by N. Mackintosh, a psychologist (as is Plotkin). This appears to be a good effort to extend or generalize Darwinian principles to the sphere of psychological ideas, and was apparently

inspired in part by Gerald Edelman's books, which in turn were inspired by recent insights into the immune system.)

33 Crow, T. J.: **A Darwinian approach to the origins of psychosis**. *British Journal of Psychiatry* 167, 12-25 (1995). <> (For brief account, see *The Scientist* 18(7):34, Apr 12, 2004; and letter in response, May 10, 2004, p10.)

34 Richerson, P. J. and Boyd, R. **Not By Genes Alone: How Culture Transformed Human Evolution**. Chicago: Univ. Chicago Press, (2004) <> (Reviewed in *Nature* 432:951, Dec 23/30, by R Dunbar, who opens saying that he "continues to be surprised by the number of educated people (some of them biologists) who think that offering explanations for human behavior in terms of culture somehow disproves the suggestion that human behavior can be explained in Darwinian evolutionary terms." On this book, she says that their "position, in a nutshell, is that culture is as good a candidate for Darwinian treatment as behavior and morphology.")

35 Smith, E. A. and Winterhalder, B. **Evolutionary Ecology and Behavior**. New York: Hawthorne Press, (1992) <> (Reviewed in *Science*, 260:1176. Begins, "The revolution in animal behavior studies that started in the 1980's is echoed in anthropology.")

36 Shennan, S. **Genes, Memes and Human History: Darwinian Archeology and Cultural Evolution**. N.Y.: Thames and Hudson, (2002) <> (Reviewed in *Nature*, 422:20-21, March 6, 2003, by R.L. Bettinger. Appears to rest heavily on the notion of "memes" – Dawkins' word for the not-very-original concept that ideas are transmitted like genes, except culturally instead of somatically.)

37 Jensen, J. V. **Thomas Henry Huxley: Communicating for Science**. Newark, NJ: Univ. of Delaware Press (Distributor: Assoc. Univ. Presses, Cranbury, NJ.), (1991) <> (Reviewed in *Science*, 256:1061, 15 May 1992, along with another book on him by Michael Collie.)

38 Collie, M. **Huxley at Work. With the Scientific Correspondence of T.H. Huxley and the Rev. Dr. George Gordon** ... N.Y.: MacMillan, (1991) <> (Reviewed in *Science*, 256:1061, 15 May 1992, by W. Montgomery. The review is favorable but notes that "Collie's book suffers from his failure to make use of Adrian Desmond's *Archetypes and Ancestors* (Univ. of Chicago Press, 1984) ...")

39 Desmond, A. **Huxley: Evolution's High Priest**. Vol. 386: Michael Joseph Press, p. 349, (1997) <> (Refers to T.H. Huxley, not other illustrious members of that family, who has been called "Darwin's bulldog.")

✧ ✧ ✧ ✧ ✧

www.ingramcontent.com/pod-product-compliance
Lightning Source LLC
Chambersburg PA
CBHW071355170526
45165CB00001B/55